Trends in Applications of
Pure Mathematics to Mechanics

Volume II

Trends in Applications of Pure Mathematics to Mechanics

Volume II

A collection of papers presented at a symposium at Kozubnik, Poland, in September 1977

Edited by

Henryk Zorski

Polish Academy of Sciences

Pitman

London San Francisco Melbourne

PITMAN PUBLISHING LIMITED
39 Parker Street, London WC2B 5PB

FEARON PITMAN PUBLISHERS INC.
6 Davis Drive, Belmont, California 94002, USA

Associated Companies
Copp Clark Pitman, Toronto
Pitman Publishing New Zealand Ltd, Wellington
Pitman Publishing Pty Ltd, Melbourne

First published 1979

AMS Subject Classifications: 34–XX, 55–XX, 73–XX, 80–XX, 81–XX, 83–XX

British Library Cataloguing in Publication Data

Trends in applications of pure mathematics to mechanics.–(Monographs and studies in mathematics; 5).
Vol. 2: A collection of papers presented at a symposium at Kozubnik, Poland, in September 1977
1. Mechanics, Analytic–Congresses
I. Zorski, Henryk II. Symposium on Trends in Applications of Pure Mathematics, *2nd, Kozubnik, 1977* III. Series
531′.01′51 QA801

ISBN 0–273–08421–6

Typeset in Northern Ireland at The Universities Press (Belfast) Ltd.
Printed in Great Britain at Biddles of Guildford.

Contents

Preface

This book contains papers presented at the 2nd Symposium on Trends in Applications of Pure Mathematics to Mechanics held at Kozubnik, Poland, in September 1977; the Proceedings of the 1st Symposium were published earlier.† The 3rd Symposium will be held in Edinburgh in 1979 and this series of symposia will be continued under the auspices of the International Society for Interaction of Mechanics and Mathematics founded during the 2nd Symposium.

A close collaboration between specialists in mechanics and mathematics is as important as it is difficult to achieve. Although the final goal is always a better understanding of nature and the development of the human mind, there is a difference in motivation and methodology between pure and applied mathematicians. Mathematics lies within the domain of purely abstract thinking, whereas mechanics is an outcome of observations of nature.‡ However, as soon as its basic laws are formulated in a system of experimentally verified hypotheses and formal axioms, i.e., a mathematical model constituting a simplified picture of natural phenomena and processes is created, the process of reasoning becomes abstract. A new language is used, but mathematics is much more: it is a language and logic together and, therefore, it provides a way of deducing exact, physically meaningful results by means of proving theorems. Of course, any mathematical proposition is a proposition about the mathematical model, not about the physical system itself; but if the whole reasoning is mathematical, a proper interpretation of the formalism will then lead to final results which agree with experiment within the same bounds as the basic laws.

Mechanics deals with quantities and relations between quantities which

† Fichera, G. (ed.), *Trends in Applications of Pure Mathematics to Mechanics, A Collection of Papers Presented at the University of Lecce, Italy, in May 1975*, Pitman Publishing, London, San Francisco, Melbourne, 1976.

‡ Incidentally, as the history of science proves, abstract concepts becomes less abstract in the course of time.

have a physical meaning, i.e., they can in principle at least, be observed and measured, and this fact bears heavily on the attitude of an applied mathematician towards abstract procedures. It is certainly convenient and important to be able to identify all the symbols and steps of a mathematical procedure with a physical fact, since this leads to a deep understanding of the model and the physical system. However, to achieve this ideal goal, a specialist in mechanics must learn new mathematics, frequently created by his own demand, no matter how pure and abstract it is; only then will he be able to get to the heart of the problem and be able to ask the pure mathematician intelligent questions in order to attract him to mechanics and to contribute to the formulation of the mathematical model. All these facts are appreciated by many people, but they are still regarded with suspicion and/or distrust by those who prefer to treat mathematics only as an instrument to get a good fit between a formula and an experimental curve.

In order to keep things in a proper balance, an old truth should be repeated here: many branches of pure mathematics arose as an answer to questions asked by physicists, and the development of many other branches was stimulated by the needs of applied sciences.

This series of symposia is organized to bring together the pure mathematicians who are interested in mechanics itself, or who enjoy seeing their abstract ideas confronted by physical facts, and specialists in mechanics interested in deep investigations of Nature. A perusal of the papers presented during the symposium shows a variety of subjects treated, some of them almost purely mathematical, some of a more practical nature. It may be hoped, therefore, that at least to some extent the aims of the symposium were achieved.

Henryk Zorski
Polish Academy of Sciences

L. Amerio

Unilateral problems for the vibrating string equation

1 Introduction

The object of the present paper is the study of the motion of a string vibrating against an obstacle.

As is well known, unilateral problems for partial differential equations of elliptic and parabolic type have been widely studied (cf., [1]) after Fichera succeeded in solving the Signorini problem for the equilibrium of an elastic body. A new theory of elliptic and parabolic inequalities has since been developed, which is supported by papers by Lions and Stampacchia, as well as by the work of Fichera. Several other authors (notably H. Lewy and Brézis) have also made wide-ranging contributions to the theory and its applications.

On the other hand, the theory of unilateral problems for hyperbolic equations is much less developed. This theory also obviously has a notable interest for mechanics: in particular, it is concerned with the study of dynamical problems arising when an elastic body, during its motion, hits an obstacle. The reactions which occur are generally of impulse type; they create waves which propagate in the medium, producing discontinuities in the first-order derivatives of the solution (and possibly for the solution too). The equations have, therefore, to be interpreted in the sense of *distributions*.

We shall discuss two very simple but typical problems. The first of these problems has been studied by Amerio and Prouse [2], the second by Amerio [3].

Let us consider the *vibrating string* equation

$$y_{tt} - y_{xx} = 2(f(P) + J) \qquad (P = (x, t)) \tag{1.1}$$

or, in the *characteristic form*,

$$y_{\xi\eta} = f(P) + J, \tag{1.2}$$

where

$$\xi = \frac{x+t}{\sqrt{2}}, \qquad \eta = \frac{-x+t}{\sqrt{2}},$$

In (1.1), (1.2), $y(x, t) = y(P)$ denotes the *displacement* in the (x, y) plane of the point of the string with coordinate x at the time t. We assume that when the string is at rest, it lies along the x axis. $f(P)$ denotes the (given) *external force*, J the (unknown) *reaction of the obstacle.*

We shall assume, for the sake of simplicity, that the displacement has to be calculated, for both problems, on a *half-strip* $\Delta = \{0 \leqslant x \leqslant l, t \geqslant 0\}$, and we set $\Delta_s = \{0 \leqslant x \leqslant l, 0 \leqslant t \leqslant s\}$, $\forall\, 0 \leqslant t \leqslant s$. The function $y(x, t)$ must, moreover, satisfy the Cauchy *initial conditions*

$$y(x, 0) = \varphi(x), \qquad y_t(x, 0) = \psi(x) \qquad (0 \leqslant x \leqslant l) \tag{1.3}$$

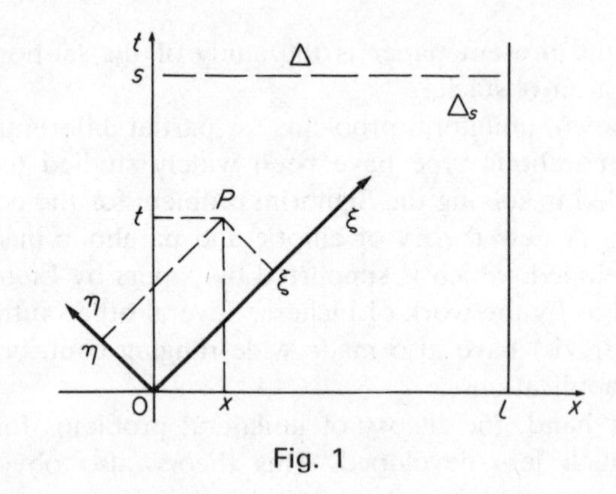

Fig. 1

and the *boundary conditions*

$$y(0, t) = y(l, t) = 0 \qquad (t \geqslant 0). \tag{1.4}$$

We consider, therefore, the motion of a string with fixed ends: more general hypotheses could, however, be considered. *Moreover,* y_{tt}, y_{xx}, $J \in \mathscr{D}'(\mathring{\Delta})$.

Problem I (Problem of the Wall) *This problem has been solved in* [1], *assuming* $f(P) \equiv 0$; *hence,*

$$\tfrac{1}{2} y_{tt} - y_{xx} = y_{\xi\eta} = J. \tag{1.5}$$

Since we have no external force, the motion is, therefore, caused only by the *initial conditions* (1.3).

Assume now that the free vibration of the string, in the (x, y) plane, is impeded by a rigid wall $y = K > 0$ that obliges the string to move in the half-plane

$$y \leq K. \tag{1.6}$$

During the motion, the string will, therefore, hit the wall and we assume that the impact is perfectly *elastic*; this fact will be interpreted by the condition that for every *impact point* $P_0(x_0, t_0)$ (in which, necessarily, $y = K$), we have

$$y_t^+(x_0, t_0) = -y_t^-(x_0, t_0) < 0. \tag{1.7}$$

We require, in other words, that during the impact against the wall, the speed does not vary in absolute value, but changes sign. Hence, the *local kinetic energy* remains unchanged:

$$\tfrac{1}{2}(y_t^+(x_0, t_0))^2 = \tfrac{1}{2}(y_t^-(x_0, t_0))^2.$$

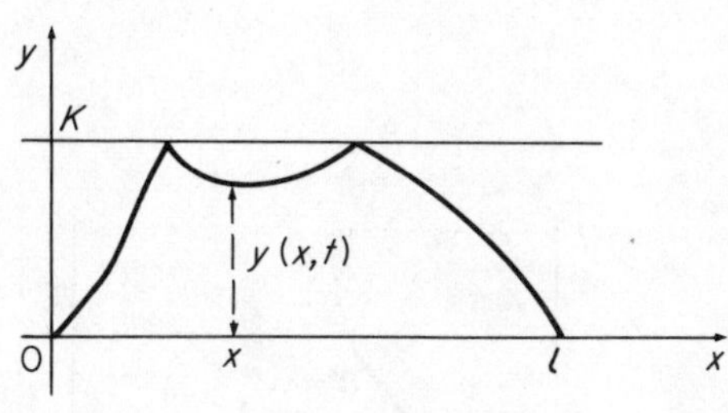

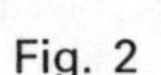
Fig. 2

Other types of impact can, obviously, be considered. Supposing, for instance, that the impact is partially *elastic* or *inelastic*, then

$$y_t^+(x_0, t_0) = -\rho y_t^-(x_0, t_0^-) < 0 \qquad (0 < \rho < 1)$$

and

$$y_t^+(x_0, t_0) = 0,$$

respectively. These cases have been studied by Citrini [3]. One can suppose, moreover, that there exist two walls, which leads to the limitations

$$-H \leq y(x, t) \leq K \qquad (H, K > 0).$$

Problem II (Problem of the Point-shaped Obstacle) *We assume that the free vibration of the string is impeded by a point G which moves in the (x, y) plane (for instance,* under *the vibrating string) with an arbitrary law:*

$$G = G(t) = (\lambda(t), \alpha(t)).$$

In this problem, we obtain the solution under very general hypotheses. The external force $f(P)$ is, in fact, an arbitrary *continuous* function; also,

$\alpha(t)$ and $\lambda(t)$ are supposed to be arbitrary *continuous* functions. However, let us make another assumption, of clear physical meaning: the function $\lambda(t)$ *must satisfy the Lipschitz condition* $|\lambda'(t)| \leq 1$, a.e., $\lambda'(t) \neq \pm 1$ on an interval. This means that the longitudinal speed of $G(t)$ cannot be greater than the speed of a wave travelling in the string. Moreover, the equality does not hold at an interval.

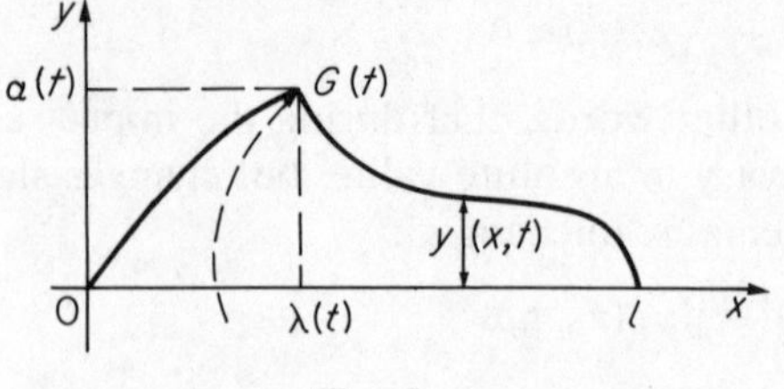

Fig. 3

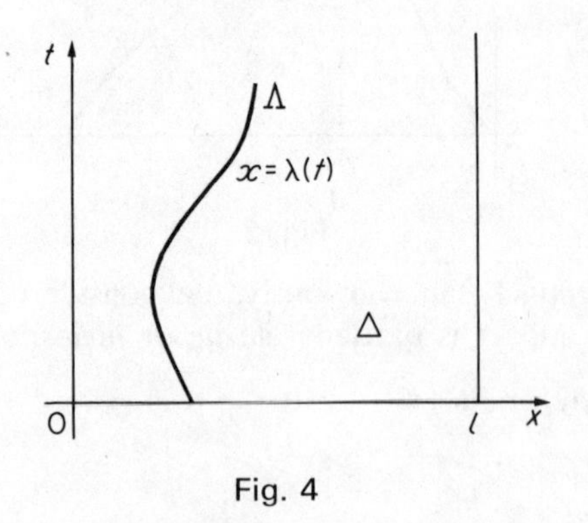

Fig. 4

The problem under scrutiny has the following analytical interpretation. We consider, in the (x, t) plane, a line

$$x = \lambda(t) \qquad (t \geq 0, 0 < \lambda(t) < l) \tag{1.9}$$

and we require that the displacement $y(P)$ satisfies the *unilateral condition*

$$y(\lambda(t), t) \geq \alpha(t) \qquad (t \geq 0). \tag{1.10}$$

Observe that we can impose (cf., [3]) a *pair of unilateral conditions:*

$$\alpha(t) \leq y(\lambda(t), t) \leq \beta(t).$$

This corresponds to the following mechanical problem. The free motion of the string is impeded by a *ring* through which the string is obliged to pass; such a ring is always orthogonal to the plane (x, y), it has its centre

at the point $G(\lambda(t), (\beta(t)-\alpha(t))/2)$ and its diameter is $\beta(t)-\alpha(t)$.
One can also consider other point-shaped obstacles or rings.

2 The problem of the wall

Assume that the *initial functions* $\varphi(x)$ and $\psi(x)$ satisfy the following conditions.

(a_1) $\varphi(x) \in \mathrm{Lip}\,[0, l], \qquad \varphi(0)=\varphi(l)=0,$ (2.1)

$\varphi(x) < K.$ (2.2)

(a_2) $\psi(x)$ and $\varphi'(x)$ are continuous on $[0, l]$ with the exception of, at most, a finite number of discontinuities of the first kind.

(a_3) The characteristic derivatives, at the initial time,

$$y_\xi\,|_{t=0} = \frac{\varphi'(x)+\psi(x)}{\sqrt{2}}, \qquad y_\eta\,|_{t=0} = \frac{-\varphi'(x)+\psi(x)}{\sqrt{2}} \tag{2.3}$$

do not vanish over the whole of $[0, l]$, with the exception, for each one, of a finite number of points and of intervals, at most.

We now introduce the *impact arcs* and the *impact points* for a function $u(P)$, with respect to Problem I. We *shall assume from now on that* $u(P) \in \mathrm{Lip}\,[\Delta]$.

We shall say that $\gamma = AB$ is an *impact arc* if it satisfies the following conditions.

(b_1) On γ, $u(P)=K$ (i.e., γ is a *line of level* K for the function $u(P)$);

(b_2) γ has an equation

$$t = g(x) \qquad (0 < x_1 \leqslant x \leqslant x_2 < l),$$

where

$$g(x) \in \mathrm{Lip}\,[x_1, x_2], \qquad g'(x) \in C^0]x_1, x_2[, \qquad |g'(x)| < 1. \tag{2.4}$$

(Hence, at no point of $\mathring{\gamma}$ does the tangent have a characteristic direction: γ is a *space-like* line for Eqn (1.5).)

(b_3) Consider the two characteristics η, ξ emanating from the endpoints A, B, and let S be the curved triangle so defined, inferiorly bounded by the union, σ, of two characteristic segments. Then, the *restriction* $u(P)\,|_S$ satisfies the following conditions:

$$\left.\begin{array}{ll} u(P)\,|_S \leqslant K, & u(P)\,|_S \in C^1[S-\sigma], \\ u_t^-(x_0, t_0) > 0, & \forall P_0 \in \mathring{\gamma}. \end{array}\right\} \tag{2.5}$$

Observe that (2.4) and (2.5) are equivalent to the assumptions

$$u_\xi^-(P_0) < 0 \quad \textit{and} \quad u_\eta^-(P_0) < 0. \tag{2.6}$$

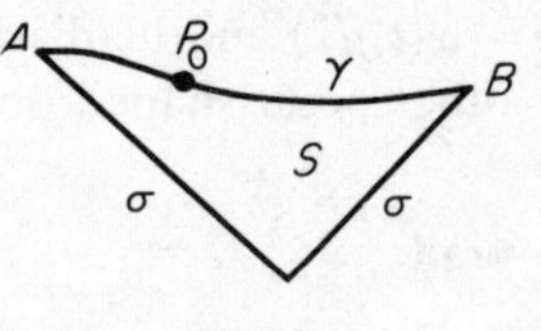

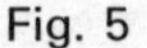
Fig. 5

We shall call P_0 an *impact point.*

We now define the *admissible functions* for Problem I. This means the set γ of all functions $y(P)$ endowed with the following *properties.*

(c_1) $y(P)$ is Lipschitz-continuous on Δ and, moreover, $y(P) \leqslant K$.

(c_2) The derivatives y_ξ, y_η are continuous on Δ with the exception of, at most, a sequence $\{\tau_n\}$ of characteristic segments and a sequence $\{\gamma_n\}$ of impact arcs (in a finite number in every rectangle Δ_s). Moreover, at every impact point P_0, the (*elastic*) *impact condition* holds:

$$y_t^+(x_0, t_0) = -y_t^-(x_0, t_0) < 0, \tag{2.7}$$

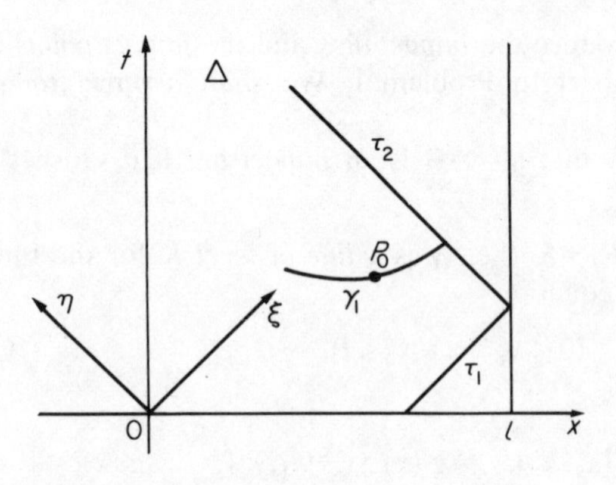

Fig. 6

equivalent, by (2.6), to the assumptions

$$\left.\begin{aligned} y_\xi^+(P_0) &= -y_\eta^-(P_0) < 0, \\ y_\eta^+(P_0) &= -y_\eta^-(P_0) < 0. \end{aligned}\right\} \tag{2.8}$$

(c_3) $y(P)$ satisfies the following *extension law.* Let us consider (with reference to the '*elementary problems*' of Cauchy, Darboux and Goursat) the situation corresponding to Figs 7, 8, 9. Moreover, let $z(x, t)$ be the solution, *on* T, of the vibrating string equation

$$z_{\xi\eta} = 0, \tag{2.9}$$

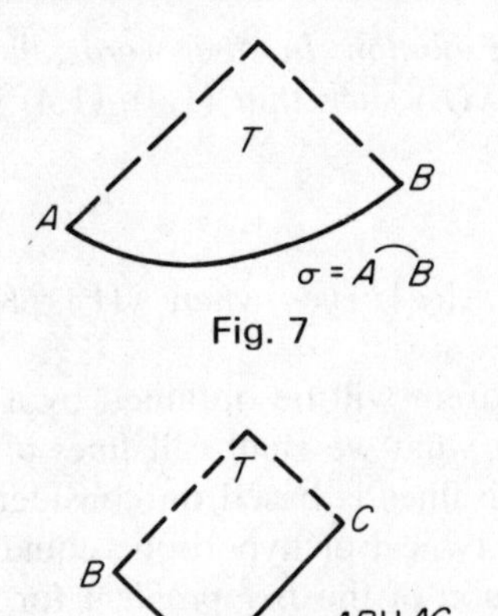

Fig. 7

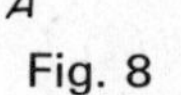

Fig. 8

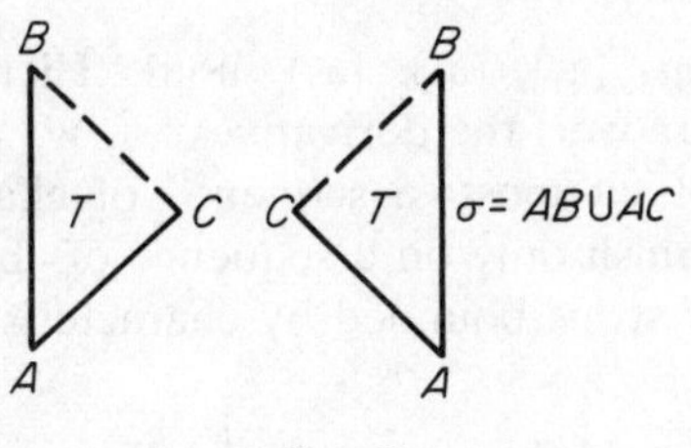

Fig. 9

corresponding to the *data*

$z(x, t)\,|_\sigma = y(x, t)\,|_\sigma, \qquad z_t(x, t)\,|_\sigma = y_t^+(x, t)\,|_\sigma$ (Cauchy);

$z(x, t)\,|_\sigma = y(x, t)\,|_\sigma$ (Darboux and Goursat).

Then, *if*

$$z(x, t) \leq K, \tag{2.10}$$

on the whole of T, we also have

$$y(x, t) = z(x, t) \tag{2.11}$$

on the whole of T. Observe that the extension law implies, essentially, that *all admissible functions must satisfy the homogeneous vibrating string equation 'wherever possible'*; in particular, this occurs on the open set, where the string does not touch the wall $(y(P) < K)$.

We can prove the following *existence and uniqueness theorem.*

Theorem *Assume that $\varphi(x)$ and $\psi(x)$ satisfy the assumptions* (a_1), (a_2), (a_3). *Then Problem I, with the initial and boundary conditions* (1.3), (1.4) *admits one and only one solution. In other words, there exists one and only one admissible function* $y(P)$ *such that* (1.3), (1.4) *hold.*

Moreover, we have

$$y_{\xi\eta}=0 \tag{2.11}$$

on the whole open set $\mathring{\Delta}-\{\gamma_n\}$, *even where* $y(P)=K$.

The proof of this theorem will be obtained by a method of *successive extensions*, starting from what we shall call *lines of influence of the wall.* The construction of such lines is based on considerations relative to the *domains of dependence*, typical of hyperbolic equations.

Let $w(P)$ be the solution of the *free problem* for Eqn (2.11), satisfying the initial and boundary conditions

$$\left.\begin{aligned} &w(x,0)=\varphi(x), \qquad w_t(x,0)=\psi(x),\\ &w(0,t)=0, \qquad w(l,t)=0, \end{aligned}\right\} \tag{2.12}$$

where the assumptions (a_1), (a_2), (a_3), hold. Then $w(P)$ is Lipschitz-continuous on Δ; moreover, the derivatives w_ξ, w_η are continuous on Δ, with the exception of, at most, a sequence of characteristic segments. Finally, w_ξ, w_η can vanish only on a sequence of characteristic segments and on a sequence of strips bounded by characteristic segments.

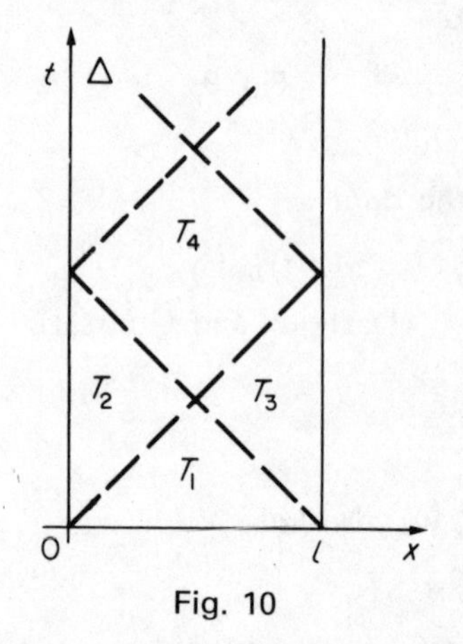

Fig. 10

We recall, also, that, by a classical procedure, $w(P)$ can be calculated by solving a sequence of Cauchy, Goursat and Darboux problems on the domains T_1, T_2, T_3, T_4, ...

If $\forall t\geqslant 0$, $w(t,x)\leqslant K$, then $w(x,t)$ gives the solution of our problem. We have, successively, by the extension law,

$$y(P)=w(P) \quad \text{on} \quad T_1, T_2, T_3, T_4, \ldots$$

and no impact arcs exist.

Assume now that there exists $P'(x', t')$ such that $w(P') > K$. We shall then define the (first) *line of influence of the wall* in the following way.

Let us fix $Q(\alpha, \beta)$ in Δ and consider the intersection Z_Q of Δ with the backward characteristic semicone with vertex in Q. We shall again call Z_Q the *backward semicone* relative to the point Q. Observe that, by (2.2), $w(P) \leq K$ in Z_Q, provided β is sufficiently small.

Now, we denote by $t(\alpha)$ the *maximum value* of β for which this inequality holds. If, therefore, $Q = (\alpha, t(\alpha))$, it follows that $w(P) \leq K\ \forall P \in Z_{Q_\alpha}$ while, choosing $t' > t(\alpha)$ arbitrarily and setting $Q' = (\alpha, t')$, there exists at least one point $P' \in Z_{Q'}$ such that $w(P') > K$. *Obviously*, $w(Q) \leq K$, where *the inequality can hold too.*

We are now able to add some considerations.

As we can prove, the value $t(\alpha)$ actually represents, for the problem with an obstacle, *the maximum time value* for which the motion of the point α of the string is *not influenced by the presence of the obstacle*, and this is true *even if* $w(Q_\alpha) < K$.

By the construction made, one may, in fact, be intuitively convinced that the following hold.

1. The function $w(P)$ has no impact points *below* the backward characteristic polygonal line σ with vertex at the point Q_α. The actual motion and the free motion coincide; in other words, on the whole of Z_{Q_α}, $y(P) = w(P)$.
2. Such impact points exist *on* or *above* the line σ, and we can find some of them as near to σ as we want.

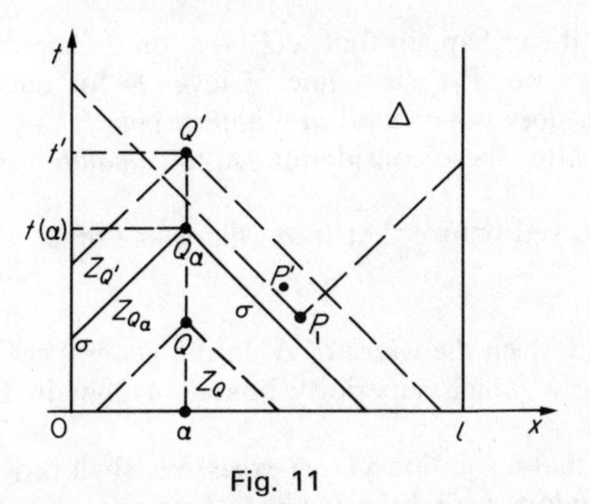

Fig. 11

Let us now consider the point $Q'(\alpha, t')$ mentioned earlier, with $t' - t(\alpha) > 0$ and *arbitrarily small.* We can then find, *on* or *above* σ, an impact point $P_1(x_1, t_1)$ such that Q' belongs to the *forward* characteristic semicone with vertex at P_1. The circumstance that $P_1(x_1, t_1)$ is an impact point means that the point x_1 of the string has hit the wall at the time t_1. An impulse has then been created, at the point P_1, because of the reaction of the wall (which produces the inversion of the speed of the point x_1). Such an impulse at P_1 will influence the law of motion at all points $P(x, t)$ which belong to the *forward* characteristic semicone emanating from P_1. This occurs, in particular, for the point $Q' \Rightarrow y(Q') \neq w(Q')$.

We are led, on the basis of this analysis, to study the line λ of equation $t = t(x)$, where x varies from 0 to l, and we shall call this line the (*first*) *line of influence of the wall.*

Let us now recall the principal properties of the line (*see* Fig. 12).

(i_1) *The function* $t(x)$ satisfies *on* $[0, l]$, *a Lipschitz condition, with constant* $= 1$; hence, $|t'(x)| \leq 1$ at all points at which the derivative exists.

(i_2) The line λ is constituted by: (1) *a finite number* (≥ 1) *of impact arcs* for the function $w(P)$ (they are, therefore, arcs of level K for the same function); (2) *a finite number* (≥ 2) of *characteristic segments*–of these, the first, M_1M_2, belongs to a characteristic η and has its first endpoint on the t axis; the last, N_1N_2, belongs to a characteristic ξ and has its second endpoint on $x = l$; moreover

$$w(M_1) = w(N_2) = 0, \qquad w(M_2) = w(N_1) = K,$$
$$w(P) \leq K \quad \text{in the remaining points.}$$

Finally, *two consecutive* impact arcs can have *an endpoint* in common; otherwise, they are connected by *one characteristic segment* or by *two consecutive characteristic segments* which belong to a *characteristic* ξ and to a *characteristic* η, respectively.

Observe that it can happen that $w(P) = K$ on a characteristic segment P_1P_2. In such a case, P_1P_2 is a line of level K for the function $w(P)$, which, however, *does not contain any impact point.*

We are able, after these considerations, *to calculate the solution* $y(x, t)$ *of Problem I.*

We have observed before that *if, on the whole of* Δ,

$$w(x, t) \leq K, \tag{2.13}$$

then $y(x, t) = w(x, t)$ *on the whole of* Δ. In the general case, there exists a line of influence λ which superiorly bounds a domain E, where (2.13) holds.

Assume now that a solution $y(x, t)$ *exists.* We shall prove that it can be obtained, in a unique way, by a *method of successive extensions*, starting from the solution $w(P)$ of the free problem.

Assume, at first, *that* $Q \in E - \lambda$. *We then have*

$$y(Q) = w(Q), \tag{2.14}$$

i.e., $w(P)$ *satisfies the vibrating string equation on* $E - \lambda$. Moreover, the value $w(P)$ is calculated by solving a finite number of problems of

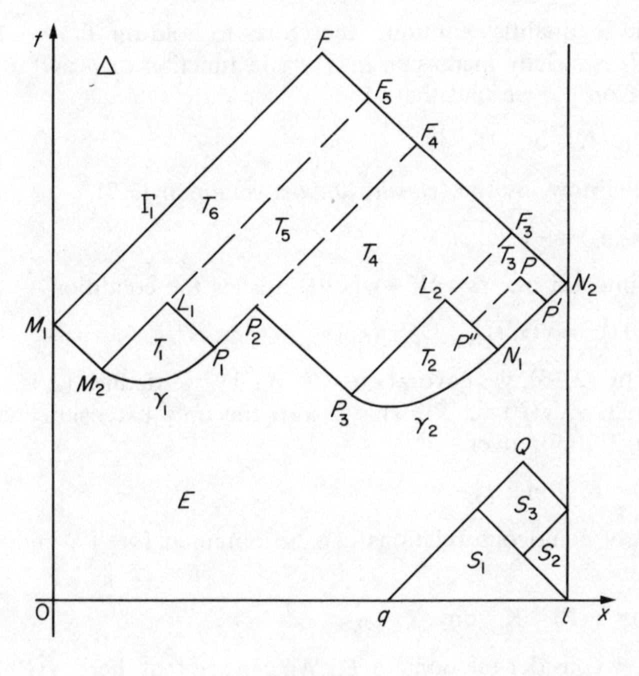

Fig. 12

Cauchy, Goursat and Darboux. Referring to Fig. 12, we have $w(P) \leq K$ in S_1, and $w(P) = y(P) = \varphi(x)$, $w_t(P) = y_t(P) = \psi(x)$ on the segment $[q, l]$. It follows, from the extension law, that $y(P) = w(P)$ on the whole of S_1. The same equality holds for the domains S_2, S_3, and (2.14) is proved. Thus, (2.14) holds in the whole domain E by the continuity of the functions $y(P)$ and $w(P)$.

We must now calculate the solution $y(P)$ above λ.

Bearing in mind the description given earlier, we can assume, without loss of generality, that λ is of the form illustrated in Fig. 12. Let $\gamma_1 = M_2P_1$ and $\gamma_2 = P_3N_1$ be the impact arcs of λ: the remaining part of λ is constituted by characteristic segments.

Let us draw from M_1 and N_2 the characteristics ξ and η and let F be their intersection. Consider then, successively, the domains T_1, T_2 bounded from below by the impact arcs γ_1, γ_2 and from above by segments of characteristics. We prove that, *necessarily*, on the whole of T_1,

$$y(P) = 2K - w(P). \tag{2.15}$$

In fact, every point $P_0 \in \mathring{\gamma}_1$ is an impact point for $w(P)$, i.e.,

$$w_\xi(P_0) > 0, \qquad w_\eta(P_0) > 0.$$

The same inequalities continue, therefore, to hold on $\mathring{T}_1 - \gamma_1$. It follows that $w(P)$ is strictly increasing in T_1, as a function of ξ and of η; since $w(P) = K$ on γ_1, we find that

$$w(P) > K \quad \text{on} \quad T_1 - \gamma_1. \tag{2.16}$$

We have now, by the (*elastic*) *impact condition* (2.7),

$$y_t^+(x_0, t_0) = -y_t^-(x_0, t_0) < 0$$

and the integral $z(x, t) = 2K - w(x, t)$ satisfies the conditions

$$z(x, t)\,|_{\gamma_2} = y(x, t)\,|_{\gamma_1}, \qquad z_t(x, t)\,|_{\gamma_2} = y_t^+(x, t)\,|_{\gamma_1}.$$

Since, by (2.16), we have $z(x, t) \leqslant K$ on T_1, we deduce (2.15) from the extension law: $y(P) = z(P)$. This is also the only extension that can be made on T_1; moreover,

$$y(P) < K \quad \text{on} \quad T_1 - \gamma_1. \tag{2.17}$$

Completely analogous relations can be obtained for T_2, and we again have

$$z(P) = y(P) < K \quad \text{on} \quad T_2 - \gamma_2. \tag{2.18}$$

Let us now consider the domain T_3. We can see that, here, $y(P)$ coincides with the integral $z(P)$, satisfying the Darboux condition:

$$z(P) = y(P) \quad \text{on} \quad N_1L_2 \cup N_1N_2.$$

In fact, $z(N_1) = y(N_1) = K$, $z(P) = y(P) < K$ on the remaining part of N_1L_2, $z(P) = y(P) \leqslant K$ on N_1N_2. It follows, on the whole of T_3, that

$$z(P) = y(P') + y(P'') - y(N_1) < K,$$

which implies

$$z(P) = y(P) \qquad (< K \text{ on } T_3 - \{N_1\}). \tag{2.19}$$

Following the same procedure, and bearing in mind that $y(P_3) = y(P_1) = y(M_2) = K$, we can calculate $y(P)$ in the remaining domains T_4, T_5, T_6 and we conclude that

$$y(P) < K \quad \text{on} \quad T_1 \cup \ldots \cup T_6 - \lambda. \tag{2.20}$$

Observe, finally, that $y(P)$ is an *integral of the homogeneous vibrating string equation in the whole domain* $U = 0lN_2FM_1$, with the sole exception of the impact arcs γ_1, γ_2.

Furthermore, we have

$$y(P) < K \quad \text{on the } \textit{polygonal line} \quad \Gamma_1 = M_1FN_2. \tag{2.21}$$

We can, therefore, repeat the preceding procedure in the domain Δ above this polygonal line, considering the integral $w_1(P)$, which vanishes on $x=0$ and $x=l$, such that

$$w_1(P)\,|_{\Gamma_1}=y(P)\,|_{\Gamma_2}.$$

A *second* line of influence is thus obtained, and so on.

One can prove that the solution $y(P)$ of Problem I can be calculated, by the given procedure, on the whole of Δ. In fact, it can be shown that there exists $\delta>0$ such that the distance between two successive lines of influence is $\geqslant\delta$.

It can be verified, in an obvious way, that the function $y(P)$ defined in Δ by (2.15), (2.17), (2.18)... is actually an admissible function: the *existence and uniqueness theorem* is therefore proved.

Remark I One can verify that the classical *energy equality holds*:

$$\int_0^l \{y_x(x,t)+y_t^2(x,t)\}\,dx=\int_0^l \{\varphi'^2(x)+\psi^2(x)\}\,dx.$$

Remark II It has been proved, by Citrini [4], that the number of lines of influence, if there are any, is *infinite*, both in the elastic and in the partially elastic case. The same number is *finite* in the inelastic case.

3 The problem of the moving point-shaped obstacle

The solution of the problem will be reduced, as in the preceding case, to that of elementary problems. To those of Cauchy, Darboux and Goursat, we must add now another problem, one which we shall call the Π problem. By solving this problem, we can calculate the *reaction of the obstacle.*

As we shall prove, the problem of the point-shaped obstacle has *one and only one solution,* without imposing any condition on the nature of the impact against the obstacle (elastic, partially elastic, inelastic). This makes a notable difference from the case of the impact against a wall.

Let us define, on the rectangle $R=0LNH=\{0\leqslant\xi\leqslant l, 0\leqslant\eta\leqslant h\}$ of the (ξ,η) plane (Fig. 13), the following *Π problem.* Let Λ be a (*time-like* oriented) line, of equation

$$\eta=g(\xi),\qquad 0\leqslant\xi\leqslant l,\tag{3.1}$$

where $g(\xi)$ is a *continuous, strictly-increasing function,* with

$$g(0)=0,\qquad g(l)=h.$$

Moreover, let $\alpha(P)$ be a *continuous function,* defined on Λ, such that

$$\alpha(0)\geqslant 0.\tag{3.2}$$

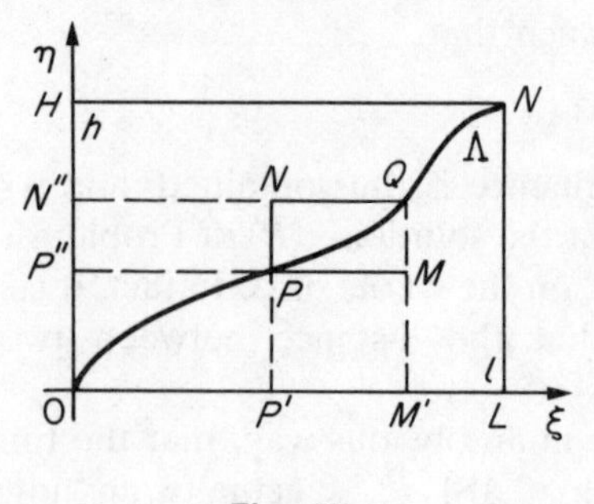

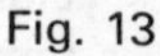
Fig. 13

We intend to calculate, on R, a function $\Gamma(P)$ which satisfies the following conditions (Π problem):

1. $\Gamma(P) \in C^0(R)$,
2. $\Gamma(P) = 0$ on $0L \cup 0H$,
3. $\Gamma(P) \geqslant \alpha(P), \qquad \forall P \in \Lambda,$ (3.3)
4. $\Gamma_{\xi\eta} \geqslant 0$,
5. $\operatorname{Supp} \Gamma_{\xi\eta} \subseteq \{P \in \Lambda \mid \Gamma(P) = \alpha(P)\}$.

In Conditions 4 and 5, the derivative $\Gamma_{\xi\eta}$ is obviously a *distribution* $\in \mathscr{D}'(\mathring{R})$; by Condition 5, $\Gamma(P)$ satisfies the homogeneous equation

$$\Gamma_{\xi\eta} = 0 \text{ on the open set } (\mathring{R} - \operatorname{Supp} \Gamma_{\xi\eta}) \supseteq \mathring{R} - \Lambda. \tag{3.4}$$

We shall prove that the Π *problem has one and only one solution.*

Let us observe, first of all, that Condition 4 is equivalent to imposing the condition

$$\Gamma(C) - \Gamma(D) - \Gamma(B) + \Gamma(A) \geqslant 0 \tag{3.5}$$

for every characteristic rectangle $S = ABCD \subseteq R$. Moreover, by Conditions 1 and 5,

$$(\operatorname{Supp} \Gamma_{\xi\eta}) \cap \mathring{S} = \emptyset \Rightarrow \Gamma(C) - \Gamma(D) - \Gamma(B) + \Gamma(A) = 0. \tag{3.6}$$

Setting $P = (\xi, \eta)$, $P' = (\xi, 0)$, $P'' = (0, \eta)$, we have, by Condition 2 and Eqn (3.5) (with $B = P'$, $D = P''$),

$$\Gamma(P) \geqslant 0. \tag{3.7}$$

Finally, $\Gamma(P)$ is an *increasing* function of P on Λ. We have, in fact, by Conditions 2, 5 and Eqn (3.5) (cf., Fig. 13),

$$0 \leqslant \Gamma(Q) - \Gamma(M) - \Gamma(N) + \Gamma(P) = \Gamma(Q) - \Gamma(P), \tag{3.8}$$

since

$$\Gamma(M) - \Gamma(P) = \Gamma(M') - \Gamma(P') = 0, \qquad \Gamma(N) - \Gamma(P) = \Gamma(N'') - \Gamma(P'') = 0.$$

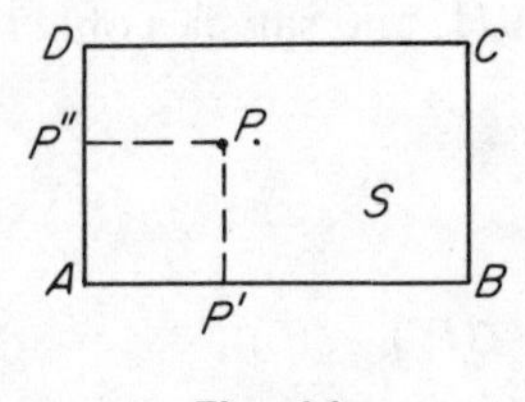

Fig. 14

Let us now prove the *uniqueness* of $\Gamma(P)$.

Assume that there exist two solutions, $\Gamma(P)$ and $\tilde{\Gamma}(P)$. By Conditions 2, 5, $\Gamma(P)=\tilde{\Gamma}(P)$ on $\Lambda \Rightarrow \Gamma(P)=\tilde{\Gamma}(P)$ on R; hence, it is sufficient to prove that $\Gamma(P)=\tilde{\Gamma}(P)$ on Λ. Assume the contrary, i.e., $\Gamma(C)>\tilde{\Gamma}(C)$, where $C \in \mathring{\Lambda}$. Since $\Gamma(0)=\tilde{\Gamma}(0)=0$, there exists an arc $\widehat{AC}$, such that $\Gamma(A)=\tilde{\Gamma}(A)$ and $\Gamma(P)>\tilde{\Gamma}(P)\geqslant \alpha(P)$ for $A<P\leqslant C$. This implies that, for $\Gamma(P)$, $\operatorname{Supp} \Gamma_{\xi\eta} \cap \mathring{S}=\emptyset$ and Eqn (3.6) holds; therefore, $\Gamma(C)-\Gamma(A)=(\Gamma(D)-\Gamma(A))+(\Gamma(B)-\Gamma(A))=0$, which is absurd, since (by Eqn (3.8))

$$\Gamma(A)=\tilde{\Gamma}(A)\leqslant \tilde{\Gamma}(C)<\Gamma(C).$$

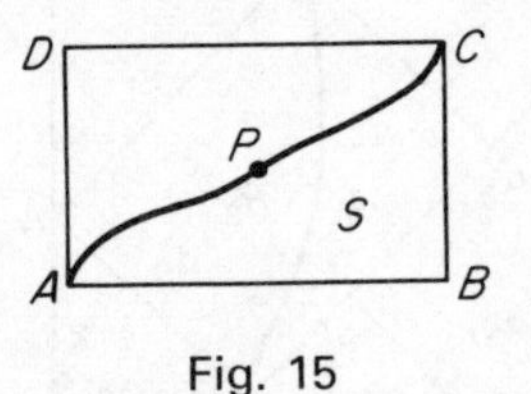

Fig. 15

The *existence* of the solution is easily proved too. Let us denote by Λ_P the arc of Λ with ends 0 and P and set

$$\alpha^+(P)=\frac{\alpha(P)+|\alpha(P)|}{2}, \qquad (P\in\Lambda), \tag{3.9}$$

$$\mathscr{U}(\xi, g(\xi))=\max_{M\in\Lambda_P} \alpha^+(M).$$

We then have, on the whole of R, the *explicit formula*

$$\Gamma(\xi,\eta)=\Gamma(P)=\begin{cases}\mathscr{U}(\xi, g(\xi)) & \text{for} \quad g(\xi)\leqslant \eta \leqslant h,\\ \mathscr{U}(g^{-1}(\eta),\eta) & \text{for} \quad g^{-1}(\eta)\leqslant \xi\leqslant l,\end{cases} \tag{3.10}$$

where g^{-1} is the inverse function of g.

Remark We can generalize the Π problem if we replace Condition 2 by the condition

$$2'.\ \Gamma(P)=\zeta(P) \quad \text{on} \quad 0L\cup 0H, \tag{3.11}$$

where $\zeta(P) \in C^0(0L \cup 0H)$ and satisfies only the (necessary) condition

$$\zeta(0) \geqslant \alpha(0). \tag{3.12}$$

Setting, $\forall P \in R$,

$$\bar{\Gamma}(P) = \Gamma(P) - \{\zeta(P') + \zeta(P'') - \zeta(0)\}, \tag{3.13}$$

we have

$$\bar{\Gamma}_{\xi\eta} = \Gamma_{\xi\eta}, \qquad \bar{\Gamma}(P) = 0 \quad \text{on} \quad 0L \cup 0H. \tag{3.14}$$

Let us now calculate $\bar{\Gamma}(P)$ by imposing Conditions 1, 2, 3, 4, 5, where $\alpha(P)$ is replaced by $\alpha(P) - \{\zeta(P') + \zeta(P'') - \zeta(0)\}$. Therefore, $\bar{\Gamma}(P)$ exists and is unique. We can easily verify that the function $\Gamma(P)$, defined by (3.13), satisfies Conditions 1, 2, 3, 4, 5.

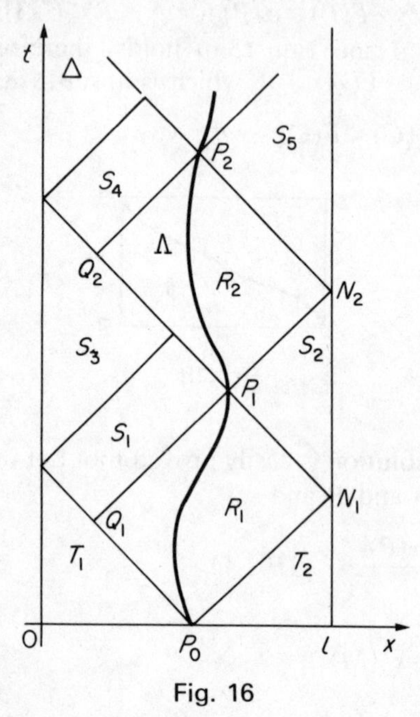

Fig. 16

We now apply the preceding result in order to solve the problem defined by (1.2), (1.3), (1.4), (1.10).

Let $w(P)$ be the solution of the *free problem*. Setting

$$y(P) = w(P) + \Gamma(P) \qquad (P \in \Delta), \tag{3.15}$$

we have to find a function $\Gamma(P)$ defined by the following conditions:

(i) $\Gamma(P) \in C^0(\Delta)$;

(ii) $\Gamma(P) = 0$ on the boundary $\partial\Delta$;

(iii) $\Gamma(P) \geqslant \alpha(P) - w(P)$ on Λ; (3.16)

(iv) $\Gamma_{\xi\eta} \geqslant 0$;
(v) $\text{Supp}\, \Gamma_{\xi\eta} \subseteq \{P \in \Lambda \mid \Gamma(P) = \alpha(P) - w(P)\}$;
(vi) $\Gamma(P) = 0$ on $T_1 \cup T_2$.

It is obvious that, if $\Gamma(P)$ satisfies Conditions (i) to (vi), then the function $y(P)$ given by (3.15) will be a solution of our problem; observe that Condition (vi) means that the impulses generated by the impact of the string against the obstacle do not influence the solution $y(P)$ in the exterior of the forward characteristic semicone with vertex at $P_0(\lambda(0), 0)$.

Let us prove that $\Gamma(P)$ *exists on all* Δ *and is unique*.

In fact, we know $\Gamma(P)$, with value zero, on the inferior edges P_0N_1 and P_0Q_1 of the rectangle R_1; hence, we can calculate $\Gamma(P)$ on R_1 by solving a Π problem. We then obtain $\Gamma(P)$ on S_1, S_2, S_3 by solving the corresponding Darboux and Goursat problems for the equation $\Gamma_{\xi\eta} = 0$. Therefore, the values of $\Gamma(P)$ are known on the edges P_1N_2 and P_1Q_2 of the rectangle R_2, and we can calculate $\Gamma(P)$ on R_2 by solving a Π problem (with non-zero values on the edges). In such a way, we obtain the function $\Gamma(P)$ on all Δ and (3.15) gives *the unique solution of the problem. The reaction of the obstacle is the distribution* $\Gamma_{\xi\eta}$.

We add, finally, that an energy equation can also be proved for the problem studied here (cf., Citrini [4]).

References

1 Lions, J.-L., *Quelques Méthodes des Résolution de Problèmes aux Limites Nonlinéaires*, Dunod, Paris, 1969.

2 Amerio, L. and Prouse, G., Study of the motion of a vibrating string against an obstacle, *Rend. di Mat. (2)*, **8**, Serie VI, 563–585, 1975.

3 Amerio, L., Su un problema di vincoli unilaterali per l'equazione non omogenea della corda vibrante, *IAC (CNR) Pubbl.*, Serie III, no 109, 1976; On the motion of a string vibrating through a moving ring with a continuously variable diameter, *Rend. Acc. Naz. dei Lincei*, **62,** Serie VIII, 134–142, 1977.

4 Citrini, C., Sull'urto parzialmente elastico o anelastico di una corda vibrante contro un' ostacolo, *Rend. Acc. Naz. dei Lincei*, **59,** Serie VIII, 368–376, 667–676, 1975; The energy theorem in the impact of a string vibrating against a point-shaped obstacle, *Rend. Acc. Naz. dei Lincei*, **62,** Serie VIII, 143–149, 1977.

Remark The results at 2, 3 can be considerably improved (Amerio, L., Continuous solutions of the problem of a string vibrating against an obstacle, *Rend. Sem. Mat. Univ. di Padova*, A unilateral problem for a nonlinear vibrating string equation, *Rend. Acc. Naz. dei Lincei*—to appear).

Professor L. Amerio,
Istituto Matematico del Politecnico di Milano,
Piazza Leonardo da Vinci 32,
Milano,
Italy

B. Barberis and D. Galletto

Foundations of Newtonian cosmology†

Dedicated to Lydia Sarti

Abstract

Contrary to the opinion of many authors, we prove that, in a strictly Newtonian context, it is possible to develop a cosmological theory not only in the bounded case but also in the case in which the Universe is considered unbounded. Moreover, contrary to all the analogous treatments developed until now, this theory (which is an organic synthesis of previous articles) is developed without resorting to the Newtonian theory of gravitation. On the contrary, we show that the foundations of the Newtonian theory of gravitation (in particular Newton's law of gravitation) and Hubble's law necessarily follow from the simple hypotheses that the fluid $\mathcal{U}$, which we assume as a model of the Universe, is homogeneous and has an isotropic behaviour with respect to any element O (namely that $\mathcal{U}$ is such that all its elements have purely radial velocities with respect to the frame of reference with origin O and which is in translational motion with respect to inertial frames). Moreover, all the frames having the elements of $\mathcal{U}$ as origins and which are in translational motion with respect to inertial frames (co-moving frames) are equivalent to one another in the sense that $\mathcal{U}$ has the same behaviour, from both the kinematical and the dynamical points of view, with respect to them.

The significance of these results is evident. Let $\mathcal{R}_O$ denote the co-moving frame with origin at the centre of mass O of our galaxy. Then from the hypotheses, verified by astronomical observation, that on a large scale the Universe is homogeneous and that its behaviour is isotropic with respect to $\mathcal{R}_O$, there necessarily follow: (a) Hubble's law; (b) Newton's law of gravitation, in the sense that the action exerted by any typical galaxy (i.e., with purely radial recessional motion) on another one is necessarily expressed by Newton's law of gravitation. With respect to any co-moving frame $\mathcal{R}_O$, the resultant acting on any galaxy P is equal to the

† Research supported in part by GNFM of the Italian Council for Research.

gravitational action exerted on P by that part of the Universe contained in the sphere with centre O and radius $|OP|$. All co-moving frames are equivalent to one another, in the sense that the Universe has, from both the kinematical and the dynamical points of view, the same behaviour with respect to them. In particular, the co-moving frame with origin at the centre of mass of our galaxy can be considered as inertial.

In the context of Newtonian mechanics, it follows from the above hypotheses, suggested by astronomical observation, that the action exerted between any two typical galaxies is necessarily expressed by Newton's law of gravitation; hence, the whole process of the expansion of the Universe is ruled by this law. This result is the first verification on an extragalactic scale of the validity of the Newtonian theory of gravitation.

We include a short critical-historical review of previous papers on Newtonian cosmology in order to emphasize the significance of our results.

1 Introduction

The first attempt to develop a cosmological theory dates back only to the end of the last century and it was made by the astronomer Seeliger.† This attempt failed not in consequence of its Newtonian origin, but because of the assumption made that, on a large scale, the Universe, believed spatially unbounded, endless in time and with a stellar density on the average constant, was static.

We find this conviction again in the analysis of the cosmological consequences of general relativity developed by Einstein [2] in 1917.

The first observation of the spiral nebulae motion dates back to 1912, when it was not yet clear if spiral nebulae were inside or outside our galaxy. During the subsequent years, astronomers discovered that almost all spiral nebulae move with recessional motion with respect to our galaxy. In 1924, E. P. Hubble proved that spiral nebulae are outside our galaxy; in 1929, he announced his famous law: the velocity of galaxies is radial and it is proportional to their distance from our galaxy.

The twenty or so years subsequent to the formulation of general relativity were spent in the development of relativistic cosmology, while Newtonian cosmology was completely neglected.

Only in 1934, after the discovery that the Universe, far from being static, is expanding, did we have a revival of Newtonian cosmology by Milne and McCrea [3, 4]. They proved that, by means of a suitable interpretation of their results, Newtonian cosmology presents a deep analogy with relativistic cosmology to such a point that in 1952 Bondi (*see*

† For an analysis of Seeliger's attempt *see* North [1], Chap. 2, Para. 1.

[5], **9.1**) wrote about Milne and McCrea's work: 'This new formulation of the old subject is highly interesting, since in spite of our present denial of many of the premises of Newtonian theory, it reveals many of the essential features of relativistic cosmology without the mathematical complexity.'

As a basis for a Newtonian formulation of cosmology, Milne and McCrea supposed the Universe $\mathcal{U}$ unbounded, in other words extended to all space. They implicitly assumed that the frame of reference $\mathcal{R}_O$ introduced by them – which has its origin at an element O of $\mathcal{U}$ and which is in translational motion with respect to inertial frames – was inertial. Moreover, considering any other element P of $\mathcal{U}$, they assumed that the part of $\mathcal{U}$ external to the material sphere S_{OP} with centre at O and radius $|OP|$ gives no contribution to the motion of P. They made both these assumptions without any justification, and, hence, without worrying whether they are valid. Then according to the Newtonian theory of gravitation and by considering $\mathcal{U}$ as if it were reduced to that part of it contained in the sphere S_{OP}, they wrote down the following equation:

$$\frac{\mathrm{d}^2 OP}{\mathrm{d}t^2} = -\frac{4}{3}\pi k \mu OP, \tag{1.1}$$

where k is the gravitational constant and μ is the average density of $\mathcal{U}$ regarded as homogeneous.

In 1954, Layzer [6] denied the possibility of formulating a Newtonian cosmological theory with $\mathcal{U}$ unbounded.† He asserted that 'the theory of Milne and McCrea is not self-consistent, being incompatible with the Newtonian conception of gravitation, which it tries to incorporate', and he made the unsustainable statement that, since the distribution of matter in the Universe defines no preferred direction in space, 'the specific gravitational force vanishes everywhere'. Consequently, 'the expansion [of the Universe] is unaccelerated'. Nevertheless, Layzer emphasized that, in the case of $\mathcal{U}$ unbounded, from the general theory of relativity, it follows that 'the motion of a particle P is given without approximation by Newton's theory if the gravitational influence of the matter outside the particle's co-moving sphere [$\cdots$]S_{OP} is neglected'.‡

The position of Layzer is quite contradictory, because he denied the possibility of formulating a Newtonian cosmological theory with $\mathcal{U}$ unbounded while stating, at the same time, the above theorem, which gives rise to Eqn (1.1), i.e., to the starting point of Milne and McCrea's theory.

This notwithstanding, under the influence of Layzer's criticism,

† This possibility was denied also by North in his fine treatise [1] (*see* Chap. 8, Para. 9) in which Layzer's point of view was fully accepted. Layzer confirmed his point of view in [7, 8].

‡ This is also the conviction of Weinberg who in [9], p. 475 wrote: 'We need general relativity to justify the neglect of all the matter outside the sphere S_{OP}'.

McCrea, in 1955 [10, 11], renounced the possibility of a rigorous formulation, within a strictly Newtonian context, of a Newtonian cosmological theory with $\mathcal{U}$ unbounded.

Contrary to the assertion made by Layzer (and also by Bondi (*see* [5], **9.3**)) and accepted by McCrea, it is, in fact, possible to develop a 'strictly' Newtonian cosmological theory not only in the bounded case but also if $\mathcal{U}$ is unbounded.†

The analysis of the foundations of the theory presented here is an organic and rigorous synthesis of most of the results contained in Galletto [16–21] and in Barberis and Galletto [15, 22, 23].

The starting point is simply given by the principles of Newtonian mechanics and by the indications given by astronomical observations. These indications say that the Universe is, on a large scale, homogeneous and that its behaviour is isotropic with respect to the frame – which we shall still indicate by $\mathcal{R}_O$ – which has the origin at the centre of mass O of our galaxy and which is in translational motion with respect to inertial frames. By saying that the behaviour of $\mathcal{U}$ is isotropic with respect to $\mathcal{R}_O$, we mean that, with respect to $\mathcal{R}_O$, the velocity of any galaxy is strictly radial.

These are the assumptions we use, besides, obviously, the principles of Newtonian mechanics. Therefore, in this treatment, we do not assume the Newtonian theory of gravitation, in particular Newton's law of gravitation, as a starting point; on the contrary we derive the foundations of this theory from the above assumptions. This is completely different from all other treatments,‡ which assume the Newtonian theory of gravitation as a starting point.

In particular, the above mentioned theorem deduced by Layzer, working within the framework of general relativity, is here deduced from the above assumptions in a strictly Newtonian context, without even having to resort to the Newtonian theory of gravitation.

2 Remarks on actions at a distance

In Newton's law of gravitation, the linear dependence on masses is postulated. Instead, it can be proved that this assumption necessarily

† For the sake of brevity, we shall not mention the attempt made in 1955–1956 by Heckmann and Schücking [12, 13] to develop a Newtonian cosmological model with $\mathcal{U}$ unbounded and to avoid Layzer's criticism. This attempt is rather complex and gives rise to criticism (*see* Layzer [7, 14]). For an analysis of Heckmann and Schücking's treatment, *see* Barberis and Galletto [15].

‡ *See*, for instance, in addition to Milne [3] and Milne and McCrea [4], Heckmann and Schücking [12, 13, 24], Bondi [5], Callan, Dicke and Peebles [25], Zeldovich [26], Heckmann [27], Davidson and Evans [28], Mavridès [29], etc., with the exception of the tautological attempt made by Landsberg [30]. For more details on Landsberg's paper [30], *see* the footnote on p. 31 of this paper.

follows from the principle of superposition of simultaneous forces and from Newton's third law of dynamics.† In other words, from these principles it follows that the action at a distance exerted by a point P_1 with mass m_1 on a point P_2 with mass m_2 is of the following type:‡

$$-m_1 m_2 f(|P_1P_2|)\frac{P_1P_2}{|P_1P_2|},$$

where the minus sign is suggested by experience, while $f(|P_1P_2|)$ is a positive function of the distance $|P_1P_2|$, which we shall assume to be continuous when $P_1 \neq P_2$.

In the case of a bounded continuous medium $\mathscr{C}$, the resultant $\mathbf{g}(P, t)$, referred to the unit of mass, of the actions at a distance exerted on any element P of $\mathscr{C}$ by all the other elements Q is expressed by

$$\mathbf{g}(P, t) = -\int_C \mu(Q, t) f(|QP|)\frac{QP}{|QP|}\,\mathrm{d}C,$$

where C is the configuration assumed by $\mathscr{C}$ at the present time t and μ is its density. We call $\mathbf{g}(P, t)$ *the resultant of the specific actions at a distance exerted on P by* $\mathscr{C}$.

In the case in which $\mathscr{C}$ is homogeneous ($\mu = \mu(t)$), we have

$$\mathbf{g}(P, t) = -\mu(t)\int_C f(|QP|)\frac{QP}{|QP|}\,\mathrm{d}C \tag{2.1}$$

and it follows that

2.1 If the continuous medium $\mathscr{C}$ *is homogeneous, the resultant* $\mathbf{g}(P, t)$ *of the specific actions at a distance exerted on P by* $\mathscr{C}$ *is a linear and homogeneous function of the density* $\mu(t)$.

3 Kinematical behaviour of the fluid $\mathscr{U}$

We shall represent the Universe by a perfect, homogeneous, inert§ and isolated fluid with negligible internal pressure. We shall still denote it by $\mathscr{U}$. It is the so-called cosmological fluid considered in cosmology. Let μ denote its density, O one of its elements, $\mathscr{R}_O$ the frame of reference having O as origin and which is in translational motion with respect to inertial frames. We shall call such a frame a *co-moving frame.* We shall assume that the behaviour of $\mathscr{U}$ with respect to $\mathscr{R}_O$ is isotropic, namely

† This assertion is mentioned in [16], 2. Its complete and rigorous proof is given in [22].

‡ *See* [22]. In [22], as also in [19], 2 the function f is allowed to depend explicitly on time in order not to exclude the possibility that the physical strength of action at a distance could change in time. In order to remain within the limits of classical mechanics, such a possibility is excluded in this treatment. Anyhow, even admitting this possibility (as is shown in [19], 2), we arrive at the same results stated in section 4.

§ Namely that it is electrically neutral, etc.

that the velocities of all elements of $\mathcal{U}$ are radial with respect to the frame $\mathcal{R}_O$.†

The hypothesis that $\mathcal{U}$ is homogeneous implies that μ is a function of time t only. Using a dot to indicate differentiation with respect to time, we define

$$h(t) = -\frac{1}{3}\frac{\dot{\mu}}{\mu}. \tag{3.1}$$

Therefore, denoting by r the distance from O of any element P of $\mathcal{U}$, the continuity equation becomes

$$\frac{\partial}{\partial r}(r^2\dot{r}) = 3r^2h.$$

From this equation it follows that

$$\dot{r} = h(t)r + \frac{f(t)}{r^2}.$$

The principle of conservation of matter implies that the function $f(t)$ must be identically zero (*see* [18], 1 and [23], 2). So we have

$$\frac{\mathrm{d}OP}{\mathrm{d}t} = h(t)OP, \tag{3.2}$$

which is Hubble's law and which is also satisfied, as is easy to see (*see*, for instance, [16], 3), with respect to any other co-moving frame.‡

Therefore, we have

3.1 Hubble's law is a kinematical consequence of the assumptions that $\mathcal{U}$ is homogeneous and that its behaviour is isotropic with respect to the frame $\mathcal{R}_O$. Hubble's law is satisfied with respect to every co-moving frame.

† With this, we obviously exclude the possibility of a rotating $\mathcal{U}$ with respect to inertial frames. A discussion about the significance of these frames will be made in a subsequent paper.

‡ We observe that Hubble's law necessarily implies that the function $h(t)$ has the form of (3.1). In fact, let us suppose O interior to $\mathcal{U}$ and that P is chosen in such a way that the spherical surface with centre O and radius $r = |OP|$ results interior to $\mathcal{U}$. As a consequence of Hubble's law, this spherical surface is a material surface and, therefore, $\frac{4}{3}\pi\mu r^3$ is a constant. By differentiating, it follows that

$$\dot{r} = -\frac{1}{3}\frac{\dot{\mu}}{\mu}r,$$

which, comparing with Hubble's law, states that $h(t)$ must be expressed by (3.1).

This result allows us to conclude that the hypotheses made about $\mathcal{U}$ – and suggested by the indications given by astronomical observations – imply that $\mathcal{U}$ satisfies the so-called cosmological principle, namely that $\mathcal{U}$ is homogeneous and isotropic. Moreover, from (3.2), it follows that

$$\frac{d^2OP}{dt^2}=(\dot{h}+h^2)OP, \tag{3.3}$$

and, as above, we have

3.II The law expressed by relation (3.3) *is satisfied not only with respect to the frame $\mathcal{R}_O$, but also with respect to every co-moving frame.*

Let us now suppose that $\mathcal{U}$ is bounded and let G be its centre of mass. It is easy to see that (3.2) (and therefore (3.3)) is also satisfied with respect to the frame $\mathcal{R}_G$, which is in translational motion with respect to $\mathcal{R}_O$. Since $\mathcal{U}$ is isolated, the frame $\mathcal{R}_G$ is inertial.

4 Consequences of Newton's second law of dynamics if $\mathcal{U}$ is bounded

If $\mathcal{U}$ is bounded, Newton's second law of dynamics – written with respect to the frame $\mathcal{R}_G$ for the element P and referred to the unit of mass – is

$$\frac{d^2GP}{dt^2}=\mathbf{g}(P,t), \tag{4.1}$$

where the assumptions made about $\mathcal{U}$ imply that $\mathbf{g}(P,t)$ is given by the resultant of the specific actions at a distance exerted on P by $\mathcal{U}$. According to the remarks made in section 2, $\mathbf{g}(P,t)$ is expressed by (2.1), where C is now the configuration assumed by $\mathcal{U}$ at the present time.

From (3.3), 3.II and (4.1) we obtain

$$\mathbf{g}(P,t)=(\dot{h}+h^2)GP \tag{4.2}$$

and, therefore, recalling the result 2.I, it follows that $\dot{h}+h^2$ must be a linear and homogeneous function of $\mu(t)$:

$$\dot{h}+h^2=\kappa(t)\mu(t). \tag{4.3}$$

In the whole time-interval during which the function $h(t)$ preserves the same sign that it has at the time under consideration (for instance, positive when $\mathcal{U}$ is expanding), t is in one to one correspondence with $\mu(t)$. Therefore, it follows that if κ were a function of t, it could be expressed as a function of μ in the above mentioned interval. But this is in contrast with the result – stated above – that $\dot{h}+h^2$ is a linear function of μ. Therefore, in such a time interval, κ is constant.

The above remark can be repeated for every interval in which h has the same sign.† Therefore we can conclude that

4.I The factor κ is constant.

(*See also* [19], 2.)

Therefore, from (2.1), (4.2) and (4.3), we obtain

$$\int_C f(|QP|)\frac{QP}{|QP|}\,\mathrm{d}C = -\kappa GP. \tag{4.4}$$

However, it is not possible to exclude *a priori* a dependence of κ upon the shape of the configuration C, a shape which is the same for all the configurations assumed by $\mathscr{U}$ in time, because they are homothetic (with pole G). In other words, the constant κ is the same for all the above mentioned configurations.

Let us now assume that $\mathscr{U}$ is convex. Let us take P to be an interior element. Finally, let us suppose, for the moment, that $\mathscr{U}$ is reduced to its part $\mathscr{U}_P$ which also has G as its centre of mass, has P on its boundary and has its configuration homothetic to C with pole G at the time under consideration. The configuration of $\mathscr{U}_P$ will be denoted by C_{GP}.

In such a case, as follows from (3.3) and (2.1), Newton's equation of dynamics, written for P, implies

$$(\dot{h}+h^2)_{C_{GP}}GP = -\mu\int_{C_{GP}} f(|QP|)\frac{QP}{|QP|}\,\mathrm{d}C, \tag{4.5}$$

with an obvious meaning for $(\dot{h}+h^2)_{C_{GP}}$.

From the same considerations which led us to relation (4.3), and taking into account the remarks made about κ, we obtain

$$(\dot{h}+h^2)_{C_{GP}} = \kappa\mu(t).$$

Therefore, by virtue of (4.4) and (4.5), we have the following relation:

$$\int_C f(|QP|)\frac{QP}{|QP|}\,\mathrm{d}C = \int_{C_{GP}} f(|QP|)\frac{QP}{|QP|}\,\mathrm{d}C, \tag{4.6}$$

which, recalling (2.1), states that

4.II Only that part of $\mathscr{U}$ whose configuration is C_{GP} contributes to the determination of $\mathbf{g}(P, t)$.

† The number of these intervals could be one which occurs only when $\mathscr{U}$ is indefinitely expanding.

In other words that part of $\mathscr{U}$ which is external to the material surface corresponding to the boundary of C_{GP} gives no contribution to the integral which appears in (4.4).

From 4.I, 4.II and (4.4), it is possible to deduce the following result (*see* [19], 3, 4):

4.III The configuration assumed by $\mathscr{U}$ is spherical.

Therefore, from (4.6) and (4.4), there follows

$$\int_C f(|QP|)\frac{QP}{|QP|}\,\mathrm{d}C=\int_{S_{GP}} f(|QP|)\frac{QP}{|QP|}\,\mathrm{d}C=-\kappa GP, \tag{4.7}$$

where S_{GP} is the sphere with centre at G and radius $|GP|$. Thus, the result 4.II becomes

4.IV Only that part of $\mathscr{U}$ which is contained in the material sphere with centre G and radius $|GP|$ contributes to the determination of $\mathbf{g}(P, t)$.

In the Newtonian theory of gravitation, this result is a consequence of Newton's law of gravitation. Instead, here we have obtained it directly from the hypotheses made about $\mathscr{U}$, without having recourse to the Newtonian theory of gravitation, in particular to Newton's law.

5 Further consequences of Newton's second law of dynamics

With respect to the frame $\mathscr{R}_G$, from (4.1), (2.1) and (4.7) we obtain the following expression for Newton's equation of motion of P:

$$\frac{\mathrm{d}^2GP}{\mathrm{d}t^2}=-\mu\int_{S_{GP}} f(|QP|)\frac{QP}{|QP|}\,\mathrm{d}C. \tag{5.1}$$

From (5.1) and (3.3) there follows the relation

$$-\mu\int_{S_{GP}} f(|QP|)\frac{QP}{|QP|}\,\mathrm{d}C=(\dot{h}+h^2)GP. \tag{5.2}$$

With respect to the frame $\mathscr{R}_O$, Newton's equation of motion is

$$\frac{\mathrm{d}^2OP}{\mathrm{d}t^2}=\mathbf{g}^{(r)}(P, t), \tag{5.3}$$

where $\mathbf{g}^{(r)}(P, t)$ now denotes the resultant of the fictitious specific force acting on P and of the specific actions at a distance exerted on P by the elements Q of $\mathscr{U}$.

Bearing in mind (3.3), there follows from (5.3) the relation

$$\mathbf{g}^{(r)}(P, t) = (\dot{h} + h^2)OP. \tag{5.4}$$

The comparison of (5.4) with (5.2) implies

$$\mathbf{g}^{(r)}(P, t) = -\mu(t) \int_{S_{OP}} f(|QP|) \frac{QP}{|QP|} \, dC. \tag{5.5}$$

In other words, we have

5.I $\mathbf{g}^{(r)}(P, t)$ *is equal to the resultant of the specific actions at a distance exerted on P by the elements of the material sphere*† S_{OP} *with centre at O and radius* $|OP|$.

This result is completely analogous to the result 4.IV, which is, in fact, a particular case of the present one.

With respect to the frame $\mathcal{R}_O$, Newton's equation of motion for P is therefore expressed by

$$\frac{d^2 OP}{dt^2} = -\mu \int_{S_{OP}} f(|QP|) \frac{QP}{|QP|} \, dC, \tag{5.6}$$

which is completely analogous to (5.1).

Therefore we conclude that

5.II *Newton's equation of motion written for any element P of* $\mathcal{U}$ *preserves its form – given by (5.6) – unchanged, with respect to every co-moving frame. This is the form that Newton's equation assumes in the inertial frame* $\mathcal{R}_G$.

(*See also* [20], 3.)

Thus we have

5.III *For the fluid* $\mathcal{U}$*, both from the kinematical and the dynamical points of view, there is no difference between the inertial frame* $\mathcal{R}_G$ *and every co-moving frame, in the sense that Hubble's law and the equation of motion for every point P of* $\mathcal{U}$ *always have, respectively, the forms (3.2) and (5.6) with respect to all these frames.*

In other words, we have,

5.IV $\mathcal{U}$ *has the same behaviour in every co-moving frame and this is the behaviour that* $\mathcal{U}$ *would have in an inertial frame.*

However, by this we do not want at all to say that two such frames are in uniform translational motion relative to each other. We shall specify

† Obviously, this sphere is considered to be homogeneous with density $\mu(t)$.

this fact more fully in section 8, where we shall also clarify the apparent paradox which seems to follow from the results we have obtained. Here, we just recall that the vector $\mathbf{g}^{(r)}(P, t)$ – which is given by the resultant of the specific actions at a distance exerted on P by the elements of the material sphere S_{OP} – gives, in reality, the resultant of the specific fictitious force acting on P and of the specific actions at a distance exerted on P by $\mathcal{U}$.

Taking (5.1) into account, and the relation

$$\frac{d^2OP}{dt^2}=\frac{d^2GP}{dt^2}-\frac{d^2GO}{dt^2} \tag{5.7}$$

we have

5.V *With respect to the frame $\mathcal{R}_O$, the specific fictitious force acting on P is given by the opposite of the resultant of the specific actions at a distance exerted on the origin O of $\mathcal{R}_O$ by $\mathcal{U}$.*

(*See also* [16], 3.)

Moreover, bearing in mind (5.1) and (5.3), we have

5.VI *$\mathbf{g}^{(r)}(P, t)$ is equal to the difference between the resultants of the specific actions at a distance exerted by $\mathcal{U}$, respectively, on P and on O.*

More generally, without resorting to the centre of mass G, from the relation

$$\frac{d^2O'P}{dt^2}=\frac{d^2OP}{dt^2}-\frac{d^2OO'}{dt^2}, \tag{5.8}$$

and in view of (5.3), it follows that

$$\mathbf{g}'^{(r)}(P, t)=\mathbf{g}^{(r)}(P, t)-\mathbf{g}^{(r)}(O', t), \tag{5.9}$$

with an obvious meaning for $\mathbf{g}'^{(r)}(P, t)$. With the help of (5.5), relation (5.8) turns into the following one:

$$\int_{S_{O'P}} f(|QP|)\frac{QP}{|QP|}\,dC=\int_{S_{OP}} f(|QP|)\frac{QP}{|QP|}\,dC-\int_{S_{OO'}} f(|QP|)\frac{QP}{|QP|}\,dC,$$

where $S_{O'P}$ and $S_{OO'}$ denote, respectively, the spheres with centres at O' and O and radii $|O'P|$ and $|OO'|$.

6 The explicit equation of motion of $\mathcal{U}$

Bearing in mind the result 4.I and introducing the constant

$$k=-\frac{3\kappa}{4\pi}, \tag{6.1}$$

we obtain from (5.4), (4.3) and (5.3) the Eqn (1.1), which is *the explicit form for the equation of motion of P with respect to the co-moving frame* $\mathcal{R}_O$ and which is the starting point of Milne and McCrea's theory and of all the treatments on Newtonian cosmology.† In the case in which the frame $\mathcal{R}_O$ is the inertial frame $\mathcal{R}_G$, the above equation is

$$\frac{d^2 GP}{dt^2} = -\frac{4}{3}\pi k\mu GP, \tag{6.2}$$

which could have been derived directly from (4.1), (4.2), (4.3) and 4.I without making use of all the subsequent results.‡

In the Newtonian theory of gravitation, Eq. (6.2) – from which Eqn (1.1) follows by virtue of relation (5.7) – is obtained by means of Newton's law of gravitation.

Introducing a function $U(P, t)$ – defined in C – such that

$$\text{grad } U(P, t) = -\mu(t)\int_C f(|QP|)\frac{QP}{|QP|}\,dC \equiv -\mu(t)\int_{S_{OP}} f(|QP|)\frac{QP}{|QP|}\,dC$$

and taking (5.4) into account, it follows from Eqn (5.5) that $U(P, t)$ satisfies *Poisson's equation*, namely

$$\Delta U = -4\pi k\mu.$$

Introducing the constant $M = \frac{4}{3}\pi\mu r^3$, which is the mass of the material sphere S_{OP} with centre at O and radius $r = |OP|$, Eqn (1.1) leads to

$$\ddot{r} = -k\frac{M}{r^2}. \tag{6.3}$$

This equation states the following well-known result of the Newtonian theory of gravitation:

6.I The element P moves as if the mass M of the material sphere S_{OP} were concentrated at the centre O and the specific force exerted on P by O were proportional to M and inversely proportional to the square of its distance from P, i.e., as if the above force were expressed by Newton's law of gravitation.

From (1.1), recalling (3.3), it follows immediately that

$$\dot{h} + h^2 = -\frac{4}{3}\pi k\mu, \tag{6.4}$$

† As we have already said, an attempt to make an exception to this way of proceeding has been made by P. T. Landsberg in [30]. For more details *see* the footnote on p. 31 of this paper.

‡ Then, from (6.2), we could have immediately obtained (1.1) by merely recalling (5.7). In other words Eqn (1.1) is independent of the results 4.II, 4.III, 4.IV, 5.I, etc., which have been deduced in order to point out the behaviour of $\mathcal{U}$ with respect to the co-moving frames and their properties.

which, recalling (3.1) and integrating, gives the law governing the behaviour of h as a function of μ (*see*, for instance, Barberis and Galletto [31], 2). A further integration gives t as a function of μ.

7 Determination of the function $f(|QP|)$

From (5.6) and (1.1) we obtain

$$\int_S f(|QP|)\frac{QP}{|QP|}\,dC=\frac{4}{3}\tau kOP \tag{7.1}$$

which, on the other hand, is implicitly contained in the second relation (4.7).

Equation (7.1) implies the existence of an element $\bar{Q}$ – situated, for reasons of symmetry, on the straight line joining O with P – in relation to which it results that

$$f(|\bar{Q}P|)=\frac{k}{|\bar{Q}P|^2}.$$

From this relation, it follows that $\bar{Q}$ coincides with O. Having chosen P and O arbitrarily, we find that the relation

$$f(|OP|)=\frac{k}{|OP|^2} \tag{7.2}$$

has a general validity. Therefore, the function $f(|QP|)$ is completely determined.

Thus, recalling the remarks made in section 2, we have

7.1 The action at a distance exerted between any two elements of $\mathcal{U}$ is necessarily given by Newton's law of gravitation.

In every previous treatment concerning Newtonian cosmology, this statement is assumed as a starting point, together with the Newtonian theory of gravitation.† In this treatment, on the contrary, this result is deduced, as well as the previous ones, as a consequence of the assumptions made about $\mathcal{U}$.

Owing to (6.1), the time-independence of κ implies the time-independence of k – the gravitational constant.

† It is necessary to remember that, in [30], P. T. Landsberg tried to deduce relation (7.2) from the law (3.2). But this attempt is tautological because, as observed by Barberis and Galletto [23, 32], the assumptions made by Landsberg necessarily imply, as a mathematical consequence independent of (3.2), Newton's law of gravitation.

Besides, from Eqn (6.4), there follows the relation

$$k = -\frac{3(\dot{h} + h^2)}{4\pi\mu}. \tag{7.3}$$

It is necessary to point out that – as shown in [33]† – it is possible to arrive directly at (7.2) starting from (4.4) and (4.6) and, afterwards, starting from (7.2), to deduce the sphericity of $\mathcal{U}$ and all the results contained in the previous sections. In our treatment, however, we have preferred to follow a different method (fully developed in [19]) which, without resorting to (7.2), allows us to show that all the above mentioned results follow directly from the hypotheses made about $\mathcal{U}$.

8 The case in which $\mathcal{U}$ is unbounded

The result 5.I, expressed by relation (5.5), does not depend on the radius or on the centre of $\mathcal{U}$ in the sense that, whatever such radius and such centre may be, $\mathbf{g}^{(r)}(P, t)$ is always given by (5.5). In other words, 5.I remains unchanged with respect to the ∞^4 possibilities which can occur, with the same density for $\mathcal{U}$. On account of this independence, 5.I *holds true also in the limit case in which the radius is infinite, namely in the case of $\mathcal{U}$ unbounded.*

Therefore we can assert that the results analogous‡ to the ones seen in section 5 hold true also in the case of $\mathcal{U}$ unbounded (*see also* [20], 5), i.e.,

8.I All the co-moving frames are equivalent to each other in the sense that $\mathcal{U}$ has the same behaviour with respect to them, from both the kinematical and the dynamical points of view.

By this, we mean that, with respect to every co-moving frame $\mathcal{R}_O$, relation (3.2) holds true and Newton's second law of dynamics – written for any element P of $\mathcal{U}$ and referred to the unit of mass – is expressed, with respect to $\mathcal{R}_O$, by Eqn (5.3) with $\mathbf{g}^{(r)}(P, t)$ given by relation (5.5), i.e., by the resultant of the specific actions at a distance exerted on P by that part of $\mathcal{U}$ contained in the sphere S_{OP}. In other words, with respect to $\mathcal{R}_O$, the equation of motion of P is still given by Eqn (5.6).

† In this paper, following [19], the hypothesis that C is convex is introduced. In [33] no hypotheses are made on C.

‡ Obviously, it is no longer possible to introduce the frame $\mathcal{R}_G$, because in this case the concept of centre of mass is meaningless.

Thus, we have

8.II The motion of P, considered with respect to any co-moving frame $\mathcal{R}_O$ is the same as if the frame $\mathcal{R}_O$ were inertial and $\mathcal{U}$ were reduced to its part contained in the sphere S_{OP}.

This result, together with relation (7.2) (which also holds true in the present case) is the same as the theorem mentioned in section 1, deduced by Layzer in [6] by resorting to the general theory of relativity.

If, together with the co-moving frame $\mathcal{R}_O$, we consider a second co-moving frame $\mathcal{R}_{O'}$, obviously these two frames are not in uniform translational motion relative to each other, because the acceleration of O' with respect to $\mathcal{R}_O$ is not zero: this acceleration is given by $(\dot{h}+h^2)OO'$, as follows from (3.3). Nevertheless, the previous result holds true in the sense that, for the motion of P with respect to the frame $\mathcal{R}_{O'}$, everything happens as if the frame $\mathcal{R}_{O'}$ were inertial and $\mathcal{U}$ were reduced to its part contained in the sphere $S_{O'P}$ with centre O' and radius $|O'P|$.

In this connection, it is necessary to bear in mind that *the fact that in every co-moving frame the equation of motion of P is written in the same way (equivalence of co-moving frames) does not mean at all that this equation is the same with respect to every frame, even if it has the same form with respect to each of these frames. If we change the frame, the term $(\dot{h}+h^2)OP$ also changes, i.e., the resultant of the specific forces acting on P changes, while if such frames were inertial* (and hence in uniform translational motion relative to each other) *this resultant would remain unchanged.* This is expressed by relation (5.9).†

The remarks now made clarify what at first sight could seem a paradox.

9 Further results for $\mathcal{U}$ unbounded

As a consequence of the result 8.II, with respect to any co-moving frame $\mathcal{R}_O$, it also follows that the considerations which led to the result 4.I hold true.

Therefore, Eqn (5.6) also takes the explicit form (1.1), where the constant κ is given by (6.1), and Eqns (6.3) and (6.4) and the result 6.I also hold true.

The considerations developed in section 7 can be repeated unchanged in the present case. Namely, relation (7.2) and the result 7.I, as well as relation (7.3), also hold true.

A detailed analysis (developed by assuming as a starting point the Newtonian theory of gravitation) of all the possible cases which can occur in the evolution of $\mathcal{U}$ is contained in [26] (*see* **I**. B, C, D), [5] (*see* **9.3**,

† For some details on the case of $\mathcal{U}$ unbounded, *see* [21].

9.4), etc.† In these treatments, the analysis contained in Milne and McCrea's papers [3, 4] of 1934 is thoroughly developed with the same assumptions, made without any justification.

10 On the Newtonian behaviour of the Universe

If we assume according to what direct astronomical observations suggest, as a model for the Universe a fluid like the one considered above,‡ the previous results allow us to state the following conclusions,§ for which it is not necessary to know if the Universe is bounded or not.

10.I *Let $\mathcal{R}_O$ denote the co-moving frame with origin at the centre of mass O of our galaxy. Then, from the hypotheses, verified by astronomical observations, that on a large scale the Universe is homogeneous and that its behaviour is isotropic with respect to $\mathcal{R}_O$, there necessarily follow:*

(a) *Hubble's law*, which is satisfied not only with respect to the frame $\mathcal{R}_O$, but also with respect to every co-moving frame;
(b) *Newton's law of gravitation*, in the sense that the action exerted by any typical galaxy on another one is necessarily expressed by Newton's law of gravitation.

10.II *With respect to any co-moving frame $\mathcal{R}_O$, the resultant acting on any galaxy P is equal to the gravitational action exerted on P by that part of the Universe contained in the sphere with centre O and radius $|OP|$.*

10.III *All the co-moving frames are equivalent to one another, in the sense that the Universe has, from both the kinematical and the dynamical points of view, the same behaviour with respect to them.*

In particular, recalling the result 8.II, we have

10.IV *The co-moving frame with the origin at the centre of mass of our galaxy can be considered as inertial.*

11 Some remarks and conclusions

In [35], McCrea definitively abandoned the case of $\mathcal{U}$ unbounded, by simply asserting, in this respect, always under the influence of Layzer's

† Such an analysis can be found also in Armellini [34], where all the previous bibliography, in particular Milne and McCrea's papers [3, 4], is completely ignored.
‡ At the present time, the internal pressure of the Universe is totally negligible.
§ For further details and remarks, *see* [18], 4, 5.

objections, that 'the treatment of an unbounded system is more satisfactorily dealt with by the methods of the theory of general relativity'.

By regarding $\mathcal{U}$ as spherical and with respect to the inertial frame $\mathcal{R}_G$, in [35], McCrea assumed as starting points Hubble's law and the Newtonian theory of gravitation. From these assumptions he deduced a result which is the same as our result 5.IV.

Without detracting anything from the result of McCrea (a result which, in the present treatment, we recall, is deduced without resorting to the Newtonian theory of gravitation), it must be said that, in [35], while emphasizing the equivalence among all co-moving frames (in the sense that $\mathcal{U}$ has the same behaviour with respect to them), it is not sufficiently emphasized that the fact that the equation of motion of P is deduced in every co-moving frame in the same way, namely by the same rule (equivalence of co-moving frames), does not mean at all that this equation is the same with respect to every co-moving frame, even if it has the same form with respect to each of them. As was extensively discussed in sections 5 and 8, by changing the frame of reference, the term $(\dot{h}+h^2)OP$ changes, i.e., the specific force acting on P (force which is the resultant of the specific fictitious force acting on P and of the resultant of the specific actions at a distance exerted on P by $\mathcal{U}$) changes. This force, on the contrary, should remain unchanged if the above frames were inertial, and, therefore, in uniform translational motion with respect to each other.

These remarks, as already stated at the end of section 8, clarify what at first sight could seem a paradox. In reality, this paradox does not exist, even if in [35] it is regarded as such.

We recall that all the results contained in the previous sections have been obtained simply by starting from the hypotheses made about $\mathcal{U}$ (i.e., from the homogeneity of $\mathcal{U}$ and from its isotropic behaviour with respect to a co-moving frame) and from the principles of Newtonian mechanics.

As we have seen, from these hypotheses, using only kinematical considerations, Hubble's law follows. In this context, therefore, Hubble's law loses the empirical aspect attributed to it by Hubble himself. Besides, from these hypotheses it necessarily follows that the action exerted by any typical galaxy on another one is expressed by Newton's law of gravitation and, therefore, the whole process of the expansion of the Universe is ruled by this law. This result gives the first verification of the validity of the Newtonian theory of gravitation on an extragalactic scale.

In the light of the above results, Newtonian mechanics reveals Newtonian cosmology as completely valid, and it removes from Newtonian cosmology the aspect of 'naive cosmology' ascribed to it by supporters of the relativistic theories. These supporters, in general, only admit that Newtonian cosmology may at best provide results which show formal analogies with the ones obtained by relativistic models. This is due to the fact that Newtonian cosmology has, until now, been developed using Newton's law of gravitation and, more generally, the foundations of the Newtonian theory of gravitation as starting points (*see* [3–5, 25, 26] etc.) instead of deducing them in a mathematical way from the indications provided by astronomical observation.

Concerning the aspect of 'naive cosmology' which has been attributed to Newtonian cosmology, Bondi's position is significant. After saying, in his fine treatise *Cosmology* of 1952, that Milne and McCrea's formulation is 'highly interesting', he concluded that 'Newtonian theory serves only as a picture, though a very useful picture, owing to the close analogy with relativistic cosmology'. And he added that 'it is hardly worthwhile nowadays to compare Newtonian theory and observation, since the Newtonian concepts are known to be untenable'.

The only exception to this position with regard to Newtonian cosmology is the one assumed in 1961 by Callan, Dicke and Peebles (*see* [25]). By considering essentially the physical meaning of Newtonian cosmology they wrote: 'It is seldom pointed out that general relativity does not explain or predict the expanding Universe any more than does Newtonian mechanics. [· · ·] There are no mysterious forces tending to disrupt clusters of galaxies. Furthermore, the mechanics of the expansion of the Universe does not even require general relativity for its discussion. [· · ·] Most important of all, this classical treatment of cosmology is neither a crude approximation to the correct relativistic calculation, nor a cooked-up montage cleverly contrived to look like the real thing. It is a completely correct discussion of certain aspects of the cosmological problem'.

Callan, Dicke and Peebles went on to say: 'Long ago McCrea and Milne showed that Newtonian mechanics was capable of describing the expansion of the Universe. However, it does not seem to have been generally recognized that such a classical treatment was not the dynamics of a crude classical model of the Universe but was rather a completely correct treatment of the real Universe'.

And without foreseeing that, in the context of Newtonian mechanics, it is possible to deduce Newtonian cosmology solely from the simple data given by astronomical observation and without having resort to the Newtonian theory of gravitation, Callan, Dicke and Peebles concluded: 'there is nothing mysterious about the motions of galaxies, [· · ·] Newtonian mechanics is adequate. The cosmological space has no centrifugal tendencies threatening to disrupt clusters of galaxies, and the expansion of the Universe is not a result of an expanding space pulling the galaxies apart.'

Obviously it must be said that the whole formulation of Newtonian cosmology – perfectly valid on the local level – has to be examined now in the light of other experimental facts, such as the ones related to the propagation of light.

References

1 North, J. D., *The Measure of the Universe,* Clarendon Press, Oxford, 1965.
2 Einstein, A., Kosmologische Betrachtungen zur allgemeinen Relativitätstheorie. *Preuss. Akad. Wiss. Berlin Sitzber.*, 142–152, 1917.
3 Milne, E. A., A Newtonian expanding universe. *Quart. J. Math.* (*Oxford Ser.*), **5,** 64–72, 1934.
4 McCrea, W. H. and Milne, E. A., Newtonian universes and the curvature of space. *Quart. J. Math.* (*Oxford Ser.*), **5,** 73–80, 1934.
5 Bondi, H., *Cosmology,* 2nd ed., Cambridge University Press, London, 1960.
6 Layzer, D., On the significance of Newtonian cosmology. *Astron. J.*, **59,** 268–270, 1954.
7 Layzer, D., Review to: 'Bemerkungen zur Newtonschen Kosmologie. I'. by O. Heckmann and E. Schücking, *Math. Rev.*, **17,** 545, 1956.
8 Layzer, D., Review to: 'On Newtonian Frames of Reference' by W. H. McCrea, *Math. Rev.*, **17,** 545, 1956.
9 Weinberg, S., *Gravitation and Cosmology: Principles and Applications of the General Theory of Relativity,* Wiley, New York, 1972.
10 McCrea, W. H., Newtonian cosmology, *Nature,* **175,** 466, 1955.
11 McCrea, W. H., On the significance of Newtonian cosmology, *Astron., J.*, **60,** 271–274, 1955.
12 Heckmann, O. and Schücking, E., Bemerkungen zur Newtonschen Kosmologie. I, *Z. Astrophys.*, **38,** 95–109, 1955.
13 Heckmann, O. and Schücking, E., Bemerkungen zur Newtonschen Kosmologie. II, *Z. Astrophys.*, **40,** 81–92, 1956.
14 Layzer, D., Review to: 'Bemerkungen zur Newtonschen Kosmologie. II'. by O. Heckmann and E. Schücking, *Math. Rev.*, **20,** 260, 1959.
15 Barberis, B. and Galletto, D., Sui fondamenti della cosmologia newtoniana. II, *Atti Acc. Sc. di Torino,* **112,** 1978, in press.
16 Galletto, D., Sulla legge di Hubble e sulla legge di gravitazione universale, *Atti Acc. Sc. di Torino,* **110,** 335–341, 1976.
17 Galletto, D., Legge di Hubble e costante di gravitazione universale, *Atti Acc. Sc. di Torino,* **110,** 401–404, 1976.
18 Galletto, D., Meccanica newtoniana ed espansione dell'Universo, *Atti Acc. Sc. di Torino,* **111,** 69–74, 1977.
19 Galletto, D., Cosmologia newtoniana e costante di gravitazione universale, *Atti Acc. Sc. di Torino,* **111,** 203–210, 1977.
20 Galletto, D., Sui fondamenti della cosmologia newtoniana. I, *Atti Acc. Sc. di Torino,* **111,** 545–554, 1977.
21 Galletto, D., Su una proprietà caratteristica del campo newtoniano, to be published.
22 Barberis, B. and Galletto, D., Remarks on Newton's law of gravitation, *Atti Acc. Sc. di Torino,* **111,** 435–439, 1977.
23 Barberis, B. and Galletto, D., Remarks on P. T. Landsberg's paper on Hubble's law and Newton's law of gravitation, *Atti Acc. Sc. di Torino,* **111,** 147–153, 1977.

24 Heckmann, O. and Schücking, E., Newtonsche und Einsteinsche Kosmologie, in *Handbuch der Physik* (S. Flügge, ed.), Springer, Berlin, 1959, Vol. LIII, pp. 489–519.

25 Callan, C., Dicke, R. H. and Peebles, P. J. E., Cosmology and Newtonian mechanics, *Amer. J. Phys.*, **33,** 105–108, 1965.

26 Zeldovich, Ya. B., Survey of modern cosmology, *Adv. Astron. Astrophys.* **3,** 241–379, 1965.

27 Heckmann, O., *Theorien der Kosmologie*, 2nd ed., Springer, Berlin, 1968.

28 Davidson, W. and Evans, A. B., Newtonian universes expanding or contracting with shear and rotation, *Int. J. Theor. Phys.*, **7,** 353–378, 1973.

29 Mavridès, S., *L'Univers relativiste*, Masson, Paris, 1973.

30 Landsberg, P. T., A deduction of the inverse square law from Newtonian cosmology, *Nature Phys. Sci.*, **244,** 66–67, 1973.

31 Barberis, B. and Galletto, D., Su una congettura in cosmologia newtoniana. *Atti Acc. Sc. di Torino*, **111,** 261–265, 1977.

32 Barberis, B. and Galletto, D., Further remarks on a paper by P. T. Landsberg, *Atti Acc. Sc. di Torino*, **112,** 103–107, 1978.

33 Galletto, D., Un teorema di unicità nella teoria del campo newtoniano e sue implicazioni comologiche, *Rend. di Matematica*, Ser. VI, **10,** 507–522, 1977.

34 Armellini, G., L'espansione dell'Universo nella meccanica classica, *Rend. Acc. Naz. Lincei*, **8,** 15–20, 1950.

35 McCrea, W. H., On Newtonian frames of reference, *Math. Gaz.*, **39,** 287–291, 1955.

Dr Bruno Barberis,
Professor Dionigi Galletto,
Istituto di Fisica Matematica,
Università di Torino,
Via C. Alberto 10,
10123 Torino,
Italy

J. Brilla

The compatible perturbation method in finite viscoelasticity

1 Introduction

In 1930, Signorini [1] proposed a formal perturbation method for the solution of traction boundary-value problems in finite elasticity leading to the solution of successive systems of equations of infinitesimal elasticity. However, Signorini himself found that for certain systems of loads, when the astatic load vanishes, the first approximation is incompatible with the finite elasticity problem and cannot be considered as its asymptotic approximation. Signorini and his associates devoted great attention to this problem of incompatibility.

Some authors regard the incompatibility for any nth approximation as an objection to classical infinitesimal theory. Therefore, attention has been directed to this problem of the incompatibility of linear and non-linear theories of elasticity.

Detailed analysis of the problem was given by Tolotti [2] and later by Grioli [3], but without a solution to the problem of incompatibility. Truesdell and Noll [4] proposed the application of surface tractions and body forces proportional to terms of the small parameter of higher degrees. This does not, in fact, solve the problem, as the surface traction and body forces are prescribed and can be given in the form considered by Signorini. Capriz and Podio Guidugli [5] use as a starting point the corresponding dynamic problem.

We have shown that the incompatibility problem can be solved directly in the case of finite elasticity (Brilla [6]), as well as in the case of finite viscoelasticity.

This paper is concerned with the compatible perturbation method in finite viscoelasticity. We assume, similarly to Signorini, that the prescribed body forces and surface tractions depend on a parameter ε. Expanding approximations of all necessary equations as power series of this parameter, we obtain successive systems of infinitesimal viscoelasticity. Solutions of these problems are determined only to within infinitesimal rotations. In contradistinction to Signorini, we determine the unknown rotations from equilibrium conditions of the deformed body or expand the rotation in terms of the parameter ε with respect to the equilibrium rotation of the deformed body. This procedure also leads to the compatibility of infinitesimal and finite viscoelasticity in the case of essential incompatibility in the sense of Signorini.

2 The initial boundary-value problem of traction

The motion of the body B may be described by a vector function $\mathbf{x}(t)=\chi(\mathbf{X}, t)$, where $\mathbf{x}(t)$ is the position of a material point $\mathbf{X}$ of the body at time t, which has the position $\mathbf{X}$ in a fixed reference configuration R. The motion of the body defines its deformation from the reference configuration.

We denote the first deformation gradient

$$\mathbf{F}=\nabla_X \boldsymbol{\chi}(\mathbf{X}, t), \qquad F^k_\alpha=\frac{\partial \chi^k}{\partial X^\alpha}=x^k_{,\alpha}. \tag{2.1}$$

It is a tensor which possesses an inverse $\mathbf{F}(t)^{-1}$. The value of $\mathbf{F}(t)$ at each point of B is determined not only by the configuration of B at time t but also by the choice of the reference configuration. We choose as the reference configuration the configuration occupied by the body at time $t=0$.

Then, in the case of an isothermal deformation, the initial boundary-value problem of traction for simple materials can be defined by

(a) the equilibrium equation

$$\operatorname{Div}\mathbf{T}+\rho\mathbf{b}=\mathbf{0}, \tag{2.2}$$

where $\mathbf{T}(\mathbf{x}, t)$ is a stress tensor, $\mathbf{b}$ is a specific body force and ρ is the mass density at time t;

(b) the constitutive equation

$$\mathbf{T}=\mathbf{F}\,\underset{s=0}{\overset{\infty}{\mathfrak{L}}}\,(\mathbf{C}^{(t)}(s))\mathbf{F}^{\mathrm{T}}, \tag{2.3}$$

where $\mathfrak{L}$ denotes a functional, $\mathbf{C}^{(t)}(s)=\mathbf{C}(t-s)$ is the history of the right Cauchy–Green tensor

$$\mathbf{C}=\mathbf{F}^{\mathrm{T}}\mathbf{F}; \tag{2.4}$$

(c) the boundary condition

$$\mathbf{t}=\mathbf{T}\mathbf{n}, \tag{2.5}$$

where $\mathbf{t}$ is the vector of the surface traction acting on the deformed boundary and $\mathbf{n}$ is the outward unit vector normal to the boundary;

(d) the initial condition

$$\mathbf{T}(\mathbf{X}, 0) = \mathbf{T}_0(\mathbf{X}). \tag{2.6}$$

The initial conditions, in fact, depend on the form of the functional $\mathfrak{L}$ and (2.5) can be one of necessary initial conditions.

When dealing with initial boundary-value problems of traction, we assume that the reference configuration is a homogeneous natural state, i.e., that the functional of the constitutive equation is a functional of $\mathbf{F}$ only, independent of $\mathbf{X}$, and that the body is free of stresses when in reference configuration. Further, we shall assume

$$\left.\begin{aligned} &\mathbf{C}^{(t)}(s) = \mathbf{I} \quad \text{for} \quad t<0,\ s>t, \\ &\mathbf{T}(\mathbf{X}, t) = \mathbf{0} \quad \text{for} \quad t<0. \end{aligned}\right\} \tag{2.7}$$

We have to take into account the fact that the domain of definition of the boundary-value problem of traction and its boundary are changing with time. Thus, a description in the entire time range in terms of a reference configuration X is useful.

Now, using a similar method to that employed by Signorini [1] for finite elasticity problems, we assume that the magnitudes of the external forces are proportional to the parameter ε.

Thus, we have

$$\mathbf{t}_{\mathrm{R}} = \varepsilon \mathbf{t}_{1\mathrm{R}}, \qquad \mathbf{b} = \varepsilon \mathbf{b}_1. \tag{2.8}$$

We also assume that the initial conditions are proportional to the parameter ε:

$$\mathbf{T}(\mathbf{X}, 0) = \varepsilon \mathbf{T}_0(\mathbf{X}). \tag{2.9}$$

Further, we assume that there exists a solution, expressed in terms of the displacement

$$\mathbf{u}(\mathbf{x}, t) = \mathbf{x}(\mathbf{X}, t) - \mathbf{X} \tag{2.10}$$

of the initial boundary-value problem, defined by Eqns (2.2) to (2.6), the loads (2.8) and the initial condition (2.9), which depends analytically on ε and vanishes when $\varepsilon = 0$. Thus, we assume that

$$\mathbf{u} = \sum_{n=1}^{\infty} \varepsilon^n \mathbf{u}_n. \tag{2.11}$$

Then, the displacement gradient

$$\nabla \mathbf{u} = \mathbf{H} = \mathbf{F} - \mathbf{I} \tag{2.12}$$

has a series expansion

$$\mathbf{H} = \sum_{n=1}^{\infty} \varepsilon^n \nabla \mathbf{u}_n = \sum_{n=1}^{\infty} \varepsilon^n \mathbf{H}_n. \tag{2.13}$$

Now it is necessary to find the expansion of the constitutive equation or its approximation in the form of power series of the parameter ε. For this purpose, we rewrite the constitutive equation (2.3) in the form

$$\tilde{\mathbf{T}} = \prod_{s=0}^{\infty} (\mathbf{C}^{(t)}(s)), \tag{2.14}$$

where

$$\tilde{\mathbf{T}} = J\mathbf{F}^{-1}\mathbf{T}(\mathbf{F}^{-1})^{\mathrm{T}} \tag{2.15}$$

is the second Piola–Kirchhoff stress tensor and $J = |\det \mathbf{F}|$. The constitutive equation (2.15) can also be written in the form

$$\tilde{\mathbf{T}} = \prod_{s=0}^{\infty} (\mathbf{E}_{\mathrm{d}}^{(t)}(s), \mathbf{E}), \tag{2.16}$$

where

$$\mathbf{E} = \tfrac{1}{2}(\mathbf{C} - \mathbf{I}) \tag{2.17}$$

is the Green–St Venant strain tensor and

$$\mathbf{E}_{\mathrm{d}}^{(t)}(s) = \mathbf{E}(t-s) - \mathbf{E}(t) \tag{2.18}$$

is the difference of its history.

In order to define an approximation of the general constitutive operator $\prod_{s=0}^{\infty}$, we have to introduce a functional space. We shall do this for materials with fading memory. We define a function $h(s)$, called an obliviator, which characterizes the rate at which the memory fades. It should satisfy the following conditions:

(a) $h(s)$ is defined for $0 \leq s < \infty$ and has real positive values $h(s) > 0$;
(b) $h(0) = 1$;
(c) $h(s)$ decreases to zero at least as

$$\lim_{s \to \infty} s^{(1/2)+\delta} h(s) \Rightarrow 0, \qquad \delta > 0 \tag{2.19}$$

monotonically for large s.

Then, we consider the Hilbert space of tensor functions $\mathbf{G}(s)$ weighted by the obliviator $h(s)$ with the norm

$$\|\mathbf{G}(s)\|_h = \left(\int_0^{\infty} [h(s)\,|\mathbf{G}(s)|]^2 \, \mathrm{d}s \right)^{1/2}, \tag{2.20}$$

where

$$|\mathbf{G}(s)| = \{\mathrm{tr}\,[\mathbf{G}(s)]^2\}^{1/2}$$

is the norm of the symmetric tensor $\mathbf{G}(s)$.

If the functional $\prod_{s=0}^{\infty}$ is n-Fréchet differentiable at $\mathbf{E}_d^{(t)}(s)=\mathbf{0}$ and the Fréchet differentiability is uniform in the tensor parameter $\mathbf{E}$, we can write

$$\tilde{\mathbf{T}}(t)=\prod_{s=0}^{\infty}(\mathbf{E}_d^{(t)}(s),\mathbf{E})$$
$$=\boldsymbol{\pi}(\mathbf{E})+\sum_{k=1}^{n}\frac{1}{k!}\delta^k\prod_{s=0}^{\infty}(\mathbf{0},\mathbf{E};\mathbf{E}_d^{(t)}(s)^k)+o(\|\mathbf{E}_d^{(t)}(s)\|^n), \tag{2.21}$$

where

$$\boldsymbol{\pi}(\mathbf{E})=\prod_{s=0}^{\infty}(\mathbf{0},\mathbf{E}) \tag{2.22}$$

and

$$\delta^k\prod_{s=0}^{\infty}(\mathbf{0},\mathbf{E};\mathbf{E}_d^{(t)}(s)^k)$$

is a bounded homogeneous polynomial functional of degree k.

If the functional

$$\prod_{s=0}^{\infty}(\mathbf{E}_d^{(t)},\mathbf{E})$$

is also n-differentiable with respect to the tensor parameter $\mathbf{E}$, we can write

$$\tilde{\mathbf{T}}(t)=\prod_{s=0}^{\infty}(\mathbf{E}_d^{(t)}(s),\mathbf{E})$$
$$=\prod_{s=0}^{\infty}(\mathbf{0},\mathbf{0})+\sum_{k=1}^{n}\frac{1}{k!}\sum_{m=0}^{k}\binom{k}{m}\delta^{k-m}\partial^m\prod_{s=0}^{\infty}(\mathbf{0},\mathbf{0};\mathbf{E}_d^{(t)}(s)^{k-m})\{\mathbf{E}^m\}$$
$$+o((\|\mathbf{E}_d^{(t)}(s)\|+\|\mathbf{E}\|)^n) \tag{2.23}$$

where

$$\delta^{k-m}\partial^m\prod_{s=0}^{\infty}(\mathbf{0},\mathbf{0};\mathbf{E}_d^{(t)}(s)^{k-m})\{\mathbf{E}^m\}$$

is a bounded homogeneous polynomial of degree $k-m$ in $\mathbf{E}_d^{(t)}(s)$ and m-linear transformation of the tensor parameter $\mathbf{E}$.

Using the representation of $\mathbf{E}_d^{(t)}(s)$ and $\mathbf{E}$ in terms of the displacement gradient, we can express the approximation of the constitutive equation (2.23) in the form of a power series of the parameter ε. In a quite straightforward way, this can be done in the case of the representation of the kth Fréchet differential by the k-tuple integral.

The conditions under which the functionals of the constitutive equation (2.23) can be represented by integral polynomials were studied by Rivlin [7].

In such a case, we can write

$$\tilde{\mathbf{T}}(t) = \boldsymbol{\pi}(\mathbf{E}) + \sum_{k=1}^{n} \int_0^\infty \dots \int_0^\infty \mathbf{K}_k(s_1, \dots, s_k, \mathbf{E}) \times \{\mathbf{E}_d^{(t)}(s_1), \dots, \mathbf{E}_d^{(t)}(s_k)\} \, ds_1 \dots ds_k \tag{2.24}$$

where $\mathbf{K}_k(s_1, \dots, s_k, \mathbf{E}) \{\cdot\}$ for each choice of $s_1, \dots, s_k$ and $\mathbf{E}$ is a multilinear tensor function of k tensor variables. The values $\mathbf{K}_k(s_1, \dots, s_k, \mathbf{E})$ may be regarded as tensors of order $2k$ and have symmetries corresponding to the symmetries of the constitutive equation. When we assume that the kernels $\mathbf{K}_k(\ )$ are n-differentiable with respect to the tensor parameter $\mathbf{E}$, the expansion in terms of the polynomial of ε is straightforward.

Now we restrict ourselves to the case of finite linear viscoelasticity. The constitutive equation assumes the form (Coleman and Noll [8])

$$\tilde{\mathbf{T}}(t) = \boldsymbol{\pi}(\mathbf{E}) + \int_0^\infty \mathbf{K}(s, \mathbf{E})\{\mathbf{E}_d^t(s)\} \, ds \tag{2.25}$$

with an error of order $o(\|\mathbf{E}_d^{(t)}(s)\|)$. Here, $\mathbf{K}(s, \mathbf{E}) \{\ \}$ for each s and each $\mathbf{E}$ is a linear transformation of the space of symmetric tensor functions into itself. $\mathbf{K}(s, \mathbf{E})$ is a tensor of the fourth order with the property

$$\int_0^\infty |\mathbf{K}(s, \mathbf{E})|^2 h(s)^{-2} \, ds < \infty \tag{2.26}$$

The constitutive equation (2.25) can be written as

$$\tilde{\mathbf{T}}(t) = \boldsymbol{\pi}(\mathbf{E}) + \int_0^\infty \mathbf{K}(s, \mathbf{E})\{\mathbf{E}^{(t)}(s) - \mathbf{E}(t)\} \, ds. \tag{2.27}$$

By setting $\tau = t - s$ and considering the assumed homogeneous natural state, we arrive at

$$\tilde{\mathbf{T}}(t) = \boldsymbol{\pi}^*(\mathbf{E}) + \int_0^t \mathbf{K}^*(t - \tau, \mathbf{E})\{\mathbf{E}(\tau)\} \, d\tau, \tag{2.28}$$

where

$$\tilde{\mathbf{T}}_e(0) = \boldsymbol{\pi}^*(\mathbf{E}) = \boldsymbol{\pi}(\mathbf{E}) - \int_0^t \mathbf{K}(t - \tau, \mathbf{E})\{\mathbf{E}(t)\} \, d\tau$$

is the instantaneous elastic stress.

In fact, when dealing with practical problems, we do not know the whole history of deformation. Also, experiments for determination of the form of material functionals can start only at a finite time. Therefore, we assume that before the time $t = 0$ the body is at rest and has forgotten its deformation history for $t < 0$. We assume the state of free equilibrium at and before $t = 0$.

Now, expanding $\boldsymbol{\pi}(\mathbf{E})$ and $\mathbf{K}(t-\tau, \mathbf{E})$ in Taylor series at $\mathbf{E}=0$, one arrives at

$$\tilde{\mathbf{T}}(t)=\sum_{k=1}^{n}\frac{1}{k!}\boldsymbol{\pi}^{*(k)}(\mathbf{0})\mathbf{E}^{k}$$
$$+\int_{0}^{t}\left[\mathbf{K}^{*}(t-\tau,\mathbf{0})+\sum_{k=1}^{n}\frac{1}{k!}\mathbf{K}^{*(k)}(t-\tau,\mathbf{0})\mathbf{E}(t)^{k}\right]\{\mathbf{E}(\tau)\}\,\mathrm{d}\tau, \quad (2.29)$$

where, according to the assumption of the state of free equilibrium, we have set $\boldsymbol{\pi}(\mathbf{0})=0$.

In order to rewrite the constitutive equation (2.29) in form of a power series of ε, we expand the Green–St Venant strain tensor:

$$2\mathbf{E}=\mathbf{H}+\mathbf{H}^{\mathrm{T}}+\mathbf{H}^{\mathrm{T}}\mathbf{H}$$
$$=2\sum_{n=1}^{\infty}\varepsilon^{n}\mathbf{E}_{n}+\sum_{n=2}^{\infty}\varepsilon^{n}\sum_{k=1}^{n-1}\mathbf{H}_{k}^{\mathrm{T}}\mathbf{H}_{n-k}, \quad (2.30)$$

where we have introduced the nth infinitesimal strain tensor

$$2\mathbf{E}_{n}=\mathbf{H}_{n}+\mathbf{H}_{n}^{\mathrm{T}}, \quad (2.31)$$

Inserting (2.30) into (2.29) and collecting terms proportional to ε^{n} yields

$$\tilde{\mathbf{T}}(t)=\sum_{n=1}^{\infty}\varepsilon^{n}\tilde{\mathbf{T}}_{n}=\sum_{n=1}^{\infty}\varepsilon^{n}\mathbf{P}_{1}(\mathbf{E}_{n})$$
$$+\sum_{n=2}^{\infty}\varepsilon^{n}\mathbf{P}_{n}(\mathbf{H}_{1},\ldots,\mathbf{H}_{n-1}), \quad (2.32)$$

where

$$\mathbf{P}_{1}(\mathbf{E}_{n})=\mathbf{L}\mathbf{E}_{n}+\int_{0}^{t}\mathbf{K}(t-\tau)\mathbf{E}_{n}(\tau)\,\mathrm{d}\tau, \quad (2.33)$$

$$\pi^{*(1)}(\mathbf{0})=\mathbf{L}, \qquad \mathbf{K}(t-\tau)=\mathbf{K}^{*}(t-\tau,\mathbf{0}) \quad (2.34)$$

and $\mathbf{P}_{k}$ is to be determined by collecting terms with like powers of ε. $\mathbf{P}_{k}$ are polynomial tensor functions of $k-1$ tensor variables and are uniquely determined by the approximation of the constitutive equation (2.29).

Using the property of homogeneity and multilinearity of terms in the approximation of the general constitutive equation (2.23), we can expand it also in the form of power series of ε.

The equilibrium equation (2.2) assumes, in terms of the second Piola–Kirchhoff stress tensor, the form

$$\mathrm{Div}\,(\mathbf{F}\tilde{\mathbf{T}})+\rho_{\mathrm{R}}\mathbf{b}=\mathbf{0}, \quad (2.35)$$

where Div is the divergence operator with respect to $\mathbf{X}$ in the reference configuration. Then

$$\text{Div}\,[(\mathbf{I}+\mathbf{H})\tilde{\mathbf{T}}]+\rho_{\mathrm{R}}\mathbf{b}=\mathbf{0} \tag{2.36}$$

or, in component form,

$$g^k_\alpha \tilde{T}^{\alpha\beta}_{,\beta}+(u^k_{,\alpha}T^{\alpha\beta})_{,\beta}+\rho_{\mathrm{R}}b^k=0, \tag{2.37}$$

where g^k_α is the contravariant shift vector. Multiplying this equation by g^α_k, we obtain

$$\tilde{T}^{\alpha\beta}_{,\beta}+(u^\alpha_{,\gamma}T^{\gamma\beta})_{,\beta}+\rho_{\mathrm{R}}b^\alpha=0. \tag{2.38}$$

Making use of the series expansion of $\tilde{\mathbf{T}}$ and $\mathbf{u}$ and collecting terms proportional to ε^n yields

$$\varepsilon(\tilde{T}^{\alpha\beta}_{(1),\beta}+\rho_{\mathrm{R}}b^\alpha_{(1)})+\sum_{n=2}^{\infty}\varepsilon^n\left[\tilde{T}^{\alpha\beta}_{(n),\beta}+\sum_{k=1}^{n-1}(u^\alpha_{(n-k),\gamma}T^{\gamma\beta}_{(k)})_{,\beta}\right]=0. \tag{2.39}$$

Finally, it is necessary to express the boundary condition in a power series of ε. In the reference configuration,

$$\mathbf{t}_{\mathrm{R}}=\mathbf{T}_{\mathrm{R}}\mathbf{n}_{\mathrm{R}}, \tag{2.40}$$

where

$$\mathbf{T}_{\mathrm{R}}=J\mathbf{T}(\mathbf{F}^{-1})^{\mathrm{T}}=\mathbf{F}\tilde{\mathbf{T}} \tag{2.41}$$

is the first Piola–Kirchhoff stress tensor.

Making use of expansions of $\tilde{\mathbf{T}}$ and $\mathbf{F}$ yields

$$\varepsilon\mathbf{t}^\alpha_{(1)\mathrm{R}}=\varepsilon\tilde{T}^{\alpha\beta}_{(1)}\mathbf{n}_{\mathrm{R}\beta}+\sum_{n=2}^{\infty}\varepsilon^n\left[\tilde{T}^{\alpha\beta}_{(n)}+\sum_{k=1}^{n-1}u^\alpha_{k,\gamma}\tilde{T}^{\gamma\beta}_{(n-k)}\right]\mathbf{n}_{\mathrm{R}\beta}. \tag{2.42}$$

Thus, we have arrived at series expansions of all equations defining the initial boundary-value problem. We assume that there exist such $\bar{\varepsilon}$ that, for any ε, which satisfies the inequality $\mathrm{O}\leqslant\varepsilon\leqslant\bar{\varepsilon}$, the problem under consideration has a solution.

Then, considering terms at like powers of ε yields the following systems of linear equations:

$$\tilde{T}^{\alpha\beta}_{(n),\beta}+\rho_{\mathrm{R}}b^\alpha_{(n)}=0, \tag{2.43a}$$

$$\tilde{T}^{\alpha\beta}_{(n)}=P^{\alpha\beta}_{(1)}(E_n)+P^{\alpha\beta}_{(n)}(\mathbf{H}_1,\ldots,\mathbf{H}_{n-1}), \tag{2.43b}$$

$$\tilde{T}^{\alpha\beta}_{(n)}n_{\mathrm{R}\beta}=t^\alpha_{(n)\mathrm{R}}, \tag{2.43c}$$

$$\mathbf{T}_n(\mathbf{X},0)=\mathbf{T}_n(X), \tag{2.43d}$$

where $\mathbf{b}_1$, $\mathbf{t}_{1\mathrm{R}}$ and $\mathbf{T}_1(\mathbf{X})$ are prescribed external forces and initial stresses, respectively, and

$$\left.\begin{aligned} \rho_R b^\alpha_{(n)} &= \sum_{k=1}^{n-1} (u^\alpha_{(k),\gamma} \tilde{T}^{\gamma\beta}_{(n-k)})_{,\beta}, \\ t^\alpha_{(n)R} &= -\sum_{k=1}^{n-1} u^\alpha_{(k),\gamma} \tilde{T}^{\gamma\beta}_{(n-k)} n_{R\beta}, \\ \mathbf{T}_n(\mathbf{X}) &= 0, \qquad n \geq 2. \end{aligned}\right\} \tag{2.44}$$

Substituting the stress components into the equilibrium equation (2.43a) and the boundary conditions (2.43c) according to the constitutive equation (2.43b), we obtain

$$P^{\alpha\beta}_{(1)}(\mathbf{E}_n)_{,\beta} + \rho_R b^{*\alpha}_{(n)} = 0, \tag{2.45a}$$

$$P^{\alpha\beta}_{(1)}(\mathbf{E}_n) n_{R\beta} = t^{*\alpha}_{(n)R}, \tag{2.45b}$$

$$\mathbf{P}_1(\mathbf{E}_n)\,|_{t=0} + \mathbf{P}_n(\mathbf{H}_1, \ldots, \mathbf{H}_{n-1})\,|_{t=0} = \mathbf{T}^*_n(X) \tag{2.45c}$$

where

$$\left.\begin{aligned} &\rho_R b^{*\alpha}_{(1)} = \rho_R b^\alpha_{(1)}, \qquad t^{*\alpha}_{(1)R} = t^\alpha_{(1)R}, \\ &\mathbf{T}^*_1(X) = \mathbf{T}_1(X), \\ &\rho_R b^{*\alpha}_{(n)} = \sum_{k=1}^{n-1} (u^\alpha_{(k),\gamma} \tilde{T}^{\gamma\beta}_{(n-k)})_{,\beta} + P^{\alpha\beta}_{(n)}(H_1, \ldots, H_{n-1})_{,\beta}, \\ &t^{*\alpha}_{(n)R} = -\left[\sum_{k=1}^{n} u^\alpha_{(k),\gamma} T^{\gamma\beta}_{(n-k)} + P^{\alpha\beta}_{(n)}(H_1, \ldots, H_{n-1})\right] n_{R\beta}. \end{aligned}\right\} \tag{2.46}$$

Thus, we have arrived at a successive system of traction initial boundary-value problems of classical infinitesimal viscoelasticity for the same material and the same boundary. For $n = 1$, we have to consider the given external forces and the initial condition; for $n \geq 2$, we must consider fictitious external forces which depend explicitly in a known way upon previously determined solutions $\mathbf{u}_1, \ldots, \mathbf{u}_{n-1}$ and, thus, also, upon material coefficients of higher orders.

3 Conditions of existence of the solution

The solution of a static initial boundary-value problem must be such that the prescribed external forces are in equilibrium in the deformed state for each value of t. Hence, the resultant force and the resultant torque acting upon a body in the configuration x vanish:

$$\left.\begin{aligned} &\int_{\partial B} \mathbf{t}\,\mathrm{d}s + \int_B \rho \mathbf{b}\,\mathrm{d}v = 0, \\ &\int_{\partial B} \mathbf{p} \times \mathbf{t}\,\mathrm{d}s + \int_B \mathbf{p} \times \rho \mathbf{b}\,\mathrm{d}v = 0, \end{aligned}\right\} \tag{3.1}$$

where $\mathbf{p}$ is the position vector of $\mathbf{x}$ from a fixed origin. With respect to the reference configuration, we have the equivalent conditions

$$\left.\begin{aligned}\int_{\partial B}\mathbf{t}_{\mathrm{R}}\,\mathrm{d}s_{\mathrm{R}}+\int_{B}\rho_{\mathrm{R}}\mathbf{b}\,\mathrm{d}v_{\mathrm{R}}=0,\\ \int_{\partial B}\mathbf{p}\times\mathbf{t}_{\mathrm{R}}\,\mathrm{d}s_{\mathrm{R}}+\int_{B}\mathbf{p}\times\rho_{\mathrm{R}}\mathbf{b}\,\mathrm{d}v_{\mathrm{R}}=0,\end{aligned}\right\}\tag{3.2}$$

where $\mathbf{p}$ is the position vector of $\mathbf{x}$, but now expressed as a function of $\mathbf{X}$.

In order that the system (2.45) has the solutions $\mathbf{u}_n$, it is necessary that the prescribed and fictitious external forces satisfy the equilibrium conditions

$$\int_{\partial B}\mathbf{t}^*_{n\mathrm{R}}\,\mathrm{d}s_{\mathrm{R}}+\int_{B}\rho_{\mathrm{R}}\mathbf{b}^*_n\,\mathrm{d}v_{\mathrm{R}}=0,\tag{3.3a}$$

$$\int_{\partial B}\mathbf{p}_{\mathrm{R}}\times t^*_{n\mathrm{R}}\,\mathrm{d}s_{\mathrm{R}}+\int_{B}\mathbf{p}_{\mathrm{R}}\times\rho_{\mathrm{R}}\mathbf{b}^*_n\,\mathrm{d}v_{\mathrm{R}}=0.\tag{3.3b}$$

It is obvious that the prescribed forces $\varepsilon\mathbf{b}_1$ and $\varepsilon\mathbf{t}_{1\mathrm{R}}$ have to satisfy (3.3). Then, applying the divergence theorem and the equilibrium equation (2.45a), we find easily that (3.3a) is automatically satisfied. Similarly, we find that (3.3b) is satisfied if

$$\int_{B}G_\alpha\times G_\beta\left[\sum_{k=1}^{n-1}u^\alpha_{(k),\gamma}\tilde{T}^{\gamma\beta}_{(n-k)}+P^{\alpha\beta}_{(n)}(\mathbf{H}_1,\ldots,\mathbf{H}_n)\right]\mathrm{d}v_{\mathrm{R}}=0,\tag{3.4}$$

where G_α, G_β are base vectors at $\mathbf{X}$. This condition is satisfied if and only if

$$\int_{B}\left[\sum_{u=1}^{n-1}u^\alpha_{(k),\gamma}\tilde{T}^{\gamma\beta}_{(n-k)}+P^{\alpha\beta}_{(n)}(\mathbf{H}_1,\ldots,\mathbf{H}_{n-1})\right]\mathrm{d}v_{\mathrm{R}}\tag{3.5}$$

is a symmetric tensor.

Since the second Piola–Kirchhoff stress tensor is symmetric, it follows from the constitutive equation (2.43b) that $P^{\alpha\beta}_{(n)}(\mathbf{H}_1,\ldots,\mathbf{H}_{n-1})$ is a symmetric tensor too. Thus, it remains to analyse the first term of (3.5). Making use of Green's theorem, we arrive at

$$\begin{aligned}S^{\alpha\beta}&=\int_{B}\sum_{k=1}^{n-1}u^\alpha_{(k),\gamma}\tilde{T}^{\gamma\beta}_{(n-k)}\,\mathrm{d}v_{\mathrm{R}}\\&=\int_{B}\sum_{k=1}^{n-1}u^\alpha_{(k)}\tilde{T}^{\gamma\beta}_{(n-k),\gamma}\,\mathrm{d}v_{\mathrm{R}}+\int_{\partial B}\sum_{k=1}^{n-1}u^\alpha_{(k)}\tilde{T}^{\gamma\beta}_{(n-k)}n_{\mathrm{R}\gamma}\,\mathrm{d}s_{\mathrm{R}}.\end{aligned}\tag{3.6}$$

Now, making use of (2.43) and (2.44), we can write (3.6) in the form

$$S^{\alpha\beta} = \int_B u^{\alpha}_{(n-1)} \rho_R b^{\beta}_{(1)} \, dv_R + \int_{\partial B} u^{\alpha}_{(n-1)} t^{\beta}_{(1)R} \, ds_R$$
$$+ \int_B \sum_{k=1}^{n-2} u^{\alpha}_{(k),\delta} \sum_{q=1}^{n-k-1} u^{\beta}_{(q),\gamma} \tilde{T}^{\gamma\delta}_{(n-k-q)} \, dv_R. \tag{3.7}$$

As the third term in (3.7) is symmetric, the condition of existence for $\mathbf{u}_n$ assumes the form

$$\int_B (u^{\alpha}_{(n-1)} \rho_R b^{\beta}_{(1)} - u^{\beta}_{(n-1)} \rho_R b^{\alpha}_{(1)}) \, dv_R$$
$$+ \int_{\partial B} (u^{\alpha}_{(n-1)} t^{\beta}_{(1)R} - u^{\beta}_{(n-1)} t^{\alpha}_{(1)R}) \, ds_R = 0, \qquad n = 2, 3, \ldots, \tag{3.8}$$

which may be expressed in the vectorial form

$$\int_B \mathbf{u}_{n-1} \times \rho_R \mathbf{b}_1 \, dv_R + \int_{\partial B} \mathbf{u}_{n-1} \times \mathbf{t}_{1R} \, ds_R = 0. \tag{3.9}$$

This equation represents the necessary conditions for the existence of the solutions of the successive systems (2.45). In fact, the approximate solution has to satisfy the equilibrium conditions in deformed states at all times; this is in agreement with (3.1). We have

Theorem *The solutions of successive systems of approximation exist only if external forces satisfy the equilibrium conditions* (3.9).

The sufficient conditions of existence of the solution can be analysed as for infinitesimal viscoelasticity. However, it is necessary also to study convergence of the solution.

In order to satisfy the equilibrium conditions (3.9), it is necessary to apply Da Silva's theorem. The solutions of the system (2.45) are solutions of the traction initial boundary-value problems of infinitesimal viscoelasticity and, thus, are determined at most to within a uniform infinitesimal rotation. This rotation can be uniquely determined from (3.9) or directly from (3.2), applying the following procedure.

The initial boundary-value problem under consideration can be defined by

$$\left.\begin{aligned} &\text{Div} \mathop{\mathfrak{F}}_{s=0}^{\infty} (\mathbf{F}) = -\rho_R \mathbf{b}, \\ &\mathop{\mathfrak{F}}_{s=0}^{\infty} (\mathbf{F}) \mathbf{n}_R = \mathbf{t}_R, \end{aligned}\right\} \tag{3.10}$$

and the initial condition $\mathbf{T}_R = \mathfrak{J}_{l=0}^{\infty}(\mathbf{F})$. We assume that the problem determines a unique deformation $\mathbf{x} = \boldsymbol{\chi}(\mathbf{X}, t)$ such that a single material point has no displacement. We take the origin at this point. Then, $\mathbf{x} = \boldsymbol{\chi}(\mathbf{0}, t) = 0$.

If we superimpose a rigid rotation $\mathbf{Q}$ about $\mathbf{0}$ upon a deformation $\boldsymbol{\chi}$, we obtain a new deformation $\boldsymbol{\chi}^* = \mathbf{Q}\boldsymbol{\chi}$. Then we can write (3.10) in the form

$$\left.\begin{aligned} \operatorname{Div} \mathop{\mathfrak{J}}_{s=0}^{\infty}(\mathbf{Q}^T\mathbf{F}^*) &= -\rho_R \mathbf{b}, \\ \mathop{\mathfrak{J}}_{s=0}^{\infty}(\mathbf{Q}^T\mathbf{F}^*)\mathbf{n}_R &= \mathbf{t}_R \end{aligned}\right\} \tag{3.11}$$

or

$$\left.\begin{aligned} \operatorname{Div} \mathop{\mathfrak{J}}_{s=0}^{\infty}(\mathbf{F}^*) &= -\rho_R \mathbf{Q}\mathbf{b}, \\ \mathop{\mathfrak{J}}_{s=0}^{\infty}(\mathbf{F}^*) &= \mathbf{Q}\mathbf{t}_R. \end{aligned}\right\} \tag{3.12}$$

If $\mathbf{F}$ is a solution of (3.10), then $\mathbf{F}^* = \mathbf{QF}$ is a solution of (3.11) or (3.12). The prescribed forces satisfy the equilibrium condition. When dealing with the infinitesimal approximation of the problem, we find out that, during an infinitesimal deformation, the moment equilibrium condition can be violated. However, according to the theorem of Da Silva, a rigid rotation $\mathbf{Q}$ can always be found such that, after the rotation, the moment equilibrium condition is satisfied. The same effect can be achieved by the rotation (3.12) of external forces. Then, the moment equilibrium condition assumes the form

$$\int_{\partial B} \mathbf{Q}\mathbf{p} \times \mathbf{t}_{1R}\, ds_R + \int_B \mathbf{Q}\mathbf{p} \times \rho_R \mathbf{b}_1\, dv_R = 0. \tag{3.13}$$

We assume a series expansion of $\mathbf{Q}$:

$$\mathbf{Q} = \mathbf{Q}_0 + \sum_{n=1}^{\infty} \varepsilon^n \mathbf{Q}_n. \tag{3.14}$$

Making use of the orthogonality condition

$$\mathbf{Q}\mathbf{Q}^T = \mathbf{I} = \mathbf{Q}_0\mathbf{Q}_0^T + \sum_{n=1}^{\infty} \varepsilon^n \sum_{k=0}^{n} \mathbf{Q}_k \mathbf{Q}_{n-k}^T \tag{3.15}$$

and equating like powers of ε yields

$$\mathbf{Q}_0\mathbf{Q}_0^T = \mathbf{I}, \qquad \sum_{k=0}^{n} \mathbf{Q}_k \mathbf{Q}_{n-k}^T = 0, \tag{3.16}$$

which are restrictions on $\mathbf{Q}_n$.

By substituting (3.14) into (3.13), we arrive at

$$\int_{\partial B} \sum_{n=0}^{\infty} \varepsilon^n \mathbf{Q}_n\left(\mathbf{p}_R + \sum_{k=1}^{\infty} \varepsilon^n \mathbf{u}_n\right) \times \mathbf{t}_{1R}\, ds_R + \int_B \sum_{n=0}^{\infty} \varepsilon^n \mathbf{Q}_n\left(\mathbf{p}_R + \sum_{k=1}^{\infty} \varepsilon^n \mathbf{u}_n\right) \times \rho_R \mathbf{b}_1\, dv_R = 0. \quad (3.17)$$

Then, collecting terms with like powers of ε, we obtain

$$\int_{\partial B} \left(\mathbf{Q}_n \mathbf{p}_R + \sum_{k=0}^{n-1} \mathbf{Q}_k \mathbf{u}_{n-k}\right) \times \mathbf{t}_{1R}\, ds_R + \int_B \left(\mathbf{Q}_n \mathbf{p}_R + \sum_{k=0}^{n-1} \mathbf{Q}_k \mathbf{u}_{n-k}\right) \times \rho_R \mathbf{b}_1\, dv_R = 0, \quad (3.18)$$

which, together with (3.16), constitutes systems sufficient for the determination of $\mathbf{Q}_n$. It is obvious that, because of (3.16), systems of equations for determination of $\mathbf{Q}_n$ are non-linear.

When looking for a linear approximation of the problem corresponding to the solution of an elastic problem due to Signorini, we set

$$\mathbf{Q}_0 \simeq \mathbf{I} + \mathbf{\Omega}_0, \qquad \mathbf{Q}_n \simeq \mathbf{\Omega}_n, \qquad n \geqslant 1, \quad (3.19)$$

where $\mathbf{\Omega}_n (n \geqslant 0)$ are skewsymmetric tensors of infinitesimal rotation. Then, we obtain

$$\int_{\partial B} \mathbf{\Omega}_0 \mathbf{p}_R \times \mathbf{t}_{1R}\, ds_R + \int_B \mathbf{\Omega}_0 \mathbf{p}_R \times \rho_R \mathbf{b}_1\, dv_R = 0 \quad (3.20)$$

and

$$\int_{\partial B} \left(\mathbf{\Omega}_n \mathbf{p}_R + \mathbf{u}_n + \sum_{k=0}^{n-1} \mathbf{\Omega}_k \mathbf{u}_{n-k}\right) \times \mathbf{t}_{1R}\, ds_R + \int_B \left(\mathbf{\Omega}_n \mathbf{p}_R + \mathbf{u}_n + \sum_{k=0}^{n-1} \mathbf{\Omega}_k \mathbf{u}_{n-k}\right) \times \rho_R \mathbf{b}_1\, dv_R = \mathbf{0}, \quad (3.21)$$

which constitute systems sufficient for the determination of $\mathbf{\Omega}_n (n \geqslant 0)$. When deriving (3.20), we have made use of the torque condition of equilibrium in the undeformed state.

Making use of the vector representation of the tensor of infinitesimal rotation, we have

$$\int_{\partial B} (\boldsymbol{\omega}_0 \times \mathbf{p}_R) \times \mathbf{t}_{1R}\, ds_R + \int_B (\boldsymbol{\omega}_0 \times \mathbf{p}_R) \times \rho_R \mathbf{b}_1\, dv_R = \mathbf{0} \quad (3.22)$$

and

$$\int_{\partial B}\left(\boldsymbol{\omega}_n \times \mathbf{p}_R + \mathbf{u}_n + \sum_{k=0}^{n-1} \boldsymbol{\omega}_k \times \mathbf{u}_{n-k}\right)\times \mathbf{t}_{1R}\, ds_R$$

$$+\int_{B}\left(\boldsymbol{\omega}_n \times \mathbf{p}_R + \mathbf{u}_n + \sum_{k=0}^{n-1} \boldsymbol{\omega}_k \times \mathbf{u}_{n-k}\right)\times \rho_R \mathbf{b}_1\, dv_R = \mathbf{0}. \quad (3.23)$$

Now it is necessary to distinguish two cases.

(a) *Problems compatible in the sense of Signorini* In this case, external forces have no axis of equilibrium and (3.22) yields

$$\boldsymbol{\omega}_0 = \mathbf{0}. \quad (3.24)$$

Equation (3.23) can be written in the form

$$\int_{\partial B} (\boldsymbol{\omega}_n \times \mathbf{p}_R)\times \mathbf{t}_{1R}\, ds_R + \int_B (\boldsymbol{\omega}_n \times \mathbf{p}_R)\times \rho_R \mathbf{b}_1\, dv_R = -\mathbf{q}_n, \quad (3.25)$$

where

$$\mathbf{q}_n = \int_{\partial B}\left(\mathbf{u}_n + \sum_{k=1}^{n-1} \boldsymbol{\omega}_k \times \mathbf{u}_{n-k}\right)\times \mathbf{t}_{1R}\, ds_R$$

$$+\int_{B}\left(\mathbf{u}_n + \sum_{k=1}^{n-1} \boldsymbol{\omega}_k \times \mathbf{u}_{n-k}\right)\times \rho_R \mathbf{b}_1\, dv_R. \quad (3.26)$$

Equation (3.25) can be expressed in the form

$$(\mathbf{A}_1 - \mathbf{I}\,\mathrm{tr}\,\mathbf{A}_1)\boldsymbol{\omega}_n = -\mathbf{q}_n, \quad (3.27)$$

where $\mathbf{A}_1$ is the astatic load corresponding to $\mathbf{t}_{1R}$ and $\rho_R \mathbf{b}_1$, given by the formula

$$\mathbf{A}_1 = \int_{\partial B} \mathbf{p}_R \otimes \mathbf{t}_{1R}\, ds_R + \int_B \mathbf{p}_R \otimes \rho_R \mathbf{b}_1\, dv_R, \quad (3.28)$$

where $\otimes$ denotes the tensor product (dyadic product) of two vectors. For $n = 1$, (3.26) corresponds to the equation given for determination of $\boldsymbol{\omega}_1$ by Grioli [3].

(b) *Problems incompatible in the sense of Signorini* In this case, external forces have at least one axis of equilibrium. Then, (3.22) admits a non-zero solution

$$\boldsymbol{\omega}_0 \neq 0. \quad (3.29)$$

However, $\boldsymbol{\omega}_0$ is limited to rotations around axes of equilibrium. For

$n=1$, one obtains

$$\int_{\partial B}(\boldsymbol{\omega}_1\times\mathbf{p}_{\mathrm{R}}+\mathbf{u}_1+\boldsymbol{\omega}_0\times\mathbf{u}_1)\times\mathbf{t}_{1\mathrm{R}}\,ds_{\mathrm{R}}+\int_B(\boldsymbol{\omega}_1\times\mathbf{p}_{\mathrm{R}}+\mathbf{u}_1+\boldsymbol{\omega}_0\times\mathbf{u}_1)\times\rho_{\mathrm{R}}\mathbf{b}_1\,dv_{\mathrm{R}}=0. \quad (3.30)$$

Restricting $\boldsymbol{\omega}_1$ to rotations around axes of equilibrium, we obtain

$$\int_{\partial B}(\boldsymbol{\omega}_0\times\mathbf{u}_1)\times\mathbf{t}_{1\mathrm{R}}\,ds_{\mathrm{R}}+\int_B(\boldsymbol{\omega}_0\times\mathbf{u}_1)\times\rho_{\mathrm{R}}\mathbf{b}_1\,dv_{\mathrm{R}}=-\int_{\partial B}\mathbf{u}_1\times\mathbf{t}_{1\mathrm{R}}\,ds_{\mathrm{R}}-\int_B\mathbf{u}_1\times\rho_{\mathrm{R}}\mathbf{b}_1\,dv_{\mathrm{R}}, \quad (3.31)$$

which can be written in the form

$$(\mathbf{B}_1-\mathbf{I}\,\mathrm{tr}\,\mathbf{B}_1)\boldsymbol{\omega}_0=-\mathbf{r}_1, \quad (3.32)$$

where

$$\mathbf{B}_1=\int_{\partial B}\mathbf{u}_1\otimes\mathbf{t}_{1\mathrm{R}}\,ds_{\mathrm{R}}+\int_B\mathbf{u}_1\otimes\rho_{\mathrm{R}}\mathbf{b}_1\,dv_{\mathrm{R}} \quad (3.33)$$

and

$$\mathbf{r}_1=\int_{\partial B}\mathbf{u}_1\times\mathbf{t}_{1\mathrm{R}}\,ds_{\mathrm{R}}+\int_B\mathbf{u}_1\times\rho_{\mathrm{R}}\mathbf{b}_1\,dv_{\mathrm{R}}. \quad (3.34)$$

Similarly, for an arbitrary n, we obtain

$$(\mathbf{B}_n-\mathbf{I}\,\mathrm{tr}\,\mathbf{B}_n)\boldsymbol{\omega}_{n-1}=-\mathbf{r}_n, \quad (3.35)$$

where

$$\mathbf{B}_n=\int_{\partial B}\mathbf{u}_n\otimes\mathbf{t}_{1\mathrm{R}}\,ds_{\mathrm{R}}+\int_B\mathbf{u}_n\otimes\rho_{\mathrm{R}}\mathbf{b}_1\,dv_{\mathrm{R}} \quad (3.36)$$

and

$$\mathbf{r}_n=\int_{\partial B}\left(\mathbf{u}_n+\sum_{k=0}^{n-2}\boldsymbol{\omega}_k\times\mathbf{u}_{n-k}\right)\times\mathbf{t}_{1\mathrm{R}}\,ds_{\mathrm{R}}+\int_B\left(\mathbf{u}_n+\sum_{k=0}^{n-2}\boldsymbol{\omega}_k\times\mathbf{u}_{n-k}\right)\times\rho_{\mathrm{R}}\mathbf{b}_1\,dv_{\mathrm{R}}, \qquad n\geqslant 2. \quad (3.37)$$

Thus, in the case of incompatibility in the sense of Signorini, the proposed procedure also yields compatible systems of equations for the determination of $\boldsymbol{\omega}_n$ and leads to compatibility between the infinitesimal theory and the finite deformation theory.

Considering the expansion

$$\mathbf{p}_R^* + \sum_{n=1}^{\infty} \varepsilon^n \mathbf{u}_n^* = \left(\mathbf{I} + \sum_{n=0}^{\infty} \varepsilon^n \boldsymbol{\Omega}_n\right)\left(\mathbf{p}_R + \sum_{k=1}^{\infty} \varepsilon^k \mathbf{u}_k\right), \tag{3.38}$$

we arrive at

$$\mathbf{u}_n^* = (\mathbf{I} + \boldsymbol{\Omega}_0)\mathbf{u}_n + \boldsymbol{\Omega}_n \mathbf{p}_R + \sum_{k=1}^{n-1} \boldsymbol{\Omega}_k \mathbf{u}_{n-k}, \tag{3.39}$$

which, in the vectorial form, can be written as

$$\mathbf{u}_n^* = \mathbf{u}_n + \boldsymbol{\omega}_0 \times \mathbf{u}_n + \boldsymbol{\omega}_n \times \mathbf{p}_R + \sum_{k=1}^{n-1} \boldsymbol{\omega}_k \times \mathbf{u}_{n-k}. \tag{3.40}$$

It is easy to show that, in each case, $\mathbf{u}_n$ satisfy the existence conditions (3.9).

We can now generalize the theorem of compatibility and uniqueness which was formulated by Signorini [1] for boundary-value problems of finite elasticity.

Theorem *If external forces $\mathbf{t}_{1R}$, $\mathbf{b}_1$ do not possess any axis of equilibrium and the corresponding traction initial boundary-value problem of infinitesimal viscoelasticity has a solution, then there exist solutions $\mathbf{u}_1, \ldots, \mathbf{u}_n, \ldots$ of the successive systems of approximation. This solution $\mathbf{u}_1, \mathbf{u}_2, \ldots$ is unique if the classical uniqueness theorem for the traction initial boundary-value problem in the infinitesimal theory holds.*

If external loads $\mathbf{t}_{1R}$, $\mathbf{b}_1$ possess an axis of equilibrium, the solution can be non-unique. However, there is no incompatibility between the non-linear theory of viscoelasticity and the infinitesimal theory of viscoelasticity. The same is true for the non-linear theory of elasticity (Brilla [6]).

References

1 Signorini, A., Sulle deformazioni termoelastiche finite, *Proc. 3rd Int. Congr. Appl. Mech.*, **2,** 80–89, 1930.

2 Tolotti, C., Orientamenti principali di un corpo elastico rispetto alla sua sollecitazione totale, *Mem. Accad. It. Cl. Sci. Mat. Nat., VII,* **13,** 1139–1162, 1943.

3 Grioli, G., *Mathematical Theory of Elastic Equilibrium* (*Recent Results*), Ergeb. angew. Math. No. 7, Springer Verlag, Berlin, Heidelberg and New York, 1962.

4 Truesdell, C. and Noll, W., The non-linear field theories of mechanics, in *Encyclopedia of Physics*, Springer Verlag, Berlin, Heidelberg and New York, 1965, Vol. III/3, pp. 223–227.

5 Capriz, G., and Podio Guidugli, P. On Signorini's perturbation method in finite elasticity, *Arch. Rat. Mech. Anal.*, **57,** 1–30, 1974.

6 Brilla, J., Compatible perturbation method in finite elasticity, *Mech. Res. Commun.*, to appear.

7 Rivlin, R. S., On the foundations of the theory of non-linear viscoelasticity, in *Mechanics of Visco-Elastic Media and Bodies* (J. Hult, ed.), Springer Verlag, Berlin, Heidelberg and New York, 1975, pp. 26–40.

8 Coleman, B. D. and Noll, W., Foundation of linear viscoelasticity, *Rev. Mod. Phys.*, **30,** 1508–1512, 1959.

Professor Jozef Brilla, D.Sc.,
Institute of Applied Mathematics and Computing Technique,
Comenius University,
Mlynská dolina,
816 31 Bratislava,
Czechoslovakia.

K. B. Broberg

Mathematical methods in fracture mechanics

Abstract

Problems of fracture mechanics are usually separated into two parts: consideration of the process region near the crack-tip, where microseparations occur, and consideration of the surrounding region. The process-region side concerns material microstructure, whereas the surrounding region can be treated as a continuum. Certain problems can be idealized by artificially extending the continuum over the process region. The justification of the separation artifice is analysed and some mathematical methods used for treating the ensuing problems are reviewed.

1 Introduction

The most important problem of fracture mechanics concerns the onset of fracture, i.e., onset of unstable crack growth. However, the preceding and succeeding events, onset and progress of stable crack growth and propagation and arrest of unstable crack growth, also receive much attention.

Fracture mechanics as a discipline differs from pure continuum mechanics in that the fracture process itself occurs in a small non-continuum region: the process region. In this region, intrinsic length dimensions of the material are essential. The fracture problem always consists of matching two different regions, the process region and the continuum, to each other. Great simplifications are often possible, due to the smallness of the process region and to its autonomy, a concept introduced by Barenblatt [1]. This concept simply tells us that, during loading of the body, the mechanical events near the crack-tip depend only on the material, not on the load distribution, the body geometry or the crack length.

The smallness of the process region and the property of autonomy often enable us to consider separate treatments of the continuum side and of the process-region side. When treating the continuum side, the domain is usually artificially extended over the process region.

2 The fracture process

The fracture process consists of three stages.

1. Loading before crack growth.
2. Stable crack growth.
3. Unstable crack growth.

Between the three stages, two critical states appear.

1. Onset of stable crack growth.
2. Onset of unstable crack growth.

3 The smallness of the process region

Examples of the intrinsic material dimensions that determine the process-region size in the static case are the distance between inclusions (at dimple-type fracture) and the interatomic distance (at ideal cleavage). In practically all cases, the process-region size is very small compared to the length of cracks, the presence of which substantially increases the risk of fracture.

4 Autonomy

In order to illustrate the concept of autonomy, a case of antiplane strain in an elastic-plastic material will be examined (*see* Fig. 1). A cylindrical

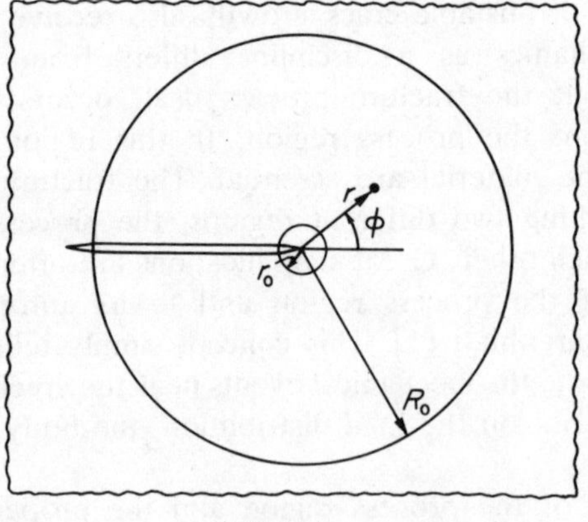

Fig. 1

coordinate system r, ϕ, z is introduced, the line $r=0$ coinciding with the right edge of the crack. Assume that an experiment gives the following results.

1. The region $r_0<r<R_0$, where

$$r_0/R_0 \ll 1, \tag{1}$$

is found to behave elastically up to the first critical state. It is understood that r_0 is the smallest and R_0 the largest radius compatible with this observation. (Moreover, R_0 should not exceed the crack length.)

2. The stress τ_{rz} at $r=R_0$ is found to be

$$\tau_{rz}=G\sum_{1,3,5,\ldots}\gamma_n \sin(n\phi/2), \tag{2}$$

where

$$|\gamma_n|<|\gamma_1| \quad \text{for} \quad n\geqslant 3, \tag{3}$$

and G is the modulus of rigidity.

3. The displacement w at $r=r_0$ is found to be

$$w(r_0,\phi)=\sum_{1,3,\ldots}\delta_n \sin(n\phi/2). \tag{4}$$

For $r=r_0$, a mathematical treatment gives

$$\tau_{rz}=\sum_{1,3,\ldots}\frac{G\sin(n\phi/2)}{1+(r_0/R_0)^n}\left\{2\gamma_n\left(\frac{r_0}{R_0}\right)^{(n/2)-1}-\frac{n\delta_n}{2r_0}\left[1-\left(\frac{r_0}{R_0}\right)^n\right]\right\}. \tag{5}$$

Recalling that $r_0/R_0\ll 1$, one obtains

$$\tau_{rz}(r_0,\phi)\approx K_{\mathrm{III}}(2\pi r_0)^{-1/2}\sin(\phi/2)-\frac{G}{2r_0}\sum_{1,3,\ldots}n\delta_n\sin(n\phi/2), \tag{6}$$

where

$$K_{\mathrm{III}}=G\gamma_1(2\pi R_0)^{1/2} \tag{7}$$

is the conventional stress intensity factor.

Equations (6) and (7) show that the action of the load at $r=R_0$ on the region $r\leqslant r_0$ consists of the action of the load component $G\gamma_1\sin(\phi/2)$ only. At monotone loading (i.e., γ_1 is monotone increasing), this action is known if the value of K_{III} is known. Hence, another experiment with the same material, but different crack length, body geometry and exterior load distribution, will give the same mechanical state of the region $r\leqslant r_0$ at the same value of K_{III}, provided that conditions (1) and (3) are fulfilled. Condition (3) simply excludes load distributions that are either 'pathological' or uninteresting in fracture problems.

The result can be described as follows.

For the class of mode III crack problems characterized by

$$R_0 > \frac{1}{p} r_0, \qquad p \ll 1, \tag{8}$$

the events taking place in the region $r < r_0$ *are, within an accuracy given by p, independent of the body geometry, the crack length and the load distribution* (excluding 'pathological' and uninteresting distributions) *during monotone loading up to the occurrence of the first critical state. This is the property of autonomy.*

The value of p in (8) can be taken to indicate the *scale of yielding*. Small-scale yielding is conventionally regarded to prevail when p is less than some few per cent.

Now, it is assumed for a moment that an annular region $r_a < r < r_b$ can be found, such that the relations $r_0/r_a \ll 1$ and $r_b/R_0 \ll 1$ are simultaneously fulfilled. Then one gets, approximately,

$$\tau_{rz} \approx \frac{K_{III}}{(2\pi r)^{1/2}} \sin(\phi/2) \quad \text{for} \quad r_a < r < r_b. \tag{9}$$

This relation holds with an accuracy represented by $p_1 \ll 1$ if

$$r_a \geqslant r_0/p_1 \quad \text{and} \quad r_b \leqslant p_1 R_0,$$

implying

$$R_0 > \frac{1}{p_1^2} r_0, \tag{10}$$

i.e., a much more restrictive condition than the one given by (8). The existence of a region in which (9) holds is, thus, not necessary for autonomy.

By idealizing the problem beforehand and taking $r_0 = 0$, relation (9), then valid for $r \ll R_0$, is easily found. It expresses the well-known square root stress singularity. K_{III} is determined correctly in this way within the accuracy imposed by (8). Thus, K_{III} can be found by searching the stress singularity of the idealized problem, even when (9) does not hold anywhere in the real case with the same expectation of accuracy as for K_{III} itself.

The results obtained can be carried over to cases of symmetric or antisymmetric in-plane strain or stress (modes I and II). They concern extremely small-scale yielding (because of the condition $r_0/R_0 \ll 1$). However, even cases of large-scale yielding can most often be included, since the region of autonomy must not necessarily cover the whole plastic region but only the process region (and, in cases where an extent of stable crack growth is considered, also the non-continuum region surrounding the path of growth). Thus, for instance, while the contour of the plastic region may be strongly influenced by the distribution of the exterior load, the influence of this distribution *may* be 'smoothed out' as the process region is approached. *This is then a consequence of ellipticity. On the other hand if a large hyperbolic domain exists around the crack-tip* (as at perfect plasticity at large-scale yielding) *then details of the load distribution can be propagated via characteristics to the very neighbourhood of the tip.*

It should also be noted that failure of ellipticity can occur in non-linear elasticity at large strains, cf. Knowles and Sternberg [2].

By analogy with the small-scale yielding case, where the stress intensity factor K_{III} uniquely specifies the state of the process region in a given material, one single quantity can be used to describe this state at large-scale yielding, provided that autonomy holds. A useful quantity is the J-integral (to be defined later on), which is path-independent in an elastic (even non-linearly) region and approximately path-independent in the plastic region around the process region.

5 Weight function representation of stress intensity factors

Bueckner [3] and Rice [4] studied a reciprocity theorem for stress intensity factors (extremely small-scale yielding and linear elasticity). The stress intensity factor at a tip of a crack in a symmetrical case can be written in the form

$$K = \int_\Gamma \overline{t(s)}\, \bar{h}(a, s)\, ds + \iint_A \overline{f(x, y)}\bar{h}(a, x, y)\, dx\, dy, \tag{11}$$

where $\bar{t}$ is the boundary traction (load per unit of length), s an arc-length along the body boundary Γ, $\bar{f}$ the volume force (force per unit of area), a the crack length, A the body area and $\bar{h}(a, x, y)$ a weight function.

The weight function is given by the formula

$$\bar{h}(a, x, y) = \frac{E}{K'} \cdot \frac{\partial \bar{u}'(a, x, y)}{\partial a}$$

for the case of in-plane stress. E is Young's modulus, u' the displacement field and K' the stress intensity factor at some load system. It turns out that $\bar{h}(a, x, y)$ does not depend on the particular choice of this load system. The determination of $\bar{h}(a, x, y)$ is, however, generally very difficult.

Cases of in-plane or antiplane strain lead to slightly modified expressions.

6 The J-integral

For elastic stress–strain fields, a surface-integral representation of a property related to the presence of a defect (a singularity or an inhomogeneity) was developed by Eshelby [5]. A specialization to plane elastostatic fields leads to the path-independent J-integral, introduced by Rice [6]. He demonstrated its usefulness for crack problems. Knowles and Sternberg [7] showed that two other path-independent integrals also exist for general plane elastostatic fields.

The J-integral is defined as

$$J=\int_{\Gamma}[W\,\mathrm{d}x_2-n_j\sigma_{ji}(\partial u_i/\partial x_1)\,\mathrm{d}s]\qquad (i,j=1,2), \tag{12}$$

which vanishes when Γ is a closed path. W is the strain energy per unit of volume, u_i the displacement, n_i the normal outwards from Γ and s an arc-length along Γ. σ_{ji} is the stress, defined by

$$\sigma_{ji}=\partial W/\partial u_{i,j}, \tag{13}$$

so that non-linear elasticity is included.

Similarly, for a plate with thickness h, traction-free on the plane surfaces $x_3=\pm h/2$, the integral

$$P=\int_{\Gamma}\left[W_{\mathrm{av}}\,\mathrm{d}x_2-\left(n_j\sigma_{ji}\frac{\partial u_i}{\partial x_1}\right)_{\mathrm{av}}\mathrm{d}s\right] \tag{14}$$

is found to vanish for a closed path Γ in the plane of the plate. The subscript av denotes the average value over the thickness h.

In elastic-plastic crack problems, *proportional loading very often prevails with fair accuracy, except in a near-tip region.* Then a total strain theory can be used, implying path-independence of the J-integral in a major part of the plastic region.

7 The separation procedure

The onset of stable crack growth is governed by a local criterion. When the process region reaches a certain state, crack growth is initiated. *Onset of unstable crack growth,* on the other hand, *is a question of global instability* [8]. Thus, for instance, grip-controlled conditions postpone the second critical state as compared to load-controlled conditions.

The fracture process could, in principle, be studied by using a suitable model of the process region, say a box of fixed dimensions with known response to loading. The line model, introduced by Barenblatt [1], is also expected to lead to fairly accurate results. A characteristic of any process-region model should be that, roughly speaking, it is extended under the action of decreasing forces. More strictly, it should be unstable under load-controlled conditions, so that the response has to be given for grip-controlled conditions. *In situ*, the conditions are something between grip- and load-control.

A few treatments with models of the process region exist. Barenblatt [1] assumed linear elasticity outside a line model. This linearization leads to a Keldysh–Sedov problem. A line model was used by Andersson [9] in a study of stable crack growth, whereas a box model was treated by Andersson and Bergkvist [10] for non-growing cracks and by Andersson [11] for a stably-growing crack. Numerical methods were used.

Incorporation of a model of the process region can generally be avoided when the first critical state is studied. For classes of problems in which autonomy holds, it is sufficient to find one single quantity – usually the J-integral – specifying the state of the process region. But, in view of the preceding discussion (sections 4 and 6), *this can be done by treating an idealized problem, where the process region is degenerated to a point and the constitutive equations are extrapolated to infinite strains.* For a given material, then, the criterion for onset of crack growth reads [12]

$$J = J_c, \tag{15}$$

where J_c is a material constant that can be experimentally determined.

Stable crack growth is governed by a *balance condition* [8, 13]:

$$\frac{dU_0}{da} = \frac{dD_0}{da}, \tag{16}$$

where dU_0/da is the energy flow per unit of crack growth from the continuum to the process region and dD_0/da is the energy dissipation per unit of crack growth in the process region. Unstable crack growth occurs when $da \to \infty$ at an infinitesimal increase of loading. For classes of problems in which autonomy holds, the right-hand side of (16) is a function of the crack growth $(a - a_0)$ only, in each given material. One intrinsic difficulty, however, is that dU_0/da is strongly coupled to the presence of the process region. Thus, while J in (15) can, in general, be accurately determined in an idealized treatment – where the process region is degenerated to a point – this is not possible for dU_0/da. In fact, such a treatment for a plane strain case without strain hardening in the near-tip region leads to the result $dU_0/da = 0$ [14].

In the dynamic case of growing cracks, the small-scale yielding case can be treated by using the balance condition [15, 16]

$$\frac{dU}{da} = \frac{dD}{da}, \tag{17}$$

where dU/da is the energy flow per unit of crack growth to the dissipative region and dD/da the energy dissipation per unit of crack growth. Often, the dissipative region is small enough for a steady-state condition to prevail in its vicinity at each moment. Then, dD/da is a material constant that depends only on the crack-tip velocity.

8 The continuum problem

The discussion in the preceding section has shown that a separate treatment of an idealized continuum problem is useful in several cases of fracture mechanics. The idealization consists of a degeneration of the process region to a point which necessitates extrapolation of the constitutive equations to infinite strains.

Analytical treatments of the (idealized) continuum problem are generally limited to relatively simple geometries and constitutive equations. A review of methods used will be given in the following sections.

9 Static problems at linear elasticity

9.1 Eigenfunction expansion

The method of eigenfunction expansion often enables calculation of K (or J) without determination of the complete stress–strain state. As an example, consider a long twisted bar with circular cross-section (radius R) and with a longitudinal crack which penetrates symmetrically from the surface to a depth $a < R$. The twist per unit of length is denoted Θ and is caused by torques at the bar ends. The problem consists in solving the equation

$$\Delta\Phi = -2, \tag{18}$$

where Φ is the potential energy for the stresses, used in the St Venant theory of torsion. The boundary condition is

$$\Phi = 0. \tag{19}$$

After introduction of bipolar coordinates ξ, η, z, a solution of (18), satisfying the condition of bounded strain energy, is found by eigenfunction expansion:

$$\Phi = -\frac{c^2 \sin^2 n}{(\cosh\xi - \cos\eta)^2} + \frac{c^2}{\pi} \sum_{1,3,\ldots} I_n(\xi_0) e^{-(n/2)(\xi-\xi_0)} \cos\frac{n\eta}{2}, \tag{20}$$

where ξ_0 is the value of ξ at the mantle surface, c a certain constant and

$$I_n(\xi_0) = \int_{-\pi}^{\pi} \frac{\sin^2\eta \cos(n\eta/2)}{(\cosh\xi_0 - \cos\eta)^2} d\eta. \tag{21}$$

For $\xi \to \infty$ (i.e., when the edge of the crack is approached), one obtains

$$\Phi \to \frac{c^2}{\pi} I_1(\xi_0) e^{-(1/2)(\xi-\xi_0)} \cos\frac{\eta}{2}, \tag{22}$$

whereupon the stress intensity factor is easily found to be

$$K_{\mathrm{III}} = \frac{G\Theta a(2R-a)^2}{4\pi(R-a)^2} \left\{ \frac{4R(R-a)+3a^2}{2a[R(R-a)]^{1/2}} \tan^{-1} \frac{2[R(R-a)]^{1/2}}{a} - 3 \right\} \cdot \left(\frac{2\pi aR}{2R-a} \right)^{1/2}. \tag{23}$$

9.2 Complex potentials

For problems of in-plane stress and strain or antiplane strain, harmonic potentials can be found. Then, one can take advantage of the simplicity and elegance connected with analytic functions.

In the case of a crack $y=0$, $b<x<c$ in a body symmetrical with respect to the plane $y=0$ and symmetrically loaded, some modifications of the usual basic equations [17] can be made so that the two unknown analytical functions are analytic inside the body contour, i.e., also at the crack tips. This is particularly useful in connection with the *boundary collocation method*, after mapping onto the unit circle $|\zeta|<|$, since approximations of the type

$$Q_1(\zeta) \approx \sum_0^N A_n \zeta^n, \qquad Q_2(\zeta) \approx \sum_0^M B_n \zeta^n \tag{24}$$

can assuredly be used to present the two unknown analytic functions Q_1 and Q_2 with optional level of accuracy if N and M are chosen large enough. At boundary collocation, the boundary condition is satisfied at sampling points along the boundary.

9.3 Integral equations

Integral equations are useful in crack problems for, among other reasons, their suitability for numerical treatments. This has been demonstrated, for instance, by Erdogan and collaborators. An excellent review is given in [18].

A simple example will show how an integral equation arises in a crack problem. A crack $y=0$, $|x|<a$ is situated in an infinite plate subjected to a constant stress $\sigma_y = \sigma_\infty$ at infinity. The lower half-plane is studied. By

studying first the load case

$$\left.\begin{aligned}\sigma_y &= q(\xi)\,\mathrm{d}\xi\,\delta(x-\xi)\\ \tau_{yx} &= 0\end{aligned}\right\} \quad \text{on} \quad y=0, \tag{25}$$

where $\delta(x)$ is the Dirac delta function, and then superposing such load cases, one obtains an integral equation

$$\int_a^\infty \frac{q(\xi)\,\mathrm{d}\xi}{\xi^2-x^2}=0 \quad \text{for} \quad |x|>a \tag{26}$$

with the integrable solution

$$q(x)=\frac{A\,|x|}{(x^2-a^2)^{1/2}} \quad \text{for} \quad |x|>a. \tag{27}$$

The condition at infinity then gives $A=\sigma_\infty$.

A variation of the technique consists in initially regarding a dislocation with Burger's vector $f(\xi)\,\mathrm{d}\xi$. Then,

$$\sigma_y=\int_{-a}^{a}\frac{f(\xi)\,\mathrm{d}\xi}{x-\xi}+\sigma_\infty \tag{28}$$

and the condition of traction-free crack surfaces implies

$$\int_{-a}^{a}\frac{f(\xi)\,\mathrm{d}\xi}{x-\xi}=-\sigma_\infty, \qquad |x|<a; \tag{29}$$

this integral equation has to be solved under the condition that the total Burger's vector is zero, i.e., the crack closes at the ends:

$$\int_{-a}^{a} f(\xi)\,\mathrm{d}\xi=0. \tag{30}$$

Closely related to integral equation methods is the formulation of a Hilbert problem. This is sometimes simpler, because the problem can be treated in one single step, avoiding complications that occasionally can arise by involving two steps in the procedure.

9.4 Integral transform methods

Integral transform methods are often used in cases where the edge of the crack is curved. A comparatively simple example will be studied, concerning a 'penny-shaped' crack in an infinite solid.

A cylindrical coordinate system r, ϕ, z is introduced. The crack occupies the plane $z=0$, $r<a$. At infinity, the stress $\sigma_z=\sigma_\infty$ is acting.

After introduction of Hankel transforms, the displacement u_z and the stress σ_z can be represented by the expressions

$$Gu_z = -\frac{1-\nu}{1-2\nu}\int_0^{\infty} \xi^2 A(\xi) J_0(\xi r)\,d\xi, \tag{31}$$

$$\sigma_z = \frac{1}{1-2\nu}\int_0^{\infty} \xi^3 A(\xi) J_0(\xi r)\,d\xi + \sigma_\infty \tag{32}$$

for $z=0$. Since

$$\sigma_z = 0 \quad \text{for} \quad r < a, \tag{33}$$

$$u_z = 0 \quad \text{for} \quad r \geqslant a, \tag{34}$$

dual integral equations are obtained for the determination of $A(\xi)$. In the example studied, they can be readily solved, cf., [19].

10 Elastic-plastic problems

10.1 General considerations

Most elastic-plastic crack problems have been solved by using a small-strain theory and concern cases where incremental and total strain theories coincide. The latter circumstance implies, among other things, that the J-integral is path-independent. It does not *per se* concern the presence or non-presence of autonomy.

The idealization of considering the process region to be point-sized and the consequential extrapolation of the constitutive equations to infinite strains does not introduce much error in the case of non-growing cracks, but for growing cracks it is, unfortunately, as has been previously discussed, detrimental.

The classical solution of an elastic-plastic crack problem was given in 1956 by Hult and McClintock [20]. The problem concerned a case of antiplane strain at perfect plasticity and small-scale yielding. Their results were extended to large-scale yielding and to strain-hardening by Rice in a series of important papers, *see* [21, 22].

Some general properties of perfectly plastic antiplane strains will first be considered, whereupon some general methods for cases of proportional loading at antiplane strain (including strain hardening) will be examined.

By writing

$$\bar{\tau} = |\tau_{xz}, \tau_{yz}|,$$

the yield criterion, valid in the plastic region, reads

$$|\bar{\tau}| = k \tag{35}$$

and the equilibrium condition has the form

$$\operatorname{div} \bar{\tau} = 0 \tag{36}$$

for elastic and plastic regions.

Curves along which $\bar{\tau}$ is constant are found to be straight lines.

Compatibility demands that

$$\operatorname{curl} \bar{\gamma} = 0, \tag{37}$$

where $\bar{\gamma} = |\gamma_{xz}, \gamma_{yz}|$ is the strain vector, derived from the displacement $w(x, y)$.

The case of *proportional loading* is now studied. It means that $\bar{\tau}$ is constant at each material point in the plastic region during loading. Then, it can be shown that

$$w = \text{constant along a characteristic line.} \tag{38}$$

Consider now two infinitesimally neighbouring characteristic lines. Denote by s the distance from the point of intersection. Then, since w is a constant along each line,

$$\operatorname{grad} w = \frac{\text{constant}}{s}, \tag{39}$$

$$\operatorname{grad} w \perp \text{to a characteristic line,} \tag{40}$$

Thus, since $\bar{\gamma} = \operatorname{grad} w$,

$$\bar{\gamma} = \frac{\bar{A}}{s} \text{ along a characteristic line.} \tag{41}$$

where $\bar{A}$ is a constant vector parallel to $\bar{\tau}$. Then, since the plastic strains are zero at each regular point of the boundary between a plastic region and a virginly elastic region, a characteristic line cannot intersect such a boundary twice, except when one intersection is a singular point.

10.2 Crack problems at antiplane strain and perfect plasticity

Figure 2 shows a possible arrangement of characteristic lines for a crack problem at antiplane strain and perfect plasticity. Then,

$$\tau_{rz} = 0, \tag{42}$$

$$\tau_{\phi z} = k, \tag{43}$$

$$\gamma_{rz} = 0, \tag{44}$$

$$\gamma_{\phi z} = \frac{k}{G} \frac{R(\phi)}{r}, \tag{45}$$

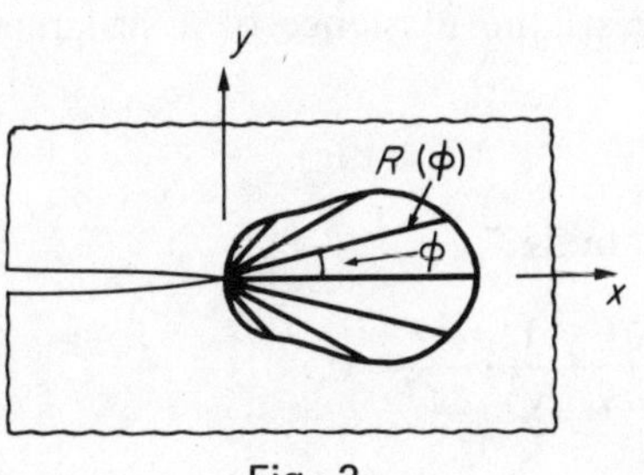

Fig. 2

where $R(\phi)$ is the radius vector of the boundary between elastic and plastic regions. The near-tip solution given by (44), (45) as $r \to 0$ is completely determined by the position of the elastic-plastic boundary. Thus, autonomy cannot be present in a class of problems where $R(\phi)$ is not a unique function of ϕ. This will be considered further in the following.

10.3 The hodograph transform

A *hodograph transform* is sometimes convenient in treatment of elastic-plastic crack problems at antiplane strain. The following treatment is due to Rice [21, 22].

For an elastic region at antiplane strain, the constitutive equation is written

$$\tau = \tau(\gamma), \tag{46}$$

where

$$\tau = |\bar{\tau}| \quad \text{and} \quad \gamma = |\bar{\gamma}|; \tag{47}$$

$\bar{\gamma}$ is parallel to $\bar{\tau}$ at each point. The relations also hold for a plastic region at proportional loading.

In the hodograph transform, the position vector $\bar{r}$ is written as a function of $\bar{\gamma}$ or of $\bar{\tau}$. Then (36) is transformed to

$$\text{div}_\tau \, \bar{r} = 0, \tag{48}$$

where subscript τ denotes that τ and ψ are regarded as independent variables (ψ is the angle between $\bar{\tau}$ and the positive x-axis). Furthermore, (37) is changed to

$$\text{curl}_\gamma \, \bar{r} = 0, \tag{49}$$

with γ and ψ as independent variables.

Equation (49) suggests the existence of a scalar potential function F such that

$$\bar{r} = \text{grad}_\gamma F. \tag{50}$$

Then, using (48), one finds

$$\frac{\tau(\gamma)}{\gamma\tau'(\gamma)} \cdot \frac{\partial^2 F}{\partial \gamma^2} + \frac{1}{\gamma} \cdot \frac{\partial F}{\partial \gamma} + \frac{1}{\gamma^2} \cdot \frac{\partial^2 F}{\partial \psi^2} = 0. \tag{51}$$

In a linearly elastic region, $\gamma\tau'(\gamma) = \tau(\gamma)$, so that (51) reduces to the Laplace equation. For power-law strain hardening,

$$\tau = \text{const} \cdot \gamma^N, \tag{52}$$

so that (51) takes the simple form

$$\frac{1}{N} \cdot \frac{\partial^2 F}{\partial \gamma^2} + \frac{1}{\gamma}\frac{\partial F}{\partial \gamma} + \frac{1}{\gamma^2} \cdot \frac{\partial^2 F}{\partial \psi^2} = 0. \tag{53}$$

A solution of (53) satisfying conditions of antisymmetric displacements and traction-free surfaces is

$$\bar{r} = \sum_{1,2,\ldots} C_{1m} \gamma^{\mu_{1m}-1} [\mu_{1m} \sin \beta_m \cdot \hat{\gamma} - (2m-1) \cos \beta_m \cdot \hat{\psi}], \tag{54}$$

where $m = 1, 2, 3 \ldots$, $\beta_m = (m - \frac{1}{2})(\pi - 2\psi)$, $C_{11} \neq 0$ and

$$\mu_{1m} = \tfrac{1}{2}\{1 - N - [(1-N)^2 + 4N(2m-1)^2]\}^{1/2}, \tag{55}$$

so that $0 > \mu_{11} > \mu_{12} > \ldots$. Other solutions exist, but either they violate the condition of bounded strain energy or they are of less interest because they are regular at the crack-tip.

Inversion of (54) gives

$$\gamma = r^{-1/(1+N)} f_1(\phi) + r^{-\mu_{12}/(1+N)-1} f_2(\phi) + \ldots, \tag{56}$$

where $f_1(\phi)$ is determined to within an amplitude factor. Equation (56) contains one singular term (negative power of r) when $2/7 \leq N \leq 1$ and two or more singular terms when $0 < N < 2/7$. As $N \to 0$ (perfect plasticity), all μ_{1m} tend towards zero, implying that the singular terms flow together so that γ becomes proportional to r^{-1} but the ϕ-dependence is undetermined if the constants C_{1m} are left undetermined. This is in agreement with the previous discussion of perfect plasticity, since the function $R(\phi)$ cannot be determined by a study of the near-tip region only.

Obviously, the larger N is, the larger is the class of problems for which autonomy holds. For $N = 0$, only small-scale yielding problems can be included in this class. It should be observed that *power-law strain hardening leads to proportional loading of the near-tip region.* In fact, it can be

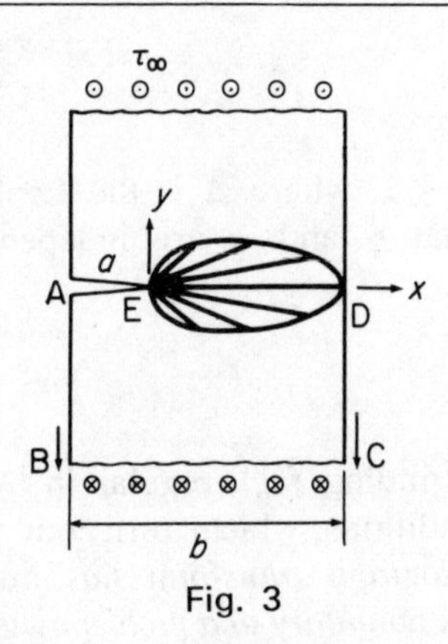

Fig. 3

shown that this property is limited to just power-law strain hardening.

Next, the matching between elastic and plastic regions will be considered for the case of perfect plasticity. As an illustration, a problem considered by Rice [22] will be very briefly discussed.

An infinite strip, $-a<x<(b-a)$ contains a crack, $y=0, -a\leqslant x<0$. It is subjected to the stress $\tau_{yz}=\tau_\infty$ at infinity. The limit load case for which $\tau_\infty = k(b-a)/b$ is studied (*see* Fig. 3).

Introduction of $\zeta=\xi+i\eta$ such that

$$\xi=(G/k)\gamma_{xz}, \tag{57}$$

$$\eta=(G/k)\gamma_{yz}, \tag{58}$$

gives the division into elastic and plastic regions along the curve $|\zeta|=1$, $\eta>0$ in the ξ-plane, the hodograph plane. Studying only the lower half of the body implies that one quarter of the ζ-plane is concerned (*see* Fig. 4). From (50), one obtains

$$x=\frac{G}{k}\cdot\frac{\partial F}{\partial \xi}, \tag{59}$$

$$y=\frac{G}{k}\cdot\frac{\partial F}{\partial \eta}, \tag{60}$$

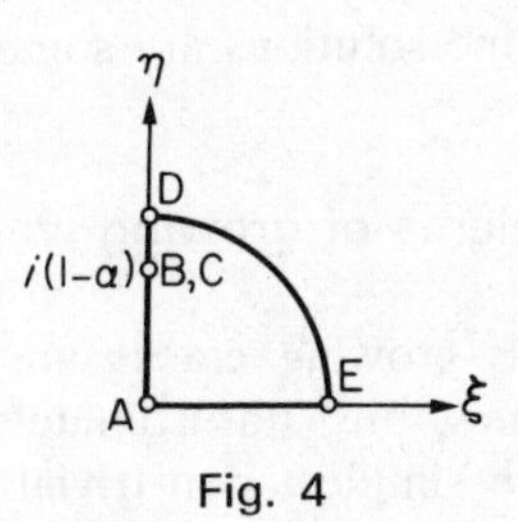

Fig. 4

whereas (51) gives

$$\Delta_\gamma F = 0 \tag{61}$$

for the elastic region $|\zeta| < 1$, where Δ is the Laplace operator and the subscript γ indicates that γ and ψ are independent variables. After writing

$$F = \frac{kb}{G} \operatorname{Re}[f(\zeta)], \tag{62}$$

the problem consists in finding $f(\zeta)$, regular in $|\zeta| < 1, \xi > 0, \eta > 0$, and satisfying boundary conditions which turn out to be conditions on $\operatorname{Im}[f(\zeta)]$. *Thus, the hodograph transform has transformed the original problem with an unknown boundary to a problem with known boundaries.*

10.4 About the uniqueness of the solution at perfect plasticity

Unfortunately, Eqns (42) to (45) do not present a unique solution of an elastic-perfectly plastic crack problem, because a displacement discontinuity along a portion of the line $\phi = 0$ is possible, cf., [23, 24]. In fact, even a mode III equivalence to the Dugdale model [25] (the plastic region degenerated to the line of discontinuity) is possible.

10.5 In-plane elastic-plastic crack problems

For in-plane elastic-plastic crack problems, a hodograph transform is unfortunately not possible. For non-moving cracks, power-law strain hardening ensures proportional loading in the near-tip region. When the strain-hardening exponent $N > 0$, the near-tip strain field is – to within a scaling factor – unique for each given material and can then be found by considering the near-tip region only. This was done for in-plane cases by Rice and Rosengren [26] (in a numerical treatment using an Airy stress function) and by Hutchinson [27]. 'Fully plastic' solutions, where power-law strain hardening is assumed for the whole body, which implies proportional loading, have also been obtained [28, 29].

For near-tip fields, slip-line solutions are sometimes useful, cf., [30].

10.6 Elastic-plastic problems of growing cracks

Problems concerning stably-growing cracks are very important in the theory of fracture mechanics, but, unfortunately, extremely difficult to treat analytically. Even the simplest non-trivial case (perfect plasticity, small-scale yielding, steady growth and antiplane strain) presents much difficulty. This problem was treated by Chitaley and McClintock [31], but, since they used the idealization of a point-size process region, a logarithmic singularity in the strains at the crack-tip was obtained. Such a singularity is too weak to allow a non-zero energy release rate, which is necessary for crack growth. Therefore, some model of a process region must be incorporated.

The inherent difficulties in analytical treatments of stably growing cracks constitute a considerable obstacle against a deeper understanding of the fracture process.

11 Dynamic problems

11.1 Growing cracks

Hitherto, the mathematical treatment of fast-growing cracks has been limited to the case of small-scale yielding. In that case, it is possible to show that a near-tip solution exists, which, to within an amplitude factor, is unique for each given material and crack-tip velocity [32, 33].

The energy flow to the crack-tip per unit of crack growth is most conveniently found by smoothing the singularity at the crack-tip by introducing a small Barenblatt region [34, 15].

An example concerning a crack extending symmetrically in an infinite plate at constant crack-tip velocity (V) and in-plane stress or strain will now be studied [35].

The equations of motion can be written

$$\operatorname{div}\operatorname{grad}\Phi=\frac{\partial^2\Phi}{\partial\tau^2},\qquad \Phi=\Phi(x,z,\tau),\tag{63}$$

$$-\operatorname{curl}\operatorname{curl}\bar{\psi}=\frac{1}{k^2}\cdot\frac{\partial^2\bar{\psi}}{\partial\tau^2},\qquad \bar{\psi}=\psi(x,z,\tau)\hat{y}\tag{64}$$

where c_d is the propagation velocity of irrotational waves, k^2 the ratio between the propagation velocities of equivoluminal waves and irrotational waves, $\tau=c_dt$, $\Phi=\operatorname{div}\bar{u}$ and $\bar{\psi}=\operatorname{curl}\bar{u}$. The half-plane $z\geqslant 0$ is studied and the boundary conditions state that

$$\tau_{zx}=0\quad\text{on}\quad z=0\tag{65}$$

$$\sigma_z=0\quad\text{on}\quad z=0,\qquad |x|<\beta_0\tau,\tag{66}$$

$$u_z=0\quad\text{on}\quad z=0,\qquad |x|>\beta_0\tau,\tag{67}$$

where $\beta_0=V/c_d$. Thus, a mixed boundary-value problem is obtained. It can be solved by establishing an integral equation, after the simple boundary-value problem of a symmetrically expanding load

$$\sigma_z = \sigma_0[U(x-\beta'\tau) - U(x+\beta'\tau)] \quad \text{on} \quad z=0, \tag{68}$$

where $U(x)$ is the unit step function, has been solved. A simpler way [36] is to treat the problem directly as a Hilbert problem. This is done by means of the Laplace transform with respect to time. It turns out that a Keldysh–Sedov problem arises. The self-similarity of the problem then implies that the resulting Laplace transforms can be immediately inverted by the Cagniard method. This technique can be easily extended to different velocity of the two tips and to non-zero loads on the crack surfaces (but such that the self-similarity is not lost). An example of the incorporation of non-zero loads on the crack surfaces is given in [36].

Another way to solve the problem in question was used by Craggs [37], who took immediate advantage of the self-similarity by introducing

$$r/t = w \tag{69}$$

so that w and ϕ are used as independent variables instead of time t and the polar coordinates r and ϕ. Then a simple boundary-value problem can be obtained.

The relative simplicity imposed by the self-similarity has a counterpart in steady-state problems, where the Galilean transformation

$$x - Vt = x' \tag{70}$$

can be used to eliminate t. Problems on infinite strips with symmetrically situated semi-infinite cracks have been studied rather extensively. Very often, the Wiener–Hopf technique is suitable [38–42], but in other cases the formulation of a Hilbert problem is advantageous [36]. Generally, the complete stress–strain field cannot be determined analytically (except by using some approximate methods [41]), but this does not prevent a determination of the stress intensity factor. Thus, for instance, when using Fourier or Laplace transforms with respect to x', the transforms need not to be inverted in full, but only for the value of the variable corresponding to the crack-tip position.

The case of non-constant crack-tip velocity presents a much more difficult problem than the constant velocity case, because neither self-similarity nor steady-state conditions prevail. Eshelby [43] treated the mode III case. The basic result obtained was that, when a crack-tip running with constant velocity suddenly stops, a certain static field radiates out (expands) from the tip. By using this result and a superposition technique (sequences of short step movements at constant velocity, followed by stops) the case of non-constant velocity can be treated. This technique, however, cannot be continued indefinitely, since the near-tip field will eventually be disturbed, for instance, by waves from the other tip. The mode I case has been treated in a similar way by Freund [44]. Various other superposition methods have been used to construct non-constant velocity crack propagation from 'simple' basic solutions, cf., [45–50].

11.2 Waves impinging on stationary cracks

For plane problems, Nilsson [51] noticed that a path-independent integral in the p-plane (p = Laplace transform variable at transformation with respect to time) can be designed by analogy with the J-integral in the static case. This integral is

$$I(p) = \int_{\Gamma} \left[(W^* + \tfrac{1}{2}\rho p^2 U_i U_i)\, \mathrm{d}x_2 - n_j S_{ji} \left(\frac{\partial U_i}{\partial x_1}\right) \mathrm{d}s \right], \qquad i, j = 1, 2, \tag{71}$$

where $W^*(U_{i,j})$ is a quadratic function such that $L^{-1}(\partial W^*/\partial U_{i,j}) = \sigma_{ji}$, $U_i = L(u_i)$, ρ = density, L is the Laplace transform operator and the other symbols are analogous to those of (32). Then, it is possible to show that the Laplace transform of the stress intensity factor is

$$K(p) = 2(1 - k^2)^{1/2} G^{1/2} [I(p)]^{1/2}. \tag{72}$$

A corresponding path-independent integral for antiplane strain can also be designed. Some problems of wave interaction with stationary cracks can be conveniently studied by means of such path-independent integrals in the p-plane [51]. One additional virtue is that stress intensity factors in viscoelastic materials can be found after appropriate modifications of the material constants appearing in (72).

Summary and comments

The problem of fracture always consists in matching a non-continuum – a microstructural region, the process region – to a continuum. In certain cases, a separate treatment of these two regions is possible, leaving as a result a very simple matching. In other cases, notably the important problem of stable crack growth, such a separate treatment is, generally, impossible. This might necessitate the use of a model of the process region, for instance, the Barenblatt line model.

The brief review of current analytical mathematical methods used for treatment of the continuum side of the problem is not complete; for instance, overlaps with the excellent reviews by Sih and Liebowitz [52] and by Rice [53] were to some extent avoided and space limitations did not allow any details. Hence, also, the reference list may appear somewhat partial.

References

1 Barenblatt, G. J., *J. Appl Math. Mech.*, **PMM 23,** 622, 1959.
2 Knowles, J. K. and Sternberg, E., On the failure of ellipticity of the equations for finite elastostatic plane strain, *CalTech Technical Report* No. 34, 1976.
3 Bueckner, H. F., *Angew. Math. Mech.*, **50,** 529, 1970.
4 Rice, J. R., *Int. J. Solids Structures*, **8,** 751, 1972.
5 Eshelby, J. D., The continuum theory of lattice defects, in *Solid State Physics* (F. Seitz and D. Turnball, eds.), Academic Press, New York, 1956, Vol. 3, pp. 79–144.
6 Rice, J. R., *J. Appl. Mech.* **35,** 379, 1968.
7 Knowles, J. and Sternberg, E., *Arch. Rat. Mech. Anal.* **44,** 211, 1972.
8 Broberg, K. B., On the treatment of the fracture problem at large scale yielding, in *Proc. International Conference on Fracture Mechanics and Technology, Hong Kong 1977* (G. C. Sih and C. L. Chow, eds.), Sijthoff and Noordhoff International Publishers, 1977, Vol II, pp. 837–859.
9 Andersson, H., *J. Mech. Phys. Solids*, **22,** 285, 1974.
10 Andersson, H. and Bergkvist, H., *J. Mech. Phys. Solids*, **18,** 1, 1970.
11 Andersson, H., 1975 Finite element methods applied to problems of moving cracks, in *Computational Fracture Mechanics* (E. F. Rybicki and S. E. Benzley, eds.), American Society of Mechanical Engineers, New York, p. 185.
12 Broberg, K. B., *J. Mech. Phys. Solids*, **19,** 407, 1971.
13 Broberg, K. B., *J. Mech. Phys. Solids*, **23,** 215, 1975.
14 Rice, J. R., An examination of the fracture mechanics energy balance from the point of view of continuum mechanics, in *Proc. of the First International Conference on Fracture, Sendai 1965* (T. Yokobori, T. Kawasaki and J. L. Swedlow, eds.), Japan Soc. for Strength and Fracture of Materials, 1967, pp. 309–340.
15 Broberg, K. B., in *Recent Progress in Applied Mechanics* (B. Broberg, J. Hult and F. Niordson, eds), Almqvist & Wiksell, Stockholm, 1967, p. 125.
16 Bergkvist, H., *J. Mech. Phys. Solids*, **21**, 229, 1972.
17 Muskhelishvili, N. J., *Some Basic Problems of the Mathematical Theory of Elasticity*, P. Noordhoff Ltd., Groningen, Holland, 1953.
18 Erdogan, F., Complex function technique, in *Continuum Physics* (A. C. Eringen, ed.), Academic Press, New York, 1975, Vol. II, p. 523.
19 Sneddon, I. N., *Fourier Transforms*, McGraw-Hill, New York, 1951.
20 Hult, A. J. and McClintock, F., *Proc. 9th Int. Congr. Appl. Mech., Brussels*, **8,** 51, 1956.
21 Rice, J. R., *J. Appl. Mech.*, **34,** 287, 1967.
22 Rice, J. R., *Int. J. Fracture Mech.*, **2,** 426, 1966.
23 Kostrov, B. V. and Nikitin, L. V., *Geophys. J. Roy. Astr. Soc.*, **14,** 101, 1967.

24 Andersson, H., *Int. J. Fracture,* **13,** 239, 1977.
25 Dugdale, D. S., *J. Mech. Phys. Solids,* **8,** 10, 1960.
26 Rice, J. R. and Rosengren, G. F., *J. Mech. Phys. Solids,* **16,** 1, 1968.
27 Hutchinson, J. W., *J. Mech. Phys. Solids,* **16,** 13, 1968.
28 Amazigo, J. C., *Int. J. Solids Structures,* **10,** 1003, 1974.
29 Goldman, N. L. and Hutchinson, J. W., *Int. J. Solids Structures* **11,** 575, 1975.
30 Rice, J. R. and Johnson, M. A., The role of large crack tip geometry changes in plane strain fracture, in *Inelastic Behaviour of Solids* (M. F. Kanninen *et al.*, eds.), McGraw-Hill, New York, 1970, pp. 641–670.
31 Chitaley, A. D. and McClintock, F. A., *J. Mech. Phys. Solids,* **19,** 147, 1971.
32 Achenbach, J. D. and Bažant, Z. P., *J. Appl. Mech.*, **42,** 183, 1975.
33 Nilsson, F., *J. Elasticity,* **4,** 73, 1974.
34 Broberg, K. B., *J. Appl. Mech.*, **31,** 546, 1964.
35 Broberg, K. B., *Arkiv för Fysik,* **18,** 159, 1960.
36 Broberg, K. B., On transient sliding motion, *Geophys. J. Roy. Astr. Soc.,* **52,** 397, 1978.
37 Craggs, J. W., Fracture criteria for use in continuum mechanics, in *Fracture of Solids,* (D. C. Drucker and J. J. Gilman, eds.), Interscience Publishers, John Wiley & Sons, New York, 1962, p. 51.
38 Nilsson, F., *Int. J. Fracture Mech.*, **9,** 403, 1972.
39 Nilsson, F., *Int. J. Fracture Mech.*, **9,** 477, 1973.
40 Kuliew, V. D., *J. Appl. Math. Mech.*, **PMM 37,** 533, 1973.
41 Kuhn, G. and Matczynski, M., *Polska Akademia Nauk,* **22,** 469, 1974.
42 Broberg, K. B., On dynamic crack propagation in elastic-plastic media, in *Proceedings of the International Conference on Dynamic Crack Propagation, Lehigh University, July 10–12, 1972, Bethlehem, Penn.*, Noordhoff, Amsterdam, 1973, pp. 461–499.
43 Eshelby, J. D., *J. Mech. Phys. Solids,* **17,** 177, 1969.
44 Freund, L. B., *J. Mech. Phys. Solids,* **20,** 129, 1972.
45 Nilsson, F., 1973 A transient crack problem for an infinite strip under antiplane shear, in *Proceedings of the International Conference on Dynamic Crack Propagation, Lehigh University, July 10–12, 1972, Bethlehem, Penn.*, Noordhoff, Amsterdam, 1973, p. 543.
46 Nilsson, F., A suddenly stopping crack in an infinite strip under tearing action, in *Fast Fracture and Crack Arrest* (G. T. Hahn and M. F. Kanninen, eds.), ASTM, Philadelphia, ASTM STP 627, 1977, pp. 77–91.
47 Nilsson, F., *Int. J. Solids Structures,* **13,** 1133, 1977.
48 Nilsson, F., 1976 Steady crack propagation followed by non-steady growth – mode I solution, Report from the Department of Strength of Materials and Solid Mechanics, The Royal Institute of Technology, S-100 44 Stockholm, Sweden.
49 Kostrov, B. V., *Int. J. Fracture,* **11,** 47, 1975.
50 Burridge, R., *Int. J. Eng. Sci.*, **14,** 725, 1976.
51 Nilsson, F., *Int. J. Solids Structures,* **9,** 1107, 1973.
52 Sih, G. C. and Liebowitz, H., Mathematical theories of brittle fracture, in *Fracture* (H. Liebowitz, ed), Academic Press, New York, 1968, vol. II, p. 67.

53 Rice, J. R., Mathematical analysis in the mechanics of fracture, in *Fracture* (H. Liebowitz, ed.), Academic Press, New York, 1968, Vol. II, p. 191.

Professor K. B. Broberg,
Division of Solid Mechanics,
Lund Institute of Technoloy,
Box 725
S-220 07 Lund,
Sweden

E. G. D. Cohen

The *N*-body problem and statistical mechanics

Abstract

The *N*-body problem in statistical mechanics, in particular as it occurs in connection with the approach to thermal equilibrium of a gas, is discussed. For the average behavior of a dilute gas, the many-body problem can be reduced to a two-body problem and the time evolution is described by the Boltzmann equation. For a dense gas, divergences occur if a straightforward generalization of the Boltzmann equation to higher densities is attempted. These difficulties are illustrated on a simple model due to the Ehrenfests. In general, a nonlocal description of the approach to equilibrium seems necessary. The incorporation of fluctuations into the theory is briefly discussed.

1 Introduction

The aim of statistical mechanics is to understand the macroscopic properties of bulk matter in terms of the microscopic properties of the particles of which matter consists. The difficulty of this endeavor is connected with the fact that bulk matter consists of very many particles and that, even if one could solve the equations of motion of all these particles, the connection with the macroscopic properties is not immediately clear.

Macroscopic properties are those observed on the system in the laboratory and involve a coarse-graining, i.e., an averaging in space and time over microscopic phases of the system. Clearly, one cannot observe in the laboratory the microscopic phase of a system, i.e., the positions and velocities of all particles [1]. Therefore, the concept of a macroscopic variable introduces, by definition, an averaging or statistical concept into physics [2]. The macroscopic laws, i.e., the laws which these macroscopic quantities obey, although a consequence of and based upon the microscopic equations of motion of the particles out of which the system consists, may be quite different from those of mechanics by the very statistical nature of macroscopic variables as compared to microscopic variables.

In this paper, we discuss the approach to equilibrium of a gas of N particles in a volume V, where the particles obey the classical equations of motion and interact with each other through a pairwise additive spherically symmetric interparticle potential of (short) range r_0. Many of the considerations, however, hold under more general conditions.

2 Approach to equilibrium in Γ-space

We shall discuss here the approach to thermodynamic equilibrium of such a system in the course of time, from a given initial state at time $t=0$. This approach itself seems already a macroscopic law in conflict with the microscopic laws of motion, since these laws predict, according to a theorem of Poincaré, a quasi-periodic motion in time for any bounded conservative system. If one represents the microscopic phase (i.e., the coordinates and velocities (or momenta) of all the N-particles) of the whole system at $t=0$ by a point in a $6N$-dimensional phase-space (Γ-space), then this point will move (quasi-periodically) in the course of time over the energy surface in Γ-space, since the energy is conserved during the motion. A macroscopic measurement at $t=0$ will at best determine a number of macroscopic variables (for instance, the density $n(\mathbf{r}, 0)$ and the temperature $T(\mathbf{r}, 0)$ at $t=0$ throughout the system, as a function of position $\mathbf{r}$ in the system). Since many microscopic phases of the whole system will correspond to one n, T field, a *region* R on the energy surface in Γ-space will correspond to any given macroscopic measurement, instead of a point for a 'microscopic measurement.' In fact, different values of the macroscopic variables will correspond to different regions in Γ-space. Although, according to Poincaré's theorem, each point inside R will describe a quasi-periodic motion, *all* the points in the region together will spread out and distribute themselves more and more evenly over the energy surface in the course of time, leading finally, for a metrically transitive system, to a uniform distribution of these points over the energy surface (the microcanonical ensemble of Gibbs). More precisely, if one associates with the macroscopic state of the system at $t=0$ an ensemble or a probability distribution function which is everywhere zero except inside the region R, where the measured macroscopic quantities confine the microscopic state of the system to be, then this distribution will, in the course of time, spread out over the energy surface and the 'ensemble fluid'–originally all contained in the region R–will stream in accordance with Liouville's equation and Liouville's theorem over the energy surface and fill more and more regions on the energy surface, which correspond to different values of the macroscopic variables than those measured at $t = 0$. Then, for metrically transitive systems after a long time, the probability of finding the system in a certain macroscopic state will be proportional to the size of the corresponding region on the energy surface. If one then assumes, in addition, that the equilibrium state of the system with respect to the macroscopic variables (for instance, in our example, n and T uniform throughout the system), occupies the overwhelmingly largest part of the energy surface, then the system will, for almost any initial macroscopic state, be found in the equilibrium state after a long time. It will be clear from this that the 'irreversible' approach to thermodynamic equilibrium, although anchored in the microscopic (time-reversal invariant) laws of mechanics, is a macroscopic law, which differs from the microscopic laws due to its statistical nature.

This picture of the approach to equilibrium is due to Gibbs and Ehrenfest. It is very general and *qualitatively* correct, but quantitatively not very precise and useful. We shall briefly come back to it at the end of section 4.

3 Approach to equilibrium in μ-space: dilute gas

A quantitative description of the approach to equilibrium was given for the first time by Maxwell and Boltzmann for a dilute gas. They considered coarse-grained quantities not in the $6N$-dimensional phase-space of the system as a whole, but in the six-dimensional phase-space of one particle (μ-space). Then, dividing μ-space into a finite number M ($M \ll N$) of small but finite cells [5], a macroscopic state of the system can be defined by giving the distribution of all N particles in the system over the M cells in μ-space, i.e., by giving the velocity and spatial distribution of the particles. Again, many microscopic phases of the system correspond to one macroscopic state, i.e., to one distribution of points over cells in μ-space. A given initial distribution of the particles over the cells should then, in the course of time, approach more and more the equilibrium, or Maxwell, distribution of the particles over the cells [6].

The μ-space picture lends itself much better to a *quantitative* description of the approach to equilibrium than the Γ-space picture. In fact, it was by studying a given μ-space distribution for a dilute gas in time that Boltzmann obtained his famous equation, over a 100 years ago. By going to the continuum limit, he introduced a distribution function $f(\mathbf{r}, \mathbf{v}, t)$ in μ-space, which gives the *average number* of particles in the system per unit volume around the point $\mathbf{r}$, $\mathbf{v}$, in μ-space, i.e., the average number of particles per unit volume around the point $\mathbf{r}$ in ordinary space and per unit volume around the velocity $\mathbf{v}$ in velocity space. Taking only binary collisions between particles [7] into account, Boltzmann obtained the following equation for f [8, 9]:

$$\frac{\partial f(\mathbf{r}, \mathbf{v}, t)}{\partial t} = -\mathbf{v} \cdot \frac{\partial f}{\partial \mathbf{r}} + O_2(ff). \tag{1}$$

Here, the streaming term $-\mathbf{v} \cdot \partial f/\partial \mathbf{r}$ contains the change of the number of particles at $\mathbf{r}$ due to the fact that they have a finite velocity, so that those that are at $\mathbf{r}$ at time t will have left $\mathbf{r}$ at time $t+\mathrm{d}t$, while others that are not yet at $\mathbf{r}$ at t will be there at time $t+\mathrm{d}t$. The (binary) collision term $O_2(ff)$ will not be written out explicitly here. The collision operator O_2 contains a statistical assumption (Stosszahl Ansatz) for the average number of binary collisions between particles, and can be written down explicitly in terms of the solution of the dynamical two-body problem in infinite space, as indicated symbolically by the subscript 2. This term contains, therefore, the essence of a statistical mechanical treatment: a *statistical* element concerning the average number of binary collisions and a *mechanical* element concerning the solution of the mechanical two-body problem. The equation is causal, in that in principle it predicts a definite later state $f(\mathbf{r}, \mathbf{v}, t)$ for every initial distribution $f(\mathbf{r}, \mathbf{v}, 0)$. In particular, it predicts a definite final state for $t \to \infty$, which should be and in fact is the Maxwell distribution function. This is the content of the H-theorem, which Boltzmann himself proved. There is *no* contradiction with the reversibility of mechanics here, since f is a coarse-grained or macroscopic quantity. We remark that, with f, all macroscopic quantities that can be expressed in terms of f, such as the pressure and the energy of the gas, will approach their thermal equilibrium values.

If one wants to say more than just that the system approaches thermodynamic equilibrium, and one also wants to discuss *how* the system approaches equilibrium, then one has to consider the characteristic relaxation times of the system in order to get some idea of the orders of magnitude of the time scales relevant in the time evolution of the system. For a dilute gas, where only binary collisions play a role, the important time scales are: the average time between two successive collisions, i.e., the average time to traverse a mean free path t_{mfp}; and the average time to traverse a macroscopic length (for instance, the length of the container) t_{macr} [10]. As these time scales are very different in magnitude under normal (STP) conditions, i.e., since $t_{\mathrm{mfp}} \sim 10^{-9}\,\mathrm{s} \ll t_{\mathrm{macr}} \sim 10^{-4}\,\mathrm{s}$, one can divide the approach to equilibrium in two stages. First, on the time scale of t_{mfp}, a kinetic stage, in which binary collisions establish–to a first approximation–a *local* velocity equilibrium and the hydrodynamical quantities are generated. Then, on the time scale of t_{macr}, a hydrodynamical stage, where the approach to spatial uniformity takes place, described by the hydrodynamical equations, and resulting finally in total equilibrium, i.e., in the Maxwell distribution function for f. The theory for the hydrodynamical stage has been worked out by Chapman and Enskog

[11]. They established the connection between the Boltzmann equation and the equations of hydrodynamics by expanding f in a systematic way in terms of the gradients of the hydrodynamic quantities (the local density $n(\mathbf{r}, t)$, velocity $\mathbf{u}(\mathbf{r}, t)$ and temperature $T(\mathbf{r}, t)$). Thus, they obtained the Euler equations for an ideal fluid (first order in the gradients) and the Navier–Stokes equation for a viscous fluid (second order in the gradients). Also higher-order hydrodynamical equations can be derived this way, the so-called Burnett equations (third order in the gradients), super-Burnett equations (fourth order in the gradients), etc. In all cases, explicit expressions for the constitutive equations in terms of the interparticle forces are obtained.

Deviations from equilibrium in the velocity distribution alone [12] decay exponentially in time, $\sim e^{-\alpha_B t}$, to equilibrium. Here, α_B is determined by the smallest eigenvalues of the linearized Boltzmann collision operator, which are directly related to the transport coefficients: the shear viscosity η and the heat conductivity λ.

The Boltzmann equation has been very extensively studied and the picture of the approach to equilibrium that has emerged from it is widely considered to be the prototype of the approach of a system to equilibrium. In the last ten years or so, it has become clear, however, that this is *not* the case, and that the picture is only typical for the approach to equilibrium of a dilute system and *not* of a dense system. Before we discuss these developments, we first have to discuss why a straightforward generalization of the Boltzmann equation to higher densities is not possible.

4 Approach to equilibrium in μ-space: dense gas

The systematic generalization of the Boltzmann equation to higher densities was first attempted by Bogolubov in 1945 [13]. Since it is not clear–not even today–how to generalize the intuitive arguments of Boltzmann in μ-space from binary collisions to ternary, quaternary, etc., collisions, Bogolubov went back to Γ-space and, by expanding the basic Liouville equation in some sense in powers of the density n, and generalizing the Stosszahl Ansatz in a systematic way, he was able to rederive the Boltzmann equation in the thermodynamic limit (i.e., in the limit $N \to \infty$, $V \to \infty N/V = n$) and, at the same time, to obtain higher-order corrections to it. This generalized Boltzmann equation was of the form [14]

$$\frac{\partial f}{\partial t} = -\mathbf{v} \cdot \frac{\partial f}{\partial \mathbf{r}} + O_2(ff) + O_3(fff) + O_4(ffff) + \dots . \tag{2}$$

Here, the l-particle collision term $O_l(f \dots f)$ contains the contributions to $\partial f/\partial t$ from l-particle collisions. The collision operator O_l contains a statistical assumption for the average number of l-particle collisions and can be written down explicitly in terms of the solution of the dynamical l-particle problem in infinite space [15]. Note that the l-particle collisions are those of l particles alone in infinite space with no other particle present. In this way, one has split, in a systematic fashion, the dynamical many-body problem involved in the computation of $\partial f/\partial t$ into contributions from one particle $(-\mathbf{v}\cdot\partial f/\partial\mathbf{r})$, two particles (O_2), three particles (O_3), etc. This implies a generalization to non-equilibrium of the idea of the virial expansions of the thermodynamic quantities of a system in thermodynamic equilibrium. In both cases, the N-particle problem is reduced to 'individual particle' problems of $1, 2, \ldots, l, \ldots$ particles, respectively, in infinite space.

Bogolubov, Choh and Uhlenbeck discussed the application of the ideas of Chapman and Enskog to the generalized Boltzmann equation (2) in analogy with the ordinary Boltzmann equation (1) [13]. A notable difficulty is that one has not been able to prove an H-theorem, i.e., the existence of the approach to equilibrium on the basis of Eqn (2), not even if one restricts oneself to O_2 and O_3 alone and to a spatially homogeneous system, where f is only a function of $\mathbf{v}$ and t and not of $\mathbf{r}$. However, the existence of the same two widely-spaced relaxation times $t_{\text{mfp}} \ll t_{\text{macr}}$ as before suggests an approach to equilibrium in two stages again and makes a discussion of the hydrodynamical stage, and, in particular, the derivation of the hydrodynamical equations, still possible. Actually, Choh and Uhlenbeck carried this out for O_2 and O_3.

Also, in a manner similar to that for the Boltzmann equation, one would expect that the approach to equilibrium of a small velocity disturbance from equilibrium would, on the basis of Eqn (2), again be exponential, $\sim e^{-\alpha t}$, where now, however, the exponent α is given by a power series in the density:

$$\alpha = \alpha_2 + n\alpha_3 + n^2\alpha_4 + \ldots. \tag{3}$$

Here, $\alpha_2 = \alpha_{\text{B}}$, the contribution of O_2, α_3 contains the contribution to α from O_3, etc. Or, also, one would expect that the transport coefficients be given by power series in the density–just as the thermodynamic quantities in thermal equilibrium–so that

$$\left.\begin{aligned} \eta &= \eta_2 + n\eta_3 + n^2\eta_4 + n^3\eta_5 + \ldots, \\ \lambda &= \lambda_2 + n\lambda_3 + n^2\lambda_4 + n^3\lambda_5 + \ldots \end{aligned}\right\} \tag{4}$$

(where η_l and λ_l contain contributions from $O_l(f\ldots f)$.

Here again one can say that the transport properties of the whole system can be expressed in terms of dynamical properties of small systems consisting of $2, 3, \ldots, l, \ldots$ particles.

Although this picture of the approach to equilibrium of a dense gas seems very plausible, it is *not* correct. In writing Eqn (2), it has been tacitly assumed that all 3-, 4-, . . ., l-, . . . particle collisions are essentially genuine 3-, 4-, . . ., l-, . . . particle collisions, where the $3, 4, \ldots, l, \ldots$ particles all collide within a region of space of a dimension of the order of the (finite) range r_0 of the interparticle forces (cf. Fig. 1). This is certainly true for the genuine l-particle collisions, but not immediately clear for extended l-particle collisions, where the particles can move temporarily outside their range of interaction. Thus, for three-particles, for instance, this implies that the extended three particle collision of the type sketched in Fig. 2 [16] can only occur if L_1, L_2 and L_3 are essentially all of the order r_0 and that otherwise it becomes 'too difficult' for particle 3 to collide with particle 2 in such a way that this particle can recollide with particle 1.

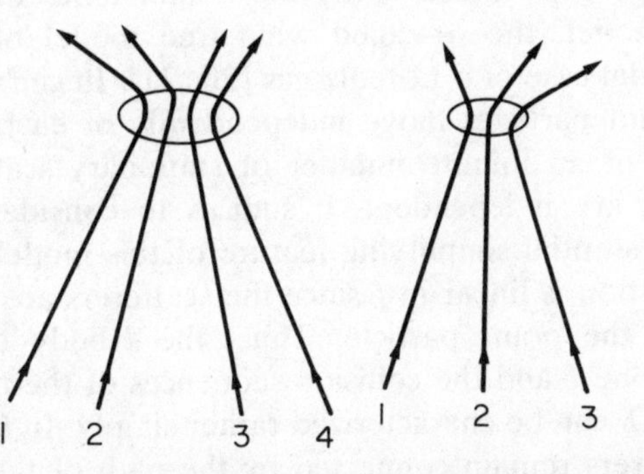

Fig. 1. Genuine ternary and quaternary collisions of three and four particles, respectively. Each of the particles is in the range of at least one other particle during the collision.

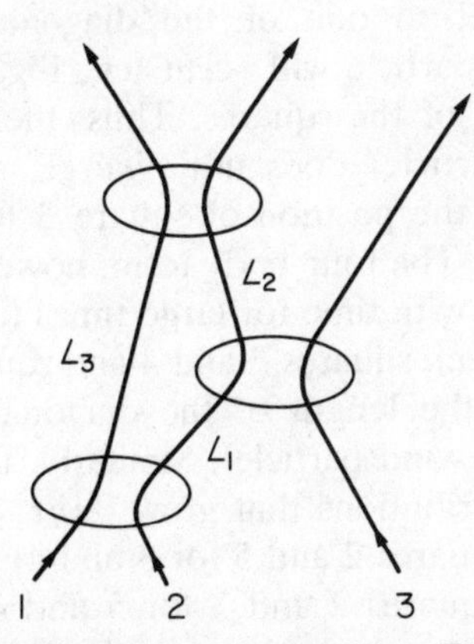

Fig. 2. Ternary collision between three particles. This collision consists of a sequence of three successive binary collisions between the particles 1 and 2, 2 and 3, and 1 and 2, respectively. The particles move freely between binary collisions over distances L_1, L_2 and L_3.

Only if this is true can one show that the operator O_3 exists, i.e., is finite. However, it has been demonstrated by studying the collision sequences between 3, 4, 5, . . . , l, . . . particles which contribute to O_3, O_4, $O_5, \ldots, O_l, \ldots$, respectively, that although O_3 indeed exists for a large class of f's, $O_4, O_5, \ldots, O_l, \ldots$ do not, and that while O_4 diverges logarithmically, $\sim\ln(T/t_{\text{coll}})$, O_l diverges $\sim(T/t_{\text{coll}})^{l-4}$, where T goes to infinity and t_{coll} is the average duration of a collision. Thus, the larger l, the worse the divergence!

We remark that, for a two-dimensional gas, O_3 diverges logarithmically, while O_4 diverges linearly and $O_l \sim (T/t_{\text{coll}})^{l-3}$. These divergences were discovered independently by J. Weinstock [17], E. A. Frieman and R. Goldman [18] and J. R. Dorfman and E. G. D. Cohen [19].

They can perhaps most easily be understood in a simple two-dimensional model, the so-called wind–tree model of the Ehrenfests, which is a special case of a Lorentz gas [20, 21]. In such a gas, an infinite number of point particles move independently of each other through a random array of an infinite number of stationary scatterers. Since the point particles are independent, it suffices to consider only one point particle. The essential simplifying feature of this model is that the basic evolution equation is linear in f since the scatterers are not affected by a collision with the point particle. Thus, the l-body operator O_l now contains only one f and the collision sequences of the point particle that contribute to O_l can be characterized rather simply. In fact, they are such that the scatterers remain connected by the path of the point particle if one cuts the path between any two successive collisions.

In the wind–tree model, the scatterers (the trees) are oriented squares with parallel diagonals and, if one starts the point (wind) particle out in a given direction parallel to one of the diagonals, only four velocity directions of the point particle will occur (cf., Fig. 3), due to the perfect orientational alignment of the squares. Thus, the three-body term (two trees and the wind particle) does not diverge, since the condition of connectedness restricts the position of square 3 to the neighborhood of square 2 (cf., Fig. 4(a)). The four-body term, however, contains contributions that grow linearly with time for large times (cf., Fig. 4(b)), since the (vertical) distance between squares 3 and 4 and square 2 can grow $\sim t/t_{\text{coll}}$, where $t_{\text{coll}} = a/v$ (a is the length of the diagonal of a tree and v the (constant) speed of the wind particle). Similarly, the five-body term (cf., Fig. 4(c)) contains contributions that grow $\sim(t/t_{\text{coll}})^2$, since the (horizontal) distance between squares 2 and 5 (or 3 and 4) as well as the (vertical) distance between the squares 2 and 3 (or 5 and 4) can grow $\sim t/t_{\text{coll}}$. In general, one expects–but no proof is available since no classification of all possible collision sequences exists–that the l-body operator $O_l(f)$ will grow $\sim(t/t_{\text{coll}})^{l-3}$ [22].

Since similar divergences occur in the case of Eqn (2), this equation

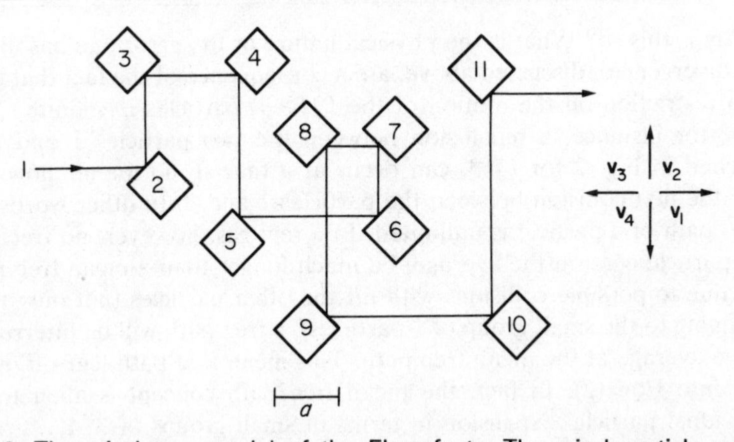

Fig. 3. The wind–tree model of the Ehrenfests. The wind particle moves between a random array of stationary square scatterers (trees) with diagonals of length a that are all oriented with parallel diagonals. If the wind particle starts out with a velocity parallel to one of the square diagonals, only four velocities will occur as a result of the perfect orientation of the squares ($\mathbf{v}_1$, $\mathbf{v}_2$, $\mathbf{v}_3$, $\mathbf{v}_4$). The speed (absolute value of the velocity) of the wind particle does not change (i.e., $|\mathbf{v}_1| = |\mathbf{v}_2| = |\mathbf{v}_3| = |\mathbf{v}_4|$), only the velocity direction changes upon collision.

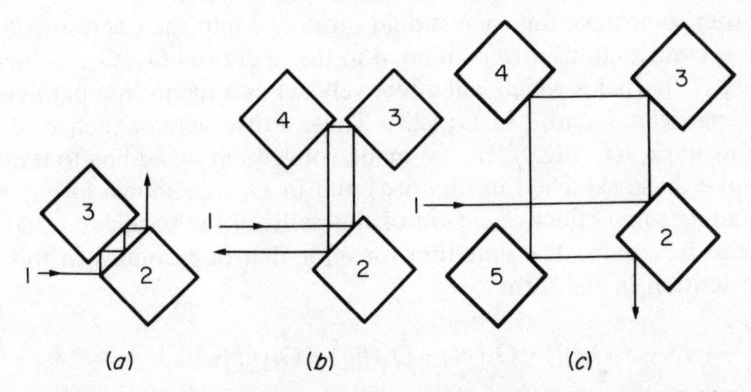

Fig. 4. (a) Three-body collision (wind particle and two trees); $O_3(f)$ is finite. (b) Four-body collision (wind particle and three trees); $O_4(f)$ diverges $\sim t/t_{\text{coll}}$. (c) Five-body collision (wind particle and four trees); $O_5(f)$ diverges $\sim (t/t_{\text{coll}})^2$.

cannot be correct, and the conclusions drawn from it (Eqns (3) to (5)) are not correct either. In fact, the coefficients η_4 and λ_4 are logarithmically infinite, η_5 and λ_5 linearly infinite, etc. This then means that a straightforward generalization of the idea of the virial expansion to nonequilibrium processes is not possible and that a straightforward generalization of the Boltzmann equation to higher densities is not possible either.

Why is this so? What is the physical nature of the error one has made? The divergences, discussed above, are a consequence of the fact that there is no restriction on the motion of the l ($l \geqslant 4$) particles in infinite space. Thus, for instance, a recollision between the two particles 1 and 2, as sketched in Fig. 2 for $l=3$, can occur at a time T no matter how long after the first collision between the particles 1 and 2. In other words, the (free) path of a particle is unlimited. In a real gas, however, no free path of a particle can, on the average, be much longer than a mean free path. For, due to possible collisions with *all* the other particles (not only those belonging to the small group of l-particles), a free path will be interrupted on the average at the mean free path. This mean free path 'cut-off' is not built into Eqn (2). In fact, the mean free path concept is alien to the 'individual particle' expansion in terms of small groups of $3, 4, \ldots, l, \ldots$ particles, since the mean free path is essentially a collective property of all the particles together and not a property of a small group of particles. Therefore, although there *is* a statistical element in each of the operators O_l, this is clearly not the correct one to obtain a meaningful equation for the change of the macroscopic quantity f with time.

One can obtain a better equation for f by rearranging the expansion (2), and thereby improving (or refining) the statistical assumption in the O_l so as to include a mean free path cut-off of particle paths.

In order to achieve this, one should introduce into the operator O_4, for instance, contributions now contained in the operators O_5, $O_6, \ldots$ involving 5, 6, ... particles which will effectively act as a mean free path cut-off for the particles 1 and 2 in O_4 [23]. Thus, a divergent sequence of four collisions in O_4 (cf., Fig. 5) can be made convergent by adding to it events in O_5 like those sketched in Fig. 6(*a*) and in O_6 like shown in Fig. 6(*b*), etc., leading to an effective cut-off of the paths of the particles 1 and 2 at the mean free path. The equation for $\partial f/\partial t$ that one obtains in this way can be written in the form

$$\frac{\partial f}{\partial t} = -\mathbf{v}\cdot\frac{\partial f}{\partial \mathbf{r}} + O_2(ff) + O_3(fff) + \bar{O}_4(ffff) + \bar{O}_5(fffff) + \ldots, \tag{5}$$

where $\bar{O}_4$ now contains part of $O_5, O_6, \ldots$. At present, Eqn (5) has only been derived up to $\bar{O}_4$.

As a consequence of the mean free path cut-off contained in the operators $\bar{O}_4$, $\bar{O}_5$, etc., instead of (4), one obtains the following density expansion of the transport coefficients:

$$\left.\begin{aligned} \eta &= \eta_2 + n\eta_3 + n^2 \ln n\eta_4'' + n^2\eta_4' + \ldots, \\ \lambda &= \lambda_2 + n\lambda_3 + n^2 \ln n\lambda_4'' + n^2\lambda_4' + \ldots. \end{aligned}\right\} \tag{6}$$

Since the mean free path is inversely proportional to the density, the prediction is thus that the transport coefficients will contain contributions

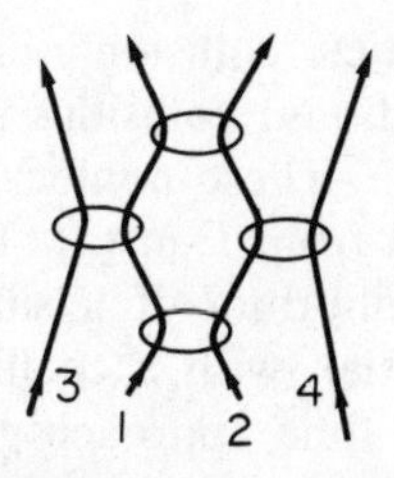

Fig. 5. Quarternary collision between four spherical particles that diverge $\sim\ln(t/t_{\text{coll}})$ in three dimensions and $\sim(t/t_{\text{coll}})$ in two dimensions.

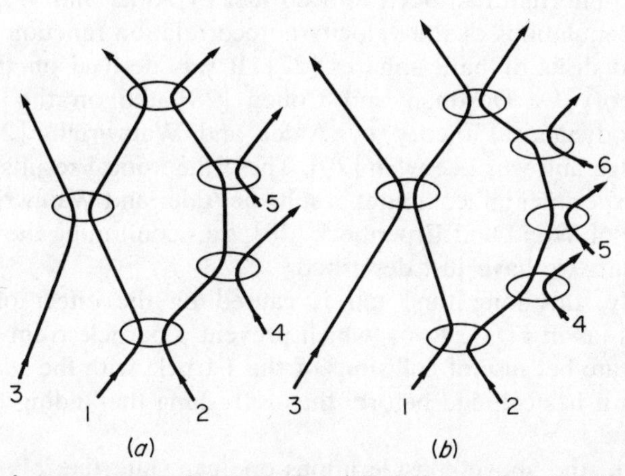

Fig. 6. Collision sequences between five (cf., (*a*)) and six (cf., (*b*)) particles in O_5 and O_6, respectively, that contribute to a free path cut-off of the path of particle 2 in O_4.

which depend logarithmically on the density. This has not yet been convincingly verified experimentally [24, 25]. We remark that the logarithmic density dependence is directly related on the one hand to the logarithmic increase of O_4 with time between first and last collision, due to dynamical events as sketched in Fig. 4, and on the other hand to the mean free path cut-off, discussed above. In order to obtain the complete density dependence of η and λ, a classification of all possible collision sequences of l ($l = 3, 4, \ldots$) particles in infinite space is needed, as well as their long time behavior. At present, this is only available for three hard spheres (or three hard disks in two dimensions) [25, 26]. Not even for as simple a model as the wind–tree model has such an analysis been made.

It is of importance to note that, in Eqn (5), $\partial f/\partial t$ at time t does *not* depend on $f(t)$ anymore at time t alone but on $f(t-\tau)$ over an effective time interval of $0<\tau<t_{\text{mfp}}$. In other words, Eqn (5) contains *nonlocal* effects in time (as well as in space) due to the influence of 'nonlocal' or

extended 4-, ..., l-, ... particle collisions, in addition to the 'local' (as, for instance, the genuine l-body) collisions present in Eqn (2) and still present in Eqn (5), of course. These nonlocal effects make the character of Eqn (5) entirely different from Eqn (2). Thus, in particular, although the initial approach to equilibrium of a small velocity deviation from equilibrium is still exponential, with a coefficient of the form given by Eqn (3) (cf., [12]), the long time approach gives rise to a so-called long time tail $\sim(t_{\mathrm{mfp}}/t)^{3/2}$ as a consequence of these nonlocal effects. We remark that a similar situation occurs in two dimensions, where the long time approach is $\sim t_{\mathrm{mfp}}/t$ as a consequence of nonlocal effects. Such a slow decay to equilibrium had been noticed first by Alder and Wainwright in computer simulations of the velocity autocorrelation function of a dense gas of hard disks or hard spheres [27]. It was derived on the basis of kinetic theory by Dorfman and Cohen [28] and on the basis of a quasi-hydrodynamical theory by Alder and Wainwright [27] and by Ernst, Hauge and van Leeuwen [29]. These theoretical results agree well with the 'experimental' computer results of Alder and Wainwright as well as of those of Wood and Erpenbeck [30], thus confirming the theoretical developments we have just described.

Physically, this long time tail is caused by the effect of extended recollisions (as in Figs 2 to 6), which prevent a particle from 'forgetting' its initial state because of collisions of this particle with the *same* particle with which it has collided before, thus refreshing the 'fading' memory of its past state.

In view of the above considerations one can state the following.

1. No matter how small the density n of the gas, there is always a time $T(n)$ such that, for all $t > T(n)$, the tail $\sim t^{-d/2}$ ($d = 2, 3$) dominates the exponential Boltzmann behavior. For such times, the Boltzmann equation is *incorrect.*
2. The long time tail makes the existence of two well-separated stages in the approach to thermodynamic equilibrium much less clear. Instead of having a kinetic stage in which the transport coefficients occurring in the hydrodynamical equations obtain their full values, followed by a hydrodynamical stage in which the time evolution of the system is described by the hydrodynamical equations, one has to take into account that, due to the long time tail, the transport coefficients may not have obtained their full values on the kinetic time scale (i.e., after a few mean free times) but still continue to grow in what one called previously the hydrodynamical stage. In the limit of infinitely slowly varying processes in time and space, this may not be a problem. In general, however, this will present a fundamental difficulty, the solution of which is not clear at the moment.
3. The long time tail also implies that the higher-order hydrodynamical

equations, such as the Burnett equations, do *not* exist in a dense gas. In fact, in two dimensions, not even the Navier–Stokes equations exist [28–32], and only for the Boltzmann equation (1) do hydrodynamical equations of all orders exist. The nonexistence of the Burnett equations, etc., in a dense gas is due to the breakdown of the constitutive relations, in that a *local* connection between fluxes and forces is no longer valid. This is reflected in a divergence of the transport coefficients if such a connection is assumed. Probably a *nonlocal* connection between fluxes and forces, reflecting perhaps the non-locality of Eqn (5), is necessary to describe the approach to equilibrium in a dense gas. In other words, the foundation and the precise range of validity and applicability of the equations of hydrodynamics, the Navier–Stokes equations, is unclear at the moment.

In spite of all these difficulties, Eqn (5) is still causal in the sense that it will predict $f(t)$ if f is given at earlier times.

We remark that a discussion of the approach to equilibrium in Γ-space that corresponds to the one given here in μ-space is lacking. Although interesting properties in the approach to equilibrium of a mechanical system in Γ-space have been discovered [33, 34], the connection between these properties and those in μ-space, discussed above, is at present completely unknown. A difficulty seems to be that the description in μ-space is for an infinite system, while that in Γ-space is for a finite system.

5 Fluctuations

Up until now, we have discussed the causal behavior of the *averages* of macroscopic quantities. However, as soon as one introduces macroscopic quantities as averages over microscopic phases, one not only has to consider the averages of the macroscopic quantities, but also the *fluctuations* around their average values. The macroscopic quantities are, in reality, *stochastic* variables. Although their average values can be predicted, they are really described by a probability distribution function, and their actual values can, therefore, only be predicted with a certain probability. In this sense, one could say that the macroscopic description is statistical and *non*causal. As we have seen, the proper description of the average of a macroscopic quantity is very difficult for a dense system. This holds *a fortiori* for the description of the fluctuations around the average, and very little is known about this to date.

Perhaps some idea can be obtained from the situation in the hydrodynamical stage. Here, the usual Navier–Stokes equations are equations for the *average* local density, velocity and energy of a fluid. Landau and Lifshitz added fluctuating terms to the linearized Navier–Stokes equations, analogous to the fluctuating term in the Langevin equation [35, 36]. The new equations for the local density, velocity and energy can be used to obtain correlations in space and time of these macroscopic quantities.

Although these fluctuating terms are not usually considered, they can be of importance, for instance, near critical points (where the fluctuations in the macroscopic variables are large) or in a discussion of multiple scattering of light by a fluid.

A derivation of the linearized Navier–Stokes equations with these Landau–Lifshitz fluctuation terms has been given for a dilute gas by Zwanzig and Bixon [37], Fox and Uhlenbeck [38], and Hinton [39] by adding a fluctuating term to the linearized Boltzmann equation (1). However, the precise nature of this fluctuating term is far from clear. In fact, a derivation of this term from the Liouville equation has not been given up until now. Recently, Logan and Kac [40] derived fluctuations for a spatially homogeneous system in μ-space from a master equation in $3N$-dimensional phase space. They showed that in order to obtain these fluctuations, terms of order $1/N^{1/2}$, if N is the number of particles in the system, had to be taken into account explicitly. These fluctuations are consistent with those derived by Fox and Uhlenbeck [38] on the basis of a fluctuating linearized Boltzmann Equation.

In conclusion, we hope to have shown that the connection between the macroscopic and the microscopic properties of matter is still far from clear, in particular as far as the approach to equilibrium is concerned, and that, therefore, a great deal remains to be investigated in the future!

Notes and references

1 Except in computer calculations, where one can, in principle, follow the microscopic behavior of the system.

2 For a discussion of the concept of macroscopic variables and requirements which they have to satisfy, *see* [3] and [4].

3 Uhlenbeck, G. E. and Ford, G. W., *Lectures in Statistical Mechanics*, American Mathematical Society, Providence, R.I., 1963, Chap. 1.

4 Uhlenbeck, G. E., in *Fundamental Problems in Statistical Mechanics* (E. G. D. Cohen, ed.), North-Holland Publishing Company, Amsterdam, 1968, Vol. II, p. 1.

5 For a given total energy of the system, the total volume in μ-space of all possible phases of the system is finite. We use here the concept of a macroscopic variable in a generalized sense. Any quantity that is sufficiently averaged over microscopic phases of the system is considered to be macroscopic, not only the 'usual' macroscopic quantities such as the density n,

temperature T, etc. There is no clear definition of 'sufficiently averaged'; $f(\mathbf{r}, \mathbf{v}, t)$, to be introduced below, contains only one averaging process less (namely, over $\mathbf{v}$) than n and T and seems, therefore, to qualify.

6 The connection with the Gibbs picture is as follows. Since to each distribution of particles over cells in μ-space there corresponds a region in Γ-space, the Maxwell distribution of the points in μ-space corresponds to the maximum-sized region on the energy surface in Γ-space mentioned above. It is this distribution in μ-space and this region in Γ-space that corresponds to the thermodynamic equilibrium state of the system.

7 I do not know of any completely satisfactory derivation of the Boltzmann equation (1). The best thing seems to be to consider Eqn (1) as *defining* $f(\mathbf{r}, \mathbf{v}, t)$.

8 *See*, for example, [3], Chap. IV.

9 Cohen, E. G. D., in *Transport Phenomena in Fluids* (H. J. M. Hanley, ed.), Marcel Dekker, New York, 1969, Chap. 6, pp. 119–156.

10 There is a third relaxation time, t_{coll}, associated with the range of the forces r_0. Under conditions of standard temperature and pressure (STP), i.e., at 0 °C and 1 atm, $t_{\text{coll}} \sim 10^{-12}$ s, so that $t_{\text{coll}} \ll t_{\text{mfp}}$.

11 *See* [3], Chap. 6, and [9], Chap. 6. These chapters give further references to the literature.

12 For instance, the thermalization of a particle is exponential. In other words, the approach to (velocity) equilibrium (the Maxwell distribution function) of a particle with a given initial velocity $\mathbf{v}_1$ (i.e., with a delta function around $\mathbf{v}_1$ as initial distribution function) is exponential.

13 *See* [3] Chap. 7 and [9], Chap. 7; also, Cohen, E. G. D., in *Fundamental Problems in Statistical Mechanics* (E. G. D. Cohen, ed.), North-Holland Publishing Company, Amsterdam, 1962, pp. 110–156.

14 Strictly speaking, the operator O_2 in Eqn (2) differs from that in Eqn (1) since it takes into account the difference in position between two colliding molecules, which O_2 in Eqn (1) does not. Since this point is of no importance here, we have ignored this difference.

15 It can characterized by similar diagrams as the l-body contributions $\sim n^l$ to the pair distribution function in thermal equilibrium (*see*, for instance, Cohen, E. G. D., *Physica*, **28,** 1025, 1962; *J. Math. Phys.*, **4,** 183, 1963).

16 From this it is clear that the Boltzmann equation takes into account a subclass of all binary collisions, viz., only those between particles that have not collided before.

17 Weinstock, J., *Phys. Rev. A*, **40,** 460, 1965.

18 Frieman, E. A. and Goldman, R., *J. Math. Phys.* **7,** 2153, 1966; **8,** 1410, 1967.

19 Dorfman, J. R. and Cohen, E. G. D., *J. Math. Phys.*, **8,** 282, 1967.

20 *See* Ehrenfest, P., *Collected Scientific Papers*, North-Holland Publishing Company, Amsterdam, 1959, p. 229. *See also*, Cohen, E. G. D., The kinetic theory of dense gases: the Lorentz gas, in *Théories cinétiques classiques et relativistes*, Colloques Internationaux CNRS No. 236, 1975, p. 269.

21 No logarithmic divergence seems to occur in this model and one expects that this would still be true if one replaced the squares by regular polygons. For *spherical* scatterers, however, logarithmic divergences do occur; in fact, for circular scatterers, $O_3(f)$ diverges logarithmically. This is related to the

occurrence here of a continuous velocity space for the point particle, rather than a discrete one as for the wind–tree model.

22 Hauge, E. H. and Cohen, E. G. D., *J. Math. Phys.*, **10,** 397, 1969; Divergences in non-equilibrium statistical mechanics and Ehrenfest's wind–tree model, *Arkiv for det Fysiske,* Seminar 1, Trondheim, No. 7, 1968.

23 Kawasaki, K. and Oppenheim, I., *Phys. Rev. A*, **136,** 1519, 1964; *see also* Dorfman, J. R., in *Lectures in Theoretical Physics,* (W. E. Brittin and A. O. Barut, eds), Gordon and Breach, London and New York, 1967, Vol. 9C, p. 443.

24 Hanley, H. J., McCarthy, R. D. and Sengers, J. V., *J. Chem. Phys.*, **50,** 857, 1969; Kestin, J., Paykoc, E. and Sengers, J. V., *Physica,* **54,** 1, 1971.

25 *See* Cohen, E. G. D. in *Lectures in Theoretical Physics,* (W. E. Brittin, ed.), University of Colorado Press, Boulder, Colorado, 1966, Vol. VIIIA, Appendix II, p. 170.

26 Sengers, J. V., Gillespie, D. T. and Hoegy, W. R., *Phys. Letters A*, **32,** 387, 1970; Hoegy, W. R. and Sengers, J. V., *Phys. Rev. A*, **2,** 2461, 1970.

27 Alder, B. J. and Wainwright, T. E., *Phys. Rev. Letters,* **18,** 1988, 1967; *J. Phys. Soc. Japan Suppl.*, **26,** 267, 1969; *Phys. Rev. A*, **1,** 18, 1970; Alder, B. J., Gass, D. M. and Wainwright, T. E., *J. Chem. Phys.*, **53,** 3813, 1970; Wainwright, T. E., Alder, B. J. and Gass, D. M., *Phys. Rev. A*, **4,** 233, 1971.

28 Dorfman, J. R. and Cohen, E. G. D., *Phys. Rev. Letters* **25,** 1257, 1970; *Phys. Rev. A*, **6,** 776, 1972; *Phys. Rev. A*, **12,** 292, 1975.

29 Ernst, M. H., Hauge, E. H. and van Leeuwen, J. M. J., *Phys. Rev. Letters,* **25,** 1254, 1970; *Phys. Rev. A*, **4,** 2055, 1971.

30 *See* Wood, W. W., in *Fundamental Problems in Statistical Mechanics,* (E. G. D. Cohen, ed.), North-Holland Publishing Company, Amsterdam, 1975, Vol. III, p. 331; Wood, W. W. and Erpenbeck, J. J., in *Statistical Mechanics,* (B. J. Berne, ed.), Plenum Press, New York, 1977, Part B, p. 1.

31 One should distinguish between hydrodynamics in two dimensions ('flatland') and in three dimensions but in two-dimensional geometry, such as the flow around an infinite cylinder. In the latter case, the Navier–Stokes equations exist, of course; difficulties encountered in treating this problem maybe connected with the non-existence of the Navier–Stokes equations in two dimensions, however [32]. This also implies that the approach to equilibrium in two dimensions is quite different from that in three dimensions and that, as a consequence, the role two-dimensional models play in understanding three-dimensional systems maybe quite different in nonequilibrium statistical mechanics than in equilibrium statistical mechanics.

32 Dorfman, J. R., Kupermann, W. A., Sengers, J. V. and McClure, C. F., *Phys. Fluids,* **16,** 2347, 1973; Gervois, A. and Pomeau, Y., *Phys. Fluids,* **17,** 2292, 1974; Pomeau, Y., *Phys. Fluids,* **18,** 2777, 1975.

33 We have in mind here the 'hierarchy' of properties that a mechanical system can have in its approach to equilibrium, each property implying the previous one and of which ergodicity is the 'lowest'. The next one, called mixing, is related to irreversibility. The 'highest' one (the so-called Bernoulli condition) makes the system behave completely randomly in a course-grained sense, although it is, of course, causal.

34 *See,* for instance, Lebowitz, J. L., Hamiltonian flows and rigorous results in non-equilibrium statistical mechanics, in *Statistical Mechanics* (S. A. Rice, K. F. Freed and J. C. Light, eds.), The University of Chicago Press, Chicago,

1971, p. 41; or Lebowitz, J. L., in *Transport Phenomena: Sitges International School of Statistical Mechanics* (G. Kirczenow and J. Morro, eds.), Lecture Notes in Physics Vol. 31, Springer Verlag, Berlin and New York, 1974, p. 202.

35 Landau, L. D. and Lifshitz, E. M., *Fluid Mechanics*, Pergamon Press, Oxford and New York, 1959, Chap. 17.

36 Fox, R. F. and Uhlenbeck, G. E., *Phys. Fluids*. **13,** 1893, 1970.

37 Bixon, M. and Zwanzig, R., *Phys. Rev.*, **187,** 1267, 1969.

38 Fox, R. F. and Uhlenbeck, G. E., *Phys. Fluids*, **13,** 2881, 1970.

39 Hinton, F. L., *Phys. Fluids*, **13,** 857, 1970.

40 Logan, J. and Kac, M., *Phys. Rev. A*, **13,** 458, 1976.

Professor E. G. D. Cohen
The Rockefeller University
1230 York Avenue
New York City
NY 10021
USA

G. Ferrarese

Intrinsic formulation of Cosserat continua dynamics

1 Introduction

It is well known that, with the development of theoretical physics and the diversification of the investigations of the internal structure of elementary particles as a result of the notion of spin, there has been a commensurate evolution in our understanding of (Cauchy-type) continuous ordinary systems–an evolution almost to the point at which the offspring has become unrecognizable to the parent.

Substantially, the concrete generalizations† have been of two types:

1. *Simple dynamical*,‡ intended to widen not the geometric aspect, but the local representation of the physical behaviour, introducing the body and contact couples. On the one hand, this extension implies a doubling of stress characteristics; from the conceptual point of view, however, it attempts to embrace three-dimensional continua and thin structures in a unique dynamical system. Indeed, for thin structures, the reduced scheme (line or deformable surface) imposes the introduction of bending and twisting couples.
2. *Geometrical-dynamical*,§ in which the geometry of the system is also improved, taking into account a possible molecular structure not reducible to the point system. This structure, which is ideally intended to be superposed over the point continuum, is described in its most natural form by a finite number of applied vectors (directors) themselves deformable–and subjected eventually to group-invariant requirements.

A continuous structure of such a type, i.e., with local directors, also finds application directly in macroscopic situations, especially in thin or

† These are sufficiently smooth, in accordance with current representation of a natural body by means of a Riemann manifold.

‡ Cf., Grioli [9, 10] and Toupin [14, 15].

§ Cf., Ericksen, [2] Green and Rivlin [7, 8, 12], Mindlin [11].

fibred continua. Consider, for instance, the variety of conditions of use of armoured concrete and the multiplicity of types of junctions in steel gridworks. Hence, generalized continua are of double interest: theoretical and practical.

This paper will be concerned with the Cosserat continuum [1], the best known microstructure and (to some extent) the simplest. This system, when constrained, includes the so-called polar continua.

Specifically, we shall try to give some details regarding the formulation of the local Cauchy problem in a hyperelastic material, assuming a time-dependent coordinate system connected with material lines (*co-moving frame of reference*). On one hand, such an assumption implies the use of local bases which are, *a priori*, unknown, e.g., the motion of the continuum; on the other hand, it permits us to operate with variables connected solely to the instantaneous characteristics of the motion, i.e., with variables of geometric-kinematic nature (such as the *metric tensor*, *the deformation rate*, the *free* and *constrained spin*, etc.) which presuppose no reference configurations–in contrast to the Piola–Kirchhoff-type Lagrangian theories.

Briefly, intrinsic formulations of this type consist of two parts: a *principal problem* and a *secondary problem*. The latter often has a simple solution.†

For a Cosserat continuum (notably that of Cauchy type), the principal problem is a first-order Cauchy problem with well-determined initial data. A precise choice of the initial data leads to the compatibility of the motion; in particular, it results in the Euclidean character of the metric tensor. Thus, we can say that the compatibility conditions and the compatibility of the constrained spin (the spatial gradient of which is uniquely related to the deformation rate gradient) are automatically satisfied by a suitable choice of the initial values.

In such a problem, the fundamental variables are: the *shifters* d^r_σ and the *free spin* $b_{\rho\sigma}$ (for the directors); the *deformation rate* $k_{\rho\sigma}$ and the *constrained spin* $w_{\rho\sigma}$ (for the point continuum).

Determination of the motion follows and constitutes the secondary aspect of the problem.

Of course, *the formulation of the problem must satisfy all the invariance properties of the mechanical theory*. In other words, for each element of the continuum, the formulation is invariant with respect to an arbitrary time-independent rotation of the Cosserat trihedron, and has tensorial character for an arbitrary transformation of the Lagrangian coordinates.

Being of general character, this paper gives only an initial approach to its subject matter. However, it opens the way to a intelligible

† Consider, for example, the role of Euler's equations in rigid-body dynamics in the cases in which they constitute an autonomous differential system (of first order).

linearized theory dependent upon the choice of the thermodynamic potential and the stability hypothesis for the initial data.

2 Macroscopic equations of a microstructure and a symbolic relationship

For a generic microstructure, the basic equations of mechanics lead, by a limit procedure, to the following macroscopic and indefinite equations:

$$\left.\begin{aligned}&\mu(\mathbf{F}-\partial_t\mathbf{v})-\frac{1}{\sqrt{g}}\partial_\sigma(\sqrt{g}\boldsymbol{\phi}^\sigma)=0\\&\mu(\mathbf{M}-\partial_t\mathbf{k})-\mathbf{e}_\sigma\times\boldsymbol{\phi}^\sigma-\frac{1}{\sqrt{g}}\partial_\sigma(\sqrt{g}\boldsymbol{\psi}^\sigma)=\mathbf{0}\ldots C\end{aligned}\right\}\tag{1}$$

with the boundary conditions

$$\mathbf{f}=n_\sigma\boldsymbol{\phi}^\sigma=0,\qquad \mathbf{m}=n_\sigma\boldsymbol{\psi}^\sigma=0\ldots\Sigma,\tag{2}$$

which are valid for every natural (dynamically possible) motion and for every time t. In these equations, $\mathbf{F}$ and $\mathbf{M}$ are the specific body forces and couples, respectively; $\boldsymbol{\phi}^\sigma$ and $\boldsymbol{\psi}^\sigma(\sigma=1,2,3)$ are the specific stress and the couple stress, respectively, in the local basis $\{\mathbf{e}_\sigma\}$; $\mathbf{f}$ and $\mathbf{m}$ are the external surface forces and couples; $\mathbf{n}=n_\sigma\mathbf{e}^\sigma$ is the inward unit normal to the boundary Σ; and g is the determinant of the metric tensor $g_{\rho\sigma}=\mathbf{e}_\rho\cdot\mathbf{e}_\sigma$:

$$g\equiv\det\|g_{\rho\sigma}\|>0.\tag{3}$$

In these general equations (which are valid in the assumed continuous representation), the microstructure appears directly only through the specific and intrinsic angular momentum $\mathbf{k}$.

It is easy to prove that the system (1), (2) can be written as a unique scalar equation (a *D'Alembert–Lagrange symbolic relation*):

$$\begin{aligned}&\int_C\mu(\mathbf{F}-\partial_t\mathbf{v})\cdot\boldsymbol{\xi}\,\mathrm{d}C+\int_C\mu(\mathbf{M}-\partial_t\mathbf{k})\cdot\boldsymbol{\eta}\,\mathrm{d}C\\&+\int_\Sigma(\mathbf{f}\cdot\boldsymbol{\xi}+\mathbf{m}\cdot\boldsymbol{\eta})\,\mathrm{d}\Sigma+\int_C(\boldsymbol{\phi}^\sigma\cdot\partial_\sigma\boldsymbol{\xi}-\mathbf{e}_\sigma\times\boldsymbol{\phi}^\sigma\cdot\boldsymbol{\eta}+\boldsymbol{\psi}^\sigma\cdot\partial_\sigma\boldsymbol{\eta})\,\mathrm{d}C\\&\qquad=0,\qquad\forall\boldsymbol{\xi},\boldsymbol{\eta}.\end{aligned}\tag{4}$$

This equation expresses the fact that *the nominal work of all the forces* (body, inertial and contact, inside and outside) *acting on the body vanishes for each natural motion of the continuum and for each time t.*

Of course, the scalar global relation (4) is equivalent to the local equations (1), (2) provided that it holds for each choice of the vector fields $\boldsymbol{\xi}$, $\boldsymbol{\eta}$ sufficiently smooth in the closure of the domain C.

3 Virtual work of the contact forces for the Cosserat continua

For a general microstructure, Eqns (1), (2) must be completed (according to the scheme employed) by the mechanical or thermomechanical equations inherent in the structure under consideration; this is in accordance with an analysis of the microforces and the dependence of $\mathbf{k}$ on the directors. Such complications do not occur directly† in the Cosserat scheme; here, the instantaneous configuration $\mathscr{C}$ is geometrically described by means of a continuum C of points and a local regular distribution of a non-deformable trihedron $\{\mathbf{d}_r\}_P$:

$$\mathscr{C} \equiv \left\{ \begin{array}{l} OP = OP(t/y) \\ \mathbf{d}_r = \mathbf{d}_r(t/y) \end{array} \qquad (r = 1, 2, 3) \right\}. \tag{5}$$

For such a structure, the virtual velocity distribution is defined (in $\mathscr{C}$) by means of the field of linear velocity $\mathbf{v}$ for the point continuum and by the field of the free spin $\mathbf{b}$,

$$\mathbf{b} = \tfrac{1}{2}\mathbf{d}^r \times \partial_t \mathbf{d}_r, \tag{6}$$

for the directors. Thus, the virtual work of the internal contact forces is given, as is Eqn (4), by the integral

$$W^{(i)} = \int_C [\boldsymbol{\phi}^\sigma \cdot (\partial_\sigma \mathbf{v} + \mathbf{e}_\sigma \times \mathbf{b}) + \boldsymbol{\psi}^\sigma \cdot \partial_\sigma \mathbf{b}]\, \mathrm{d}C, \tag{7}$$

which depends on $\mathbf{b}$ and on its spatial gradient $\partial_\sigma \mathbf{b}$.

It is clear that $W^{(i)}$ is invariant: (a) with respect to an arbitrary rotation of the $\{\mathbf{d}_r\}_P$ trihedron, eventually dependent on the 'particle', but independent of time (*principle of objectivity*); (b) with respect to a rigid motion superposition, namely for transformation of the vector fields $\mathbf{v}$ and $\mathbf{b}$ of the type;

$$\mathbf{v} \to \mathbf{v}' = \mathbf{v} + \mathbf{H} + \mathbf{K} \times OP \qquad \mathbf{b} \to \mathbf{b}' = \mathbf{b} + \mathbf{K},$$

with $\mathbf{H}$ and $\mathbf{K}$ independent of the point $P \in C$ (*principle of material indifference*).

From the thermodynamical point of view, Eqn (7) is remarkable. It suggests the most natural choice of the stress variables and of the parameters of the free states. More precisely, with regard to the stress, it

† But only because of the specification of the local inertia tensor of the microstructure. Hereafter, we are particularly interested in such continua.

is natural to assume as scalar variables the components ϕ_r^σ and ψ_r^σ of stress and couple stress, respectively, in the trihedron $\{\mathbf{d}_r\}_P$; therefore, we consider the following decomposition

$$\boldsymbol{\phi}^\sigma = \phi_r^\sigma \mathbf{d}^r, \qquad \boldsymbol{\psi}^\sigma = \psi_r^\sigma \mathbf{d}^r \qquad (\sigma = 1, 2, 3). \tag{8}$$

In the same way, with regard to the free-state parameters, including the shifters d_σ^r which appear in the decomposition

$$\mathbf{d}^r = d_\sigma^r \mathbf{e}^\sigma, \quad \text{with} \quad \mathbf{e}_\rho = \partial_\rho OP, \qquad \mathbf{e}_\rho \cdot \mathbf{e}^\sigma = \delta_\rho^\sigma, \tag{9}$$

it is natural to consider, in addition to the free spin $\mathbf{b}$, the analogous vector fields $\mathbf{b}_\sigma$:

$$\mathbf{b}_\sigma = \tfrac{1}{2}\mathbf{d}^r \times \partial_\sigma \mathbf{d}_r \qquad (\sigma = 1, 2, 3), \tag{10}$$

with their components

$$b_\sigma^r = \mathbf{d}^r \cdot \mathbf{b}_\sigma, \qquad \sim \mathbf{b}_\sigma = d_\sigma^r \mathbf{d}_r. \tag{11}$$

With such a choice, we obtain the following synthetic expression for the work of the internal contact forces (cf., [3, 4]):

$$W^{(i)} = \int_C (\phi_r^\sigma \partial_t d_\sigma^r + \psi_r^\sigma \partial_t b_\sigma^r)\, \mathrm{d}C.$$

We note that the 18 variables d_σ^r and b_σ^r† define the geometric state of the body, i.e., they generalize what is defined, for an ordinary continuum, by the metric tensor $g_{\rho\sigma}$; similarly, they have a precise geometric meaning in relation to the instantaneous configuration $\mathscr{C}$.

Indeed, besides the direct meaning of the shifters d_σ^r, *the variables* b_σ^r *are in a one to one relation with the tensor* $b_{\sigma\mu\nu}$ *associated with the Ricci rotation coefficients of the anholonomic basis* $\{\mathbf{d}^r\}_P$:

$$b_{\sigma\mu\nu} = (\partial\Gamma_{\sigma\mu,\nu} - d_{r\nu}\partial_\sigma d_\mu^r \equiv -d_{r\nu}\nabla_\sigma d_\mu^r, \tag{12}$$

and also with the anholonomic tensor $a_{\sigma\mu,\nu}$ *of the same basis:*

$$a_{\sigma\mu,\nu} = (\partial_\mu d_\sigma^r - \partial_\sigma d_\mu^r) d_{r\nu}. \tag{13}$$

In fact, the following invertible relations hold:

$$\left.\begin{aligned} b_\sigma^r &= \tfrac{1}{2} d_\rho^r \eta^{\rho\mu\nu} b_{\sigma\mu\nu} \sim b_{\sigma\mu\nu} = d_r^\rho \eta_{\rho\mu\nu} b_\sigma^r, \\ b_{\sigma\mu\nu} &= \tfrac{1}{2}(a_{\nu\sigma,\mu} + a_{\sigma\mu,\nu} - a_{\mu\nu,\sigma}) \sim a_{\sigma\mu,\nu} = b_{\sigma\mu\nu} - b_{\mu\sigma\nu}. \end{aligned}\right\} \tag{14}$$

Of course Eqn (12) leads to the formula

$$b_{\sigma\mu\nu} = -\nabla_\sigma d_{r[\mu} d_{\nu]}^r \tag{15}$$

† They are *free* variables; having chosen a reference configuration $\mathscr{C}^*$ and denoted the local rotation of the Cosserat trihedron and the local displacement by $\mathscr{R}$ and $\mathbf{u}$, respectively, the variables d_σ^r depend on $\mathscr{R}$ and on the gradient of $\mathbf{u}$ while the b_σ^r's are independent of $\mathbf{u}$.

and to the identity

$$\nabla_\sigma d_{r(\mu} d^r_{\nu)} = 0 \sim \Gamma_{\sigma(\mu,\nu)} = \partial_\sigma d_{r(\mu} d^r_{\nu)}, \tag{16}$$

which corresponds to Ricci's theorem. Indeed, since the metric tensor $g_{\mu\nu}$ is uniquely determined by the d^r_σ (but not *vice versa*):

$$g_{\mu\nu} = \delta_{rs} d^r_\mu d^s_\nu \sim \delta^s_r = d^\mu_r d^s_\mu; \tag{17}$$

Eqn (16) does not differ from

$$\nabla_\sigma g_{\mu\nu} = 0 \sim \Gamma_{\sigma(\mu,\nu)} = \tfrac{1}{2}\partial_\sigma g_{\mu\nu}.$$

Thus, Eqns (14) express the variables b^r_σ in terms of the shifters d^r_σ and their first spatial derivatives; the first implicitly, through the Christoffel symbols, as in Eqns (12) and (17) and the other explicitly, as in Eqn (13).

In other words, the geometrical state of a Cosserat continuum can be defined either by the free variables d^r_σ and b^r_σ or by the shifters d^r_σ and their first derivatives. However, the latter variables are not free, but are subject to the *eighteen* constraints of Eqn (16).

4 Hyperelastic Cosserat continua

For simplicity, we restrict ourselves to the Cosserat continua of hyperelastic type, i.e., those subjected to reversible transformations only (for instance, isothermal or adiabatic). This assumption allows us to remain in a purely mechanical state.

In such a case, the thermodynamic principles require the specific work of the internal contact forces to be an exact differential form for each continuum motion $\mathcal{M}$, i.e., in accordance with Eqn (7′),

$$\phi^\sigma_r\, \partial_t d^r_\sigma + \psi^\sigma_r\, \partial_t b^r_\sigma = -\mu\, \partial_t w, \qquad \forall \mathcal{M},\, t. \tag{18}$$

In other words, the following constitutive relations hold:

$$\phi^\sigma_r = -\mu \frac{\partial w}{\partial d^r_\sigma}, \qquad \psi^\sigma_r = -\mu \frac{\partial w}{\partial b^r_\sigma}, \tag{18′}$$

where $w(y^\sigma/d^r_\sigma/b^r_\sigma)$ is the isothermal (or adiabatic) *specific potential* (free energy), i.e., a function of the state, characteristic of the material.

It is clear that the constitutive relations (18′), which in tensorial form do not differ from

$$\phi^\sigma_\rho = -\mu d^r_\rho \frac{\partial w}{\partial d^r_\sigma}, \qquad \psi^\sigma_\rho = -\mu d^r_\rho \frac{\partial w}{\partial b^r_\sigma}, \tag{18″}$$

determine the mechanical problem.

More precisely, just as in the classical case, two types of approach arise in the non-linear theories. The more usual one implies for the continuum the choice of *a reference configuration*, which has a very important role. Starting from this configuration, we can evaluate the local displacement and the rotation of the Cosserat trihedron in terms of six scalar parameters, which become the fundamental variables of the theory.

From this point of view, at first, we must express the variables d^r_σ and b^r_σ by the foregoing parameters and their first spatial derivatives, and we must transport the stress and the couple-stress (and hence the body and the external surface forces) to the reference configuration, introducing in their place the analogous Piola–Kirchhoff characteristics. Then, projecting Eqns (1), (2) on the local basis of the reference configuration, and taking into account the constitutive relations (18″), we finally obtain the fundamental equations of the theory: six equations for the unknown variables, and similarly for the boundary conditions. Of course, we assume for **k** the classical expression of the angular momentum:

$$\mathbf{k} = \sigma_{rs} b^s \mathbf{d}^r, \quad \text{with} \quad b^s \equiv \mathbf{b} \cdot \mathbf{d}^s, \tag{19}$$

where the inertia tensor σ_{rs} and the mass density satisfy the Lagrangian conservative equations

$$\mu = \mu^0 \sqrt{(g^0/g)}, \qquad \sigma_{rs} = \sigma^0_{rs}. \tag{20}$$

A second approach, more natural, is the one known as *intrinsic*, in the sense that it involves the actual configuration only and is independent of the choice of a reference configuration. This point of view agrees with a general formulation that would be quite independent of the existence of a privileged state (natural, isotropic, hereditary, etc.) and also with the meaning of a differential evolution problem, which is directly connected with the instantaneous configuration of the system and its neighbourhood. Thus, it appears more natural to project the fundamental equations (1), (2) on the local basis $\{\mathbf{e}_\rho\}$, referring to the domain C and not to the reference one. It is clear that such a basis, referring to material lines of the continuum, has the same local role as the solid trihedron in the rigid-body dynamics.

This projection, connected with a basis deformable with the 'particle', and, like this unknown, yields equations of the type

$$\left.\begin{aligned} a_\rho &= F_\rho - \frac{1}{\mu} \nabla_\sigma \phi^\sigma_\rho, \\ x_\rho &= M_\rho + \frac{1}{\mu} (\eta^\nu_{\sigma\rho} \phi^\sigma_\nu - \nabla_\sigma \psi^\sigma_\rho), \\ x_\rho &\equiv \sigma_{\rho\mu} (\partial_t b^\mu + h^\mu_\nu b^\nu) + \sigma_{\mu\nu} b^\mu b^{*\nu}_\rho. \end{aligned}\right\} \tag{21}$$

We note that Eqns (21) have *tensorial character*, i.e., they are invariant with respect to arbitrary transformations of the Lagrangian coordinates:

$$y^\sigma = y^\sigma(y'^1, y'^2, y'^3) \quad \text{with} \quad \det \left\| \frac{\partial y^\sigma}{\partial y'^\rho} \right\| > 0,$$

and contain the *free spin* b^μ or its dual $b^*_{\rho\sigma}$:

$$b^\mu \equiv \mathbf{b} \cdot \mathbf{e}^\mu, \qquad b^*_{\rho\sigma} = \eta_{\mu\rho\sigma} b^\mu \sim b^\mu = \tfrac{1}{2}\eta^{\mu\rho\sigma} b^*_{\rho\sigma}, \tag{22}$$

and the tensor $h_{\rho\sigma}$, which includes the *constrained spin*† and the *deformation rate*:

$$h_{\rho\sigma} = \omega_{\rho\sigma} + k_{\rho\sigma} \tag{23}$$

5 Cauchy's problem

Let us consider Eqns (21), taking into account the constitutive relations (18″), the continuity equation (20), and Eqns (14) and (13), which express the variables b^r_σ in terms of the shifters d^r_σ and their first spatial derivatives. It is clear that, finally, Eqn (21) leads to relationships of the following type:‡

$$\left.\begin{aligned} a_\rho &= f_\rho(y^\sigma \mid h_{\mu\sigma} \mid b_\mu \mid d^r_\sigma \mid \partial_\nu d^r_\sigma \mid \partial_{\mu\nu} d^r_\sigma), \\ \partial_t b_\rho &= g_\rho(y^\sigma \mid h_{\mu\sigma} \mid b_\mu \mid d^r_\sigma \mid \partial_\nu d^r_\sigma \mid \partial_{\mu\nu} d^r_\sigma), \end{aligned}\right\} \tag{24}$$

where f_ρ and g_ρ are *tensorial functions* of the indicated variables, which are known as soon as the laws of the body forces and the potential w are specified. Of course, it is necessary to add to the fundamental equations (24) the geometrical-kinematical compatibility conditions of the motion, which do not differ from (cf., [6])

$$\left.\begin{aligned} &\partial_t d^r_\sigma = d^{r\rho}(h_{\sigma\rho} - b^*_{\sigma\rho}), \\ &\partial_t h_{\rho\sigma} = \nabla_\rho a_\sigma + h^\nu_\rho h_{\sigma\nu}, \\ &A_{\alpha\rho\sigma} \equiv \nabla_\alpha \omega_{\rho\sigma} - \nabla_\rho k_{\alpha\sigma} + \nabla_\sigma k_{\alpha\rho} = 0, \\ &R^\sigma_{\alpha\beta\rho} \equiv \partial_\alpha \Gamma^\sigma_{\beta\rho} - \partial_\beta \Gamma^\sigma_{\alpha\rho} + \Gamma^\mu_{\beta\rho}\Gamma^\sigma_{\alpha\mu} - \Gamma^\mu_{\alpha\rho}\Gamma^\sigma_{\beta\mu} = 0 \end{aligned}\right\} \tag{25}$$

where $R^\sigma_{\alpha\beta\rho}$ is the *curvature tensor* associated with the metric tensor (17). Equations (24), (25) allow us to express the static problem directly in terms of the 'deformation', i.e., by means of the nine variables $d^r_\sigma(y)$. In a *congruent static configuration*, such variables should satisfy the following system of equations containing partial derivatives up to the second order:

$$f_\rho = 0, \qquad g_\rho = 0, \qquad R^\sigma_{\alpha\beta\rho} = 0. \tag{26}$$

† It corresponds to the linear velocity $\mathbf{v}$, and, therefore, it is independent of the motion of the Cosserat trihedron: $\boldsymbol{\omega} = \frac{1}{2}\,\mathrm{rot}\,\mathbf{v} \equiv \frac{1}{2}\mathbf{e}^\sigma \times \partial_\sigma \mathbf{v}$.

‡ The inertia tensor $\sigma_{\rho\sigma} = d^r_\rho d^s_\sigma \sigma_{rs}$ is *singular* for a rectilinear microstructure only.

It is clear that, from the knowledge of d^r_σ, we obtain the metric $g_{\rho\sigma}$ of the static domain C and then $g_{\rho\sigma}$ define the effective position of C up to a global rigid displacement.

Of course, a static formulation in terms of stress is possible only if the constitutive equations (18″) are invertible, and, therefore, they allow us to obtain the variables d^r_σ in terms of the stress characteristics ϕ^σ_ρ and ψ^σ_ρ.

However, for the *dynamical problem*, which is more important than the static one, the previous procedure suggests in the most natural way that we consider the following differential system in the variables d^r_σ, $h_{\rho\sigma} \equiv \omega_{\rho\sigma} + k_{\rho\sigma}$ and b_ρ:

$$\left.\begin{aligned} &\partial_t d^r_\sigma = d^{r\rho}(h_{\sigma\rho} - b^*_{\sigma\rho}),\\ &\partial_t h_{\rho\sigma} = \nabla_\rho f_\sigma + h^\nu_\rho h_{\sigma\nu},\\ &\partial_t b_\rho = g_\rho, \end{aligned}\right\} \tag{27}$$

where f_ρ and g_ρ are *known functions* defined by the conditions (24).

The differential system (27) does not include all the required conditions, because it does not account for Eqn $(25)_{3,4}$. However (and this appears to be essential), it *implies*, as in the ordinary case (cf., [5] pp. 246–247), *the following differential conditions of the first order for the two tensors* $A_{\alpha\rho\sigma}$ and $R^\sigma_{\alpha\beta\rho}$:

$$\left.\begin{aligned} &\partial_t A_{\alpha\rho\sigma} = R^\nu_{\rho\sigma\alpha} f_\nu + 2h^\nu_{[\rho} A_{\sigma]\alpha\nu} - 2h^\mu_\alpha A_{[\rho\sigma]\nu},\\ &\partial_t R^\sigma_{\alpha\beta\rho} = -R^\nu_{\alpha\beta\rho} h^\sigma_\nu - R^\sigma_{\alpha\beta\nu} h^\nu_\rho - 2\nabla_{[\alpha} A^\sigma_{\beta]\rho}. \end{aligned}\right\} \tag{28}$$

They constitute a system of *linear and homogeneous* equations for the tensors under consideration, subject to (27) in the explicit way, through the coefficients f_ρ, $h_{\rho\sigma}$ and $g_{\rho\sigma}$; thus, *any smooth solution of the system* (27) *which satisfies the conditions* $A_{\alpha\rho\sigma} = 0$ and $R^\sigma_{\alpha\beta\rho} = 0$ *initially* $(t = 0)$ *conserves such conditions for every time t.*

In other words, the compatibility of the secondary problem (determination of the motion through the basis $\{\mathbf{e}_\rho\}$ and the velocity $\mathbf{v}$, starting from the $h_{\rho\sigma}$) is a consequence of system (27), as soon as the initial data satisfy the conditions $(25)_{3,4}$ for $t = 0$.

This means that, at least from the local point of view, *the dynamics of the Cosserat continua* of hyperelastic type, or with given constitutive relations, *can be translated into a well-determined first-order Cauchy problem*, constituted by the system (27). However, in *the Euclidean case, the initial data* $\mathring{d}^r_\sigma(y)$, $\mathring{h}_{\rho\sigma}(y)$ and $\mathring{b}_\rho(y)$ cannot all be arbitrary, but *must satisfy the following differential conditions*:

$$\mathring{A}_{\alpha\rho\sigma} = 0, \qquad \mathring{R}^\sigma_{\alpha\beta\rho} = 0. \tag{29}$$

We note that such conditions evidently limit the choice of the initial data in the following way. Only three of the nine $\mathring{d}^r_\sigma$ are independent, because the six combinations $\mathring{g}_{\rho\sigma} = \delta_{rs}\mathring{d}^r_\rho \mathring{d}^s_\sigma$ must characterize the Euclidean metric of initial configuration C_0; the $\mathring{k}_{\rho\sigma}(y)$ are subject to the St Venant compatibility conditions:

$$\mathring{\eta}^{\mu\alpha\beta}\mathring{\eta}^{\nu\rho\sigma}\mathring{\nabla}_\alpha \mathring{\nabla}_\rho \mathring{k}_{\beta\sigma} = 0.$$

Finally the constrained spins $\mathring{\omega}_{\rho\sigma}(y)$ are determined by the values at a point C_0, through the resolution of a total differential system.

A concrete case that could be developed in detail, on the basis of the preceding equations, is the one in which the stress tensors ϕ^σ_ρ and ψ^σ_ρ are functions of the metric tensor $g_{\rho\sigma}$ and its tensorial products associated with d^r_σ and b^r_σ. We have in mind the case in which the stress depends on the variables d^r_σ and b^r_σ through the metric $g_{\rho\sigma}$ and the tensors

$$l_{\rho\sigma} = d^r_\rho b_{r\sigma}, \qquad \bar{l}_{\rho\sigma} = d^r_\sigma b_{r\rho}, \qquad m_{\rho\sigma} = b^r_\rho b_{r\sigma} \tag{30}$$

(for instance, polynomial functions restricted by the existence of integrity bases, or isotropic type, etc.). In particular, a *linear dependence* such as

$$\left.\begin{aligned} \phi^\sigma_\rho &= p\delta^\sigma_\rho + p_1 l^\sigma_\rho + p_2 \bar{l}^\sigma_\rho + p_3 m^\sigma_\rho \\ \psi^\sigma_\rho &= q\delta^\sigma_\rho + q_1 l^\sigma_\rho + q_2 \bar{l}^\sigma_\rho + q_3 m^\sigma_\rho \end{aligned} \quad (\rho, \sigma = 1, 2, 3), \right\} \tag{31}$$

where p, p_1, p_2, p_3, q, q_1, q_2, q_3 are *scalar functions of the invariants of the tensors* (30) with respect to the metric $g_{\rho\sigma}$ (through the potential w in the hyperelastic case). Of course, such cases, excluding fluids, are not present in the classical mechanics of continua, simply because of the absence of a supplementary geometric structure such as directors.

On the contrary, the stress response of the material, in a generic configuration C, could depend on a privileged configuration, as happens, for instance, in the ordinary isotropic case. In such a case, the variables d^r_σ and b^r_σ can be usefully replaced by differences of mixed type,

$$2E^r_\sigma = d^r_\sigma - D^r_\sigma, \qquad 2F^r_\sigma = b^r_\sigma - B^r_\sigma, \tag{32}$$

where the capital letters are usually related to the privileged configuration; or by the characteristics of direct deformation (*Green tensors*)

$$2E^\rho_\sigma = D^\rho_r d^r_\sigma - \delta^\rho_\sigma, \qquad 2F^\rho_\sigma = D^\rho_r b^r_\sigma - L^\rho_\sigma. \tag{32'}$$

6 An alternative expression for the internal work

Another expression for the internal work can be used in (7′); its efficiency has already been mentioned. More precisely, introducing the Lagrangian components of stress and couple-stress,

$$\phi^{\rho\sigma} = \boldsymbol{\phi}^\rho \cdot \mathbf{e}^\sigma \equiv \phi^\rho_r d^{r\sigma}, \qquad \psi^{\rho\sigma} = \boldsymbol{\psi}^\rho \cdot \mathbf{e}^\sigma \equiv \psi^\rho_r d^{r\sigma}, \tag{33}$$

and decomposing $\phi^{\rho\sigma}$ into its symmetric and antisymmetric parts $\phi^{(\rho\sigma)}$

and $\phi^{[\rho\sigma]}$, respectively, the internal work can be expressed by the following Lagrangian form (cf., [3, 4]):

$$W^{(i)} = \frac{1}{2}\int_C (\phi^{(\rho\sigma)}\,\partial_t g_{\rho\sigma} + \psi^{\rho\sigma}\eta^{\mu}_{\nu\sigma}\,\partial_t\Gamma^{\nu}_{\rho\mu})\,dC$$

$$+\int_C (\phi^{[\rho\sigma]}c_{\sigma\rho} + \psi^{\rho\sigma}\nabla_\rho c_\sigma)\,dC, \tag{34}$$

where $c_{\rho\sigma}$ is the difference between the free spin and the constrained spin:

$$c_{\rho\sigma} = b^{*}_{\rho\sigma} \sim \omega_{\rho\sigma} \equiv d_{r[\rho}\partial_t d^r_{\sigma]}, \qquad c^{\mu} = \tfrac{1}{2}\eta^{\mu\rho\sigma}c_{\rho\sigma}. \tag{35}$$

Expression (34), such as (7′), clearly has *tensorial character*† and allows us to find, directly, the well-known Eulerian form, taking into account the kinematic meaning of the time derivative of the Christoffel symbols of the second kind:

$$\partial_t\Gamma^{\nu}_{\rho\mu} = g^{\nu\sigma}(\nabla_\rho k_{\mu\sigma} + \nabla_\mu k_{\sigma\rho} - \nabla_\sigma k_{\rho\mu}). \tag{36}$$

In fact, (34) suggests that we introduce the coefficients $\Gamma_{\rho\sigma}$:

$$\Gamma_{\rho\sigma} = \tfrac{1}{2}\Gamma^{\nu}_{\rho\mu}\eta^{\mu}_{\nu\sigma} \equiv \tfrac{1}{2}\Gamma_{\rho[\mu,\nu]}\eta^{\mu\nu}\sigma, \tag{37}$$

i.e., the *affine tensor* that is obtained by the duality law, antisymmetrizing the last two indices of the Christoffel symbols. Such coefficients, which are connected directly with the affine system

$$\mathbf{\Gamma}_\rho = \tfrac{1}{2}\mathbf{e}^\sigma \times \partial_\rho \mathbf{e}_\sigma \equiv \Gamma_{\rho\sigma}\mathbf{e}^\sigma, \tag{37′}$$

analogous to the free spin $\mathbf{b}$, *are not all independent*, unlike b^r_σ or the equivalent variables $b_{\rho\sigma} = b^r_\sigma d_{r\rho}$.

Actually, the definition (37) makes clear that the $\Gamma_{\rho\sigma}$ variables satisfy the restriction

$$g^{\rho\sigma}\Gamma_{\rho\sigma} = 0, \tag{38}$$

i.e., *the first invariant of the affine tensor $\Gamma_{\rho\sigma}$ vanishes.*

We note, explicitly, that the differences (35) agree, as in (15), with the analogous differences

$$b_{\rho\mu\nu} - \Gamma_{\rho[\mu,\nu]} = d_{r[\mu}\partial_\rho d^r_{\nu]} \tag{39}$$

or, in an equivalent form,

$$b_{\rho\sigma} - \Gamma_{\rho\sigma} = \tfrac{1}{2}d_{r[\mu}\partial_\rho d^r_{\nu]}\eta^{\mu\nu}_{\sigma}. \tag{39′}$$

Indeed, introducing the affine tensor (37), which is a function of the metric tensor $g_{\rho\sigma}$ and its spatial derivatives, the internal work (34)

† As is well known, the derivatives $\partial_t\Gamma^{\nu}_{\rho\mu}$ have tensorial character, unlike the Christoffel symbols.

assumes the *mixed form* (affine and tensorial)

$$W^{(i)} = \int_C \left(\tfrac{1}{2}\tilde{\phi}^{\rho\sigma}\, \partial_t g_{\rho\sigma} + \psi^{\rho\sigma}\, \partial_t \Gamma_{\rho\sigma} - \phi^{[\rho\sigma]} c_{\rho\sigma} + \psi^{\rho\sigma} \nabla_\rho c_\sigma\right) \mathrm{d}C, \tag{40}$$

where, for brevity, we have set

$$\tilde{\phi}^{\rho\sigma} = \phi^{(\rho\sigma)} - \psi^{\mu\nu}(\Gamma^\rho_\mu \delta^\sigma_\nu + \Gamma^{(\rho\varepsilon)}_\mu \eta^\sigma_{\nu\varepsilon}). \tag{41}$$

We note that the affine tensor $\tilde{\phi}^{\rho\sigma}$, which is a linear combination of the stress and couple-stress, depends on *all* partial derivatives of the metric tensor $g_{\rho\sigma}$; i.e., it depends on all the Christoffel symbols and not on the combinations $\Gamma_{\rho\sigma}$ alone, since

$$-2\Gamma^{(\rho,\varepsilon)}_\mu = \partial_\mu g^{\rho\varepsilon}. \tag{42}$$

The property (38) allows us to further transform (40), in the sense that the dependence on the isotropic part of the couple-stress tensor $\psi^{\rho\sigma}$ can be confined to the last term only. More precisely, with the usual decomposition

$$\psi^{\rho\sigma} = \psi g^{\rho\sigma} + S^{\rho\sigma}, \quad \text{with} \quad g_{\rho\sigma} S^{\rho\sigma} = 0 \tag{43}$$

(40) assumes the form

$$W^{(i)} = \int_C \left(\tfrac{1}{2}N^{\rho\sigma}\, \partial_t g_{\rho\sigma} + S^{\rho\sigma}\, \partial_t \Gamma_{\rho\sigma} - \phi^{[\rho\sigma]} c_{\rho\sigma} \right.$$
$$\left. + S^{\rho\sigma} \nabla_\rho c_\sigma + \psi \nabla_\rho c^\rho\right) \mathrm{d}C, \tag{44}$$

where the symmetric tensor $N^{\rho\sigma}$ is defined by the symmetric part of the stress tensor $\phi^{(\rho\sigma)}$ and by the couple-stress deviator $S^{\rho\sigma}$:

$$N^{\rho\sigma} = \phi^{(\rho\sigma)} - S^{\mu\nu}(\Gamma^{(\rho}_\mu \delta^{\sigma)}_\nu + \tfrac{1}{2}\partial_\mu g^{\varepsilon(\rho} \eta^{\sigma)}_{\varepsilon\nu}). \tag{45}$$

Of course, the mixed expression (44) does not represent a remarkable improvement with respect to (7′), at least in the general case, unless the class of possible transformations of the dynamical system are restricted, admitting, for instance, the existence of such a vectorial function of the d^r_σ, say φ_ρ, that $c_\rho = \partial_t \varphi_\rho$.

7 Polar ordinary continua

As opposed to the situation in the general case, expression (44) becomes more significant for an ordinary polar continuum, i.e., for a Cosserat continuum constrained by the condition

$$c_{\rho\sigma} \equiv 0 \sim b^*_{\rho\sigma} \equiv \omega_{\rho\sigma} \qquad \forall \mathcal{M},\ t,\ P. \tag{46}$$

This constraint is a restriction for the motion $\mathcal{M}$ of the continuum, which involves the position variables d_σ^r and their first time derivatives; briefly, it is a *linear and bilateral anholonomic constraint.*

Of course, we assume that such a constraint is *ideal*, and supported by the internal contact forces only, i.e., by the stress and couple-stress tensors.† In other words, for such a constraint, the *virtual work* of the reaction forces must vanish or, in view of (44), the following local condition must hold:

$$\tfrac{1}{2}N_{(r)}^{\rho\sigma}\,\partial_t g_{\rho\sigma} + S_{(r)}^{\rho\sigma}\,\partial_t \Gamma_{\rho\sigma} = 0, \qquad \forall \mathscr{C}, \partial_t \mathscr{C}, \tag{47}$$

at each time and for each choice of the variables $\partial_t g_{\rho\sigma}$ and $\partial_t \Gamma_{\rho\sigma}$ compatible with the restriction (38):

$$\Gamma^{(\rho\sigma)}\,\partial_t g_{\rho\sigma} - g^{\rho\sigma}\,\partial_t \Gamma_{(\rho\sigma)} = 0. \tag{48}$$

Therefore, (47) leads to the conditions $S_{(r)}^{[\rho\sigma]} = 0$, $N_{(r)}^{\rho\sigma} = 2\lambda\Gamma^{(\rho\sigma)}$, $S_{(r)}^{(\rho\sigma)} = -\lambda g^{\rho\sigma}$, where λ is a multiplicative parameter, *a priori* arbitrary. However, the further condition $g_{\rho\sigma}S_{(r)}^{\rho\sigma} = 0$ implies $\lambda = 0$.

Then, the '*virtual work principle*' (47) is exactly equivalent to the conditions

$$S_{(r)}^{\rho\sigma} = 0, \qquad \phi_{(r)}^{(\rho\sigma)} = 0 \Rightarrow N_{(r)}^{\rho\sigma} = 0. \tag{47'}$$

In other words, for an ordinary polar continuum, the stress variables $\phi^{(\rho\sigma)}$ and $S^{\rho\sigma}$ have an active character, i.e., they can be related to the 'deformation'; the $\phi^{[\rho\sigma]}$ have a passive character, and hence are unknown variables to be determined *a posteriori*.

The case is different for the scalar ψ in the decomposition (43) since it is not determined. Thus, we obtain the well-known result that *the theory of the ordinary polar continua has one degree of indeterminacy.*‡

This result is also confirmed by the local equations (1). In fact, the second equation $(1)_2$ allows us to determine, *a posteriori*, the antisymmetric part of the stress tensor $\phi^{[\rho\sigma]}$, i.e., the reaction to the constraint (46), where $\mathbf{k}$ is related to the motion $\mathscr{M}$ by (46) itself.

On the other hand, the first equation $(1)_1$ assumes the following form (cf., [4], p. 171):

$$\mu(\mathbf{p} - \partial_t \mathbf{v}) - \frac{1}{\sqrt{g}}\,\partial_\sigma(\sqrt{g}\mathbf{P}^\sigma) = 0, \tag{49}$$

where

$$\left.\begin{aligned} \mathbf{p} &\equiv \mathbf{F} - \frac{1}{2\mu\sqrt{g}}\,\partial_\nu[\mu\sqrt{g}(\mathbf{M} - \partial_t \mathbf{k}) \times \mathbf{e}^\nu], \\ \mathbf{P}^\sigma &\equiv \mathbf{Y}^\sigma - \frac{1}{2\sqrt{g}}\,\partial_\nu(\sqrt{g}\mathbf{S}^\nu) \times \mathbf{e}^\sigma, \end{aligned}\right\} \tag{50}$$

† If, more generally, the internal constraint (46) is supported also by the body forces, cf. [16]

‡ More precisely, $\psi^{\rho\sigma}$ is determined up to the scalar ψ, while $\phi^{\rho\sigma}$ is determined up to the gradient of ψ. We note that an analogous situation occurs in the Kirchhoff theory of shells.

and also

$$\mathbf{Y}^\sigma = \phi^{(\rho\sigma)}\mathbf{e}_\rho, \qquad \mathbf{S}^\sigma = S^{\rho\sigma}\mathbf{e}_\rho. \tag{51}$$

Equation (49) is a typical form of a conservative equation. Of course, the source $\mathbf{p}$ is related to the body forces and couples $\mathbf{F}$ and $\mathbf{M}$, respectively, and to the intrinsic angular momentum $\mathbf{k}$. In contrast to this, the vectorial system $\mathbf{P}^\sigma$ includes the active stress $\phi^{(\rho\sigma)}$ and $S^{\rho\sigma}$, and does not contain the parameter ψ, i.e., that confirms its indeterminacy.

From the intrinsic point of view, the formulation of the dynamical problem follows from the system (27), at least if $\mathbf{k}$ can be neglected as compared to $\mathbf{M}$.† Obviously, we must take into account the different constitutive relations, where, as in (44) and (46), we have

$$W^{(i)} = \int_C (\tfrac{1}{2}N^{\rho\sigma}\,\partial_t g_{\rho\sigma} + S^{\rho\sigma}\,\partial_t \Gamma_{\rho\sigma})\,\mathrm{d}C: \tag{52}$$

the $g_{\rho\sigma}$ and $\Gamma_{\rho\sigma}$ appear as variables of the geometric states. However, only *fourteen* of these fifteen variables are *free*, because restriction (38) holds.

For instance, in the *hyperelastic case*, the specific potential w (isothermal or adiabatic) is a proper function of the y^σ and the foregoing variables that must satisfy the condition

$$\tfrac{1}{2}N^{\rho\sigma}\,\partial_t g_{\rho\sigma} + S^{\rho\sigma}\,\partial_t \Gamma_{\rho\sigma} = -\mu\,\partial_t w$$

for each choice of the variables $\partial_t g_{\rho\sigma}$ and $\partial_t \Gamma_{\rho\sigma}$ constrained by the restriction

$$\Gamma^{(\rho\sigma)}\,\partial_t g_{\rho\sigma} - g^{\rho\sigma}\,\partial_t \Gamma_{(\rho\sigma)} = 0.$$

It follows that the free energy w is of the type

$$w = w(y^\sigma \mid g_{\rho\sigma} \mid \Gamma_{(\rho\sigma)} \mid \Gamma_{[\rho\sigma]}) \tag{53}$$

and that the constitutive relations

$$\tfrac{1}{2}N^{\rho\sigma} + \lambda\Gamma^{(\rho\sigma)} = -\mu\frac{\partial w}{\partial g_{\rho\sigma}},$$

$$S^{(\rho\sigma)} - \lambda g^{\rho\sigma} = -\mu\frac{\partial w}{\partial \Gamma_{(\rho\sigma)}}, \qquad S^{[\rho\sigma]} = -\mu\frac{\partial w}{\partial \Gamma_{[\rho\sigma]}},$$

hold, where λ is a multiplicative parameter determined by the condition $g_{\rho\sigma}S^{\rho\sigma} = 0$:

$$\lambda = \mu q, \qquad q \equiv \tfrac{1}{3}g_{\rho\sigma}\frac{\partial w}{\partial \Gamma_{(\rho\sigma)}}.$$

† More precisely, as in expression (19), if the restriction $|\sigma_{rs}\dot{b}^r\mathbf{d}^s| \ll |\sigma_{rs}b^r\dot{\mathbf{d}}^s|$ holds (this condition is analogous to that appearing systematically in the dynamics of gyroscopes).

Finally, taking into account (45), we have the following equations:

$$\left.\begin{aligned}\phi^{(\rho\sigma)} &= -\mu\left[2\left(\frac{\partial w}{\partial g_{\rho\sigma}} + q\Gamma^{(\rho\sigma)}\right) + \left(\frac{\partial w}{\partial \Gamma_{[\mu\nu]}} + \frac{\partial w}{\partial \Gamma_{(\mu\nu)}} - qg^{\mu\nu}\right)\right.\\ &\qquad \left.(\Gamma^{(\rho}_{\mu}\delta^{\sigma)}_{\nu} + \tfrac{1}{2}\,\partial_{\mu}g^{\varepsilon(\rho}\eta^{\sigma)}_{\varepsilon\nu})\right],\\ S^{(\rho\sigma)} &= -\mu\left(\frac{\partial w}{\partial \Gamma_{(\rho\sigma)}} - qg^{\rho\sigma}\right), \qquad S^{[\rho\sigma]} = -\mu\,\frac{\partial w}{\partial \Gamma_{[\rho\sigma]}}.\end{aligned}\right\} \tag{54}$$

On the other hand, projecting the indefinite equation (49) on the local basis $\{\mathbf{e}_\rho\}$, we obtain the expression for the acceleration a_ρ:

$$a_\rho = p_\rho - \frac{1}{\mu}\,\nabla_\sigma P^{\sigma}_{\ \rho}, \tag{55}$$

where, as in (50),

$$\left.\begin{aligned}p_\rho &\equiv F_\rho - \frac{1}{2\mu}\,\nabla_\sigma[\mu(M_\nu - x_\nu)]\eta^{\nu\sigma}_{\rho},\\ P^{\rho\sigma} &\equiv \phi^{(\rho\sigma)} - \tfrac{1}{2}\nabla_\mu S^{\nu\mu}\eta^{\sigma\rho}_{\nu},\end{aligned}\right\} \tag{56}$$

and, hence, in view of $(21)_3$, taking into account the constraint (46),

$$x_\nu \equiv \sigma_{\nu\mu}(\partial_t\omega^\mu + h^\mu_\varepsilon\omega^\varepsilon) + \sigma_{\varepsilon\mu}\omega^\mu\omega^\varepsilon_\nu. \tag{57}$$

Therefore, if the laws of the body forces and the potential (53) are known, then (55) allows us to express the acceleration a_ρ as a well-determined function of a certain number of variables, through the fictitious forces p_ρ and the stress tensor $P^{\rho\sigma}$.

More precisely, the tensor $P^{\rho\sigma}$, which represents the active stress and couple-stress, depends, in view of the constitutive relations (54), on the metric tensor $g_{\rho\sigma}$ and its first and second spatial derivatives; thus, concerning the stress dependence, *the acceleration a_ρ is a function of the metric tensor and its first, second and third spatial derivatives.*

Moreover, in the case of the fictitious forces p_ρ, in addition to the dependence on **F** and **M** through the x_ν tensor, there appear the shifters d^r_σ, the tensor $h_{\rho\sigma}$ and its spatial gradient and, *finally*, *$\partial_t\omega^\mu$ and its spatial derivatives as well.* Thus, *the differential system* $(27)_{1,2}$ *does not express a first-order Cauchy problem*, at least in general, unless the term $\partial_t\omega^\mu$ in (57), or x_ν directly (the usual hypothesis), can be neglected as compared to M_ν. In the latter case, a_ρ is a known function of the type

$$a_\rho = f_\rho(\varphi^\sigma \mid g_{\mu\sigma} \mid \partial_\nu g_{\mu\sigma} \mid \partial_{\alpha\nu} g_{\mu\sigma} \mid \partial_{\alpha\beta\nu} g_{\mu\sigma} \mid h_{\rho\sigma} \mid \partial_\nu h_{\rho\sigma}), \tag{58}$$

and the Cauchy problem assumes essentially the form

$$\left.\begin{aligned}\partial_t g_{\rho\sigma} &= 2h_{(\rho\sigma)},\\ \partial_t h_{\rho\sigma} &= \nabla_\rho f_\sigma + h^\nu_\rho h_{\sigma\nu}.\end{aligned}\right\} \quad (59)$$

The shifters d^r_σ can be determined, *a posteriori*, by the equations

$$\partial_t d^r_\sigma = d^{r\rho} h_{(\sigma\rho)}. \quad (60)$$

Finally, as regards the ordinary continua, the Cauchy problem is of the type (59),† with the difference that, in the expression of the acceleration (58), the second and third derivatives of the metric tensor disappear.

References

1 Cosserat, E. and Cosserat, F., *Théorie des Corps Déformables*, Paris, Herman, 1909.

2 Ericksen, J. L., Anisotropic fluids, *Arch. Rat. Mech. Anal.*, **4,** 231–237, 1959/60.

3 Ferrarese, G., *Lezioni di Meccanica Superiore*, Veschi, Roma, 1968.

4 Ferrarese, G., Sulla compatibilità dei continui alla Cosserat, *Rend. Matem. Roma*, **4,** 151–174, 1971.

5 Ferrarese, G., Sulla formulazione intrinseca della meccanica dei continui iperelastici, *Rend. Matem. Roma*, **8,** 235–249, 1975.

6 Ferrarese, G., Sulla formulazione intrinseca della dinamica dei continui alla Cosserat, *Ann. Matem. Pura Appl.*, **108,** 109–124, 1976.

7 Green, A. E. and Rivlin, R. S., Simple force and stress multipoles, *Arch. Rat. Mech. Anal.*, **16,** 325–353, 1964.

8 Green, A. E. and Rivlin, R. S., Multipolar continuum mechanics, *Arch. Rat. Mech. Anal.*, **17,** 113–147, 1964.

9 Grioli, G., Elasticità asimmetrica, *Ann. Matem. Pura Appl.*, **50,** 389–417, 1960.

10 Grioli, G., *Mathematical Theory of Elastic Equilibrium* (*Recent Results*), Ergeb. Angew. Math. no. 7, Springer Verlag, Berlin, Heidelberg and New York, 1962.

11 Mindlin, R. D., Microstructure in linear elasticity, *Arch. Rat. Mech. Anal.*, **16,** 51–78, 1964.

12 Rivlin, R. S., The formulation of theories in generalized continuum mechanics and their physical significance, in *Symposia Mathematica, Istituto Nazionalie di Alta Matematica*, Academic Press, London and New York, Vol. I, pp. 357–373, 1969.

13 Stazi, L., Sulla meccanica intrinseca dei continui iperelastici, *Rend. Circ. Matem. Palermo*, in press.

14 Toupin, R. A., Elastic materials with couple-stress, *Arch. Rat. Mech. Anal.*, **11,** 385–414, 1962.

† It is clear that an analogous formulation can be given in terms of the variables $g_{\rho\sigma}$ and v_ρ (cf., [13]) and therefore in the case of (59).

15 Toupin, R. A., Theories of elasticity with couple-stress, *Arch. Rat. Mech. Anal.*, **17,** 85–112, 1964.

16 Wozniak, C., *Constrained continuous media*, I, II, III, *Bull. Acad. Polon. Sci. Ser. Sci. Techn.*, **21,** 109–116, 167–173, 175–182, 1973.

Professor Giorgio Ferrarese,
Istituto Matematico 'G. Castelnuovo',
Università di Roma,
Roma,
Italy

I. Herrera

On the variational principles of mechanics

Abstract

Recently, the author has developed a general formulation of variational principles using functional-valued operators which simplifies the treatment of problems of continuum mechanics and its partial differential equations because it is applicable in any linear space, not necessarily normed, or with an inner product, or complete. This method has been applied to obtain dual extremum principles for initial-value problems and, more generally, for non-negative asymmetric operators. More recently, it has been used to obtain general variational principles applicable to diffraction problems and to formulate a general theory of connectivity for formally symmetric operators, which is basic in the formulation of the finite element method. Here, a systematic presentation of this method and these results is given.

1 Introduction

The numerical methods of mechanics and, more generally, of mathematical physics make extensive use of variational formulations. The modern approach to such methods is based on functional analysis. However, in many applications to mechanics, functional analysis is not used systematically, in spite of the fact that it permits the achievement of greater generality, rigour and clarity; to a large extent this is due to the fact that the applicability of functional analysis is frequently hindered by the complicated structures which are assumed in many of its theories.

The complexity of structures limits, in at least two ways, the usefulness of functional analysis:

(a) it makes it difficult (or impossible) to treat complicated situations;
(b) it diminishes the number of people able to apply it efficiently.

This paper is devoted to a presentation of a general framework which simplifies the formulation of variational principles of mechanics and, more generally, of mathematical physics [1]. It is based on the systematic use of functional-valued operators. Advantages of this approach [1–3] are as follows:

1. Problems are formulated in the most general kind of linear spaces, which are not necessarily normed, or with an inner product, or complete. Most work in this field has been done either in inner product spaces [4] or in Hilbert spaces [5, 6] and it is generally thought that this is desirable, if not essential, for the results to hold. In many applications, the introduction of the Hilbert-space structure leads to unwarranted complications which do not occur when functional-valued operators are used.
2. The removal of superfluous hypotheses in the development of a theory is always convenient, because it enlarges its applicability.
3. The symmetry condition for the potentialness of an operator can be extended to linear spaces for which no inner product or norm have to be defined [1]. This fact makes it possible to formulate a theory which is rigorous and at the same time not complicated.
4. Error bounds for approximate solutions are among the most important results that the theory yields [7]. They depend, however, on simple properties which are independent of any Hilbert space structure, and, therefore, can be obtained in the simple setting that the author has developed [3].
5. The generality achieved is greater. This fact has been especially useful in the development of variational principles for diffraction problems and the theory of connectivity to be presented in this paper.

This method has been used to obtain variational [8] and truly extremum principles [1–3] for initial-value problems,† and, more generally, for non-negative asymmetric operators [3]. A procedure which permits extending variational principles, derived by the mirror method into truly extremal principles, has also been constructed [3]. The notion of formally symmetric operators, usually applicable to differential operators only, is extended to functional-valued operators; for such operators, general variational principles have been developed for diffraction problems [9] and for a theory of connectivity [10] basic to the formulation of the finite-element method.

The description of the framework and of the main developments is brief; therefore, many details have been left out. This has required that, in some cases, the results are not presented in their greatest generality. Thus, for example, diffraction and related problems with prescribed

† These results were obtained in 1974 [1, 2]. Later, in a less general setting, Brezis and Ekeland [6], obtained related results.

jumps are not considered, although they have been treated in this more general form in the corresponding references [10]. Similarly, problems defined in affine subspaces are not considered here.

2 Preliminary notions and notations

All linear spaces to be considered will be defined on the field of real or complex numbers $\mathscr{F}$. The outer sum of two such spaces D_1 and D_2 will be represented by $D_1 \oplus D_2$. On the other hand, if a linear space D is spanned by two linearly independent subspaces D_1 and D_2, the space D is isomorphic to $D_1 \oplus D_2$ and we will write $D = D_1 \oplus D_2$; the subspaces D_1, D_2 are called a decomposition of D. In this case, given any $u \in D$, there is a unique pair of elements (u_1, u_2) such that $u_i \in D_i$ $(i = 1, 2)$ and $u = u_1 + u_2$; the elements of this pair are called projections of u on D_1 and D_2, respectively. The notation D^n will be used for the outer sum $D_1 \oplus \cdots \oplus D_n$ when $D_i = D$ $(i = 1, \ldots, n)$.

The value of an n-linear† functional $\alpha : D^n \rightarrow \mathscr{F}$ at an element $u = (u_1, \ldots, u_n) \in D^n$ will be represented by $\langle \alpha, u_1, \ldots, u_n \rangle$.‡ The notation D^{n*} will be used for the linear space of all n-linear functionals. Alternatively, D^{1*} will be written as D^*; D^{0*} is defined as $\mathscr{F}$. Notice that $D^{n*} \neq (D^n)^*$.

Functional-valued operators $P : D \rightarrow D^{n*}$ are considered in this work. Special attention will be given to the case $n = 1$, i.e., to operators of the form $P : D \rightarrow D^*$. When P is linear, its adjoint $P^* : D \rightarrow D^*$ is defined by $\langle P^* u, v \rangle = \langle Pv, u \rangle$, which holds for every $u, v \in D$; thus, the adjoint of such linear operators always exists.

For operators of the type $P : D \rightarrow D^{n*}$, the notion of continuity can be introduced without a topology in D. The operator $P : D \rightarrow D^{n*}$ is said to be bidimensionally continuous at $u \in D$ if, for every v, w, $\xi^{(1)}, \ldots, \xi^{(n)} \in D$, the function $f(\eta, \lambda) = \langle P(u + \eta v + \lambda w), \xi^{(1)}, \ldots, \xi^{(n)} \rangle$ is continuous at $\eta = \lambda = 0$.

The concept of a derivative of an operator will be used in the sense of additive Gateaux variation [11]. More precisely, an element $P'(u) \in D^{(n+1)*}$ will be called the derivative of P at $u \in D$, if, for every $v, \xi^{(1)}, \ldots, \xi^{(n)} \in D$,

$$g'(0) = \langle P'(u), v, \xi^{(1)}, \ldots, \xi^{(n)} \rangle \tag{1}$$

whenever the function $g(t) = \langle P(u + tv), \xi^{(1)}, \ldots, \xi^{(n)} \rangle$. Partial derivatives $P_{,1}(u), P_{,2}(u) \in D^{(n+1)*}$ will be considered when $D = D_1 \oplus D_2$. Using the

† Russian authors usually include continuity in the definition of linearity [12]. On the other hand, for most American authors, this concept includes only additivity and homogeneity. In this paper, we follow the latter terminology.

‡ This notation does not imply the existence of an inner product.

unique representation (v_1, v_2), $v_i \in D_i$ $(i=1,2)$, of every $v \in D$, they are defined by

$$\langle P,_i(u), v, \xi^{(1)}, \ldots, \xi^{(n)}\rangle = \langle P'(u), v_i, \xi^{(1)}, \ldots, \xi^{(n)}\rangle \qquad (i=1,2), \tag{2}$$

which holds for every $\xi^{(1)}, \ldots, \xi^{(n)} \in D$.

An operator $P: D \to D^*$ is said to be potential if there exists a functional $\psi: D \to \mathcal{F} = D^{0*}$ such that $\psi'(u) = P(u)$ for every $u \in D$. It is well known [12] that a sufficient condition for potentiality is that $P'(u)$ be symmetric for every $u \in D$. This result remains valid for the class of operators $P: D \to D^*$ considered here, if $P'(u)$ is assumed to be bidimensionally continuous at each $u \in D$, as has been shown [1] in a manner related to that suggested by Vainberg [12]. For a linear operator P, this requirement reduces to the condition that P be symmetric. Such P will be said to be non-negative if $\langle Pu, u\rangle \geqslant 0$ for every $u \in D$ and positive if, in addition, $\langle Pu, u\rangle = 0$ only when $u = 0$. Non-positive and negative operators are defined similarly.

The general problem to be considered consists in finding solutions to an equation

$$P(u) = f, \tag{3}$$

where P is a functional-valued operator $P: D \to D^*$ and $f \in D^*$. The solutions u will be restricted to be in a subset $\hat{E} \subset D$. In applications [2, 3], the elements belonging to $\hat{E}$ satisfy some additional boundary or initial conditions and frequently constitute an affine subspace; i.e., there is a subspace $E \subset D$ and an element $w \in D$, $\hat{E} = w + E$. The results to be presented can be easily extended to the case $\hat{E} \neq D$ [2, 3]; however, for simplicity, it will be assumed $\hat{E} = D$, which corresponds to taking $w \in E = D$.

Following Noble and Sewell [4], given any two elements $u_-, u_+ \in D$, define

$$\Delta X = X(u_+) - X(u_-), \tag{4a}$$

$$\Delta u = u_+ - u_-. \tag{4b}$$

If a decomposition D_1, D_2 of D is available, let $\Delta_i u \in D_i$ $(i=1,2)$ be the unique representation of Δu in terms of an element of D_1 plus an element of D_2.

When the functional X is differentiable, it is said to be convex if

$$\Delta X - \langle X'(u_-), \Delta u\rangle \geqslant 0 \tag{5}$$

or, equivalently,

$$\Delta X - \langle X'(u_+), \Delta u\rangle \leqslant 0 \tag{6}$$

for every $u_+, u_- \in D$. It is strictly convex if the strict inequality holds whenever $u_+ \neq u_-$. In addition, X is concave or strictly concave if $-X$ is convex or strictly convex, respectively.

Furthermore, X is saddle, convex on D_1 and concave on D_2, if

$$\Delta X - \langle X'(u_-), \Delta_1 u\rangle - \langle X'(u_+), \Delta_2 u\rangle \geqslant 0 \tag{7}$$

for every $u_+, u_- \in D$. It is strictly saddle if the inequality is strict whenever $u_+ \neq u_-$.

3 Variational principles

A variational principle is understood to be an assertion stating that the derivative $X'(u)$ of a functional X vanishes at a point $u \in D$ if and only if u is a solution of (3). A sufficient condition for the construction of variational principles is that the operator P be potential, because if $\psi : D \to \mathscr{F}$ is a potential of P, then the functional $X(u) = \psi(u) - \langle f, u\rangle$ will have the required property.

Theorem 1 *Assume $P : D \to D^*$ to be potential. Then $u \in D$ is a solution of (3) if and only if $X'(u) = 0$. Here, $X : D \to \mathscr{F}$ is defined as above.*

The theory of dual extremal principles for non-linear functional-valued operators has been developed in complete generality elsewhere [1]. Here, a summary of this theory for the linear case is presented. Therefore, in what follows, the linearity of the operator $P : D \to D^*$ will be assumed. In addition, the field $\mathscr{F}$ is taken as R^1.

Given a decomposition $D_1 \subset D$, $D_2 \subset D$ of D, every element $u \in D$ can be written as $u = u_1 + u_2$, where u_1, u_2 are the projections of u on D_1 and D_2, respectively. Projections of linear functionals and of functional-valued operators which are linear can be defined similarly. Given $f \in D^*$ and $P : D \to D^*$, the projections $f_1 \in D^*$ and $f_2 \in D^*$ are such that

$$\langle f_1, u\rangle = \langle f, u_1\rangle; \qquad \langle f_2, u\rangle = \langle f, u_2\rangle \tag{8}$$

for every $u \in D$, while the projections $P_1 : D \to D^*$ and $P_2 : D \to D^*$ satisfy the conditions

$$P_1 u = (Pu)_1, \qquad P_2 u = (Pu)_2. \tag{9}$$

From these definitions, it follows that

$$f = f_1 + f_2, \tag{10}$$

$$P = P_1 + P_2. \tag{11}$$

When P is non-negative, the non-negative square root of $\langle Pu, u\rangle$ will be denoted by $\|u\|_P$ whenever $u \in E$. The set

$$N_P = \{u \in D \mid \|u\|_P = 0\} \tag{12}$$

may contain non-zero vectors. However, N_P is always a linear subspace of D [3].

Definition *Let* $P: D \to D^*$ *be non-negative. Then* $u, v \in D$ *are said to be* P-equivalent *if and only if* $u - v \in N_P$. *In this case, one writes*

$$u \approx_P v. \tag{13}$$

This is an equivalence relation because N_P is a linear subspace. For non-positive and negative operators there are definitions and results corresponding to those already given for non-negative and positive operators. It is worth noticing that the relation $\approx_P$ becomes equality when P is either positive or negative.

Definition *Let* D_1, D_2 *be a decomposition of* D *and* $P: D \to D^*$ *be a self-adjoint operator. Then,* P *is said to be* saddle with respect to D_1, D_2 *if* P *is non-negative on* D_1 *and non-positive on* D_2. *It is* strictly saddle *if* P *is positive on* D_1 *and negative on* D_2.

It must be observed that, in this definition, the subspaces can not be interchanged. However, this manner of introducing them simplifies many propositions of the theory. For saddle operators, the bilinear functional $\langle Pu_1, v_1\rangle - \langle Pu_2, v_2\rangle$ is symmetric and possesses a non-negative quadratic form. The non-negative square root of $\langle Pu_1, u_1\rangle - \langle Pu_2, u_2\rangle$ will be denoted by $\|u\|_P$; when P is strictly saddle, $\|u\|_P$ is positive and therefore $\|u\|_P$ is a norm. In this case, the bilinear functional $\langle Pu_1, v_1\rangle - \langle Pu_2, v_2\rangle$ is an inner product.

Definition *When* $P: D \to D^*$ *is saddle, any pair* $u, v \in D$ *are said to be* P-equivalent (*i.e.,* $u \approx_P v$) *if and only if* $\|u - v\|_P = 0$.

Again the set (12) is a linear subspace and $\approx_P$ is an equivalence relation.

Let

$$D_{\text{I}} = \{u \in D \mid P_1 u = 0\}, \tag{14a}$$

$$D_{\text{II}} = \{u \in D \mid P_2 u = 0\}. \tag{14b}$$

Then, if $P: D \to D^*$ is saddle with respect to the decomposition D_1, D_2, the operator P is non-positive on D_{I} and non-negative on D_{II}. Furthermore [3],

$$\langle Pu, u\rangle = -\|u\|_P^2 \quad \text{when} \quad u \in D_{\text{I}}, \tag{15a}$$

$$\langle Pu, u\rangle = \|u\|_P \quad \text{when} \quad u \in D_{\text{II}} \tag{15b}$$

The dual extremum principles are closely related to Eqns (15). In view of Theorem 1, for symmetric operators, Eqn (3) is equivalent to

$$X,_1(u) = P_1 u - f_1 = 0, \tag{16a}$$

$$X,_2(u) = P_2 u - f_2 = 0. \tag{16b}$$

Theorem 2 *Let $P: D \rightarrow D^*$ be saddle with respect to the decomposition D_1, D_2 of D. Define the affine subspaces*

$$\hat{\mathscr{E}}_a = \{u \in D \mid (16a) \text{ holds}\}, \tag{17a}$$

$$\hat{\mathscr{E}}_b = \{u \in D \mid (16b) \text{ holds}\}; \tag{17b}$$

then, for every $u_a \in \hat{\mathscr{E}}_a$ and $u_b \in \hat{\mathscr{E}}_b$,

$$2[X(u_b) - X(u_a)] = \|u_b - u_a\|_P^2 \geqslant 0. \tag{18}$$

When a solution $u_0 \in D$ of (3) exists,

$$\|u_b - u_a\|^2 = \|u_b - u_0\|^2 + \|u_0 - u_a\|^2. \tag{19}$$

Proof The proof of (18), when a solution $u_0 \in D$ exists, follows immediately from Eqns (15), because

$$u_b - u_a = (u_b - u_0) + (u_0 - u_a), \tag{20}$$

$u_b - u_0 \in D_{\text{II}}$ and $u_0 - u_a \in D_{\text{I}}$. When the existence of u_0 is not assumed, the proof is slightly more complicated and is given in [3].

4 The mirror method

Let $P: D \rightarrow D^*$ be a non-negative, but not necessarily symmetric, operator. The mirror method consists in embedding the equation $Pu = f$ into the system

$$Pu = f \quad \text{and} \quad P^*u' = g, \tag{21}$$

where $u, u' \in D$ and $f, g \in D^*$. This system of equations may be written in terms of a symmetric operator $\hat{P}: \hat{D} \rightarrow \hat{D}^*$, where $\hat{D} = D \oplus D$ is the outer sum of D with itself; thus, the elements $\hat{u} = (u, u') \in \hat{D}$ are ordered pairs of members $u, u' \in D$. To express (21) in the form $\hat{P}\hat{u} = \hat{f}$ with $\hat{P}$ symmetric, it is enough to define $\hat{P}: \hat{D} \rightarrow \hat{D}^*$ and $\hat{f} \in \hat{D}^*$ by

$$\langle \hat{P}\hat{u}, \hat{v} \rangle = \langle Pu, v' \rangle + \langle Pv, u' \rangle \tag{22}$$

and

$$\langle \hat{f}, \hat{v} \rangle = \langle f, v' \rangle + \langle g, v \rangle. \tag{23}$$

Due to this fact, it is possible to formulate variational principles for the system (21). In this case, the functional $X: D \rightarrow \mathscr{F}$ of Theorem 1 is

$$X(\hat{u}) = X(u, u') = \langle Pu, u' \rangle - \langle f, u' \rangle - \langle g, u \rangle. \tag{24}$$

In general, this functional is not saddle. However, when P is non-negative, it is easy to construct decompositions $\hat{D}_1$, $\hat{D}_2$ of $\hat{D}$ for which $\hat{P}: \hat{D} \rightarrow \hat{D}^*$ is saddle. Indeed, let α, β, γ, δ be real numbers such that

$\alpha\delta - \beta\gamma \neq 0$, $\alpha\beta < 0$ and $\gamma\delta > 0$. Then, $\hat{P}$ is saddle with respect to $\hat{D}_1, \hat{D}_2$ if

$$\hat{D}_1 = \{\hat{u} \in \hat{D} \mid \alpha u + \beta u' = 0\}, \tag{25a}$$

$$\hat{D}_2 = \{\hat{u} \in \hat{D} \mid \gamma u + \delta u' = 0\} \tag{25b}$$

and Theorem 2 is applicable.

5 Variational formulations of diffraction problems

Consider a linear operator $P : D \to D^*$. Given such operator, let

$$\mathcal{N} = \{u \in D \mid Au = 0\} \tag{26}$$

be the null subspace of the antisymmetric part $A = (P - P^*)/2$ of P, which is well defined because the adjoint P^* always exists. The operator P is said to be formally symmetric if

$$\langle Pu, v \rangle = 0 \qquad \forall v \in \mathcal{N} \Rightarrow Pu = 0. \tag{27}$$

It is worth noticing that any functional-valued operator $P : D \to D^*$ satisfies the proposition that is obtained by replacing $\mathcal{N}$ with D in (27). Attention will be restricted in this section to formally symmetric operators.

Definition *A linear subspace $\mathcal{I} \subset D$ is said to be a connectivity condition if*

$$(a)\ \mathcal{N} \subset \mathcal{I}, \tag{29}$$

$$(b)\ \langle Au, v \rangle = 0, \qquad \forall u, v \in \mathcal{I}. \tag{30}$$

The connectivity condition is said to be complete *if, in addition,*

(c) *for any* $w \in D$,

$$\langle Aw, v \rangle = 0, \quad \forall v \in \mathcal{I} \Rightarrow w \in \mathcal{I}. \tag{31}$$

The set $\mathcal{E} \subset D$ is defined by

$$\mathcal{E} = \{u \in D \mid Pu = 0\}. \tag{32}$$

The relation

$$\{Au, v\rangle = 0, \qquad \forall u, v \in \mathcal{E} \tag{33}$$

is clear, because $2\langle Au, v \rangle = \langle Pu, v \rangle - \langle Pv, u \rangle$. In what follows of this section, it will be assumed that there is in D a connectivity condition $\mathcal{I}$; however, the completeness of $\mathcal{I}$ will not be taken for granted.

The problem of diffraction associated with P will now be presented.

Definition *Given $U \in D$ and $V \in D^*$, a motion $u \in D$ is said to be* a solution of the problem of diffraction *if*

$$Pu = f \quad \text{and} \quad V - u = v \in \mathscr{I}, \tag{34}$$

where $f = PU$.

There are several alternative variational formulations of this diffraction problem [9, 10]. Here, only the two which are especially relevant for numerical applications are quoted.

Theorem 3 *Define the functional*

$$X(u, u') = \tfrac{1}{2}\langle Pu, u\rangle + \langle A[u], \bar{u}\rangle + \langle AV, u'\rangle - \langle f, u\rangle \tag{35}$$

for every couple $u, u' \in D$. Here,

$$[u] = u' - u, \qquad \bar{u} = \tfrac{1}{2}(u' + u). \tag{36}$$

Then, when $\mathscr{I}$ is complete or, alternatively, when $u' - V \in \mathscr{I}$,

$$\langle X'(u, u'), (v, v')\rangle = 0 \qquad \forall (v, v'), \qquad v \in D \quad \text{and} \quad v' \in \mathscr{I} \tag{37}$$

if and only if u is a solution of the diffraction problem and $u' - u \in \mathscr{N}$.

Proof This theorem contains a slight modification of results proved in [9, 10].

6 Characterization of complete connectivity conditions

The main result to be presented in this section is that when the diffraction problem associated with a connectivity condition $\mathscr{I} \subset D$, is well posed, then $\mathscr{I}$ is complete.

For every symmetric operator $S : D \to D^*$, the operator $B = S + A/2$ satisfies

$$A = B - B^*. \tag{38}$$

It will be assumed that operators $B : D \to D^*$ (to be considered in what follows) satisfy (38). When such B is available, two problems are defined.

Definition *Given $U \in D$, the motion $u \in D$ is said to be* a solution of the external boundary-value problem (e.b.v.p.) *if $Bu = BU$ and $u \in \mathscr{I}$. The* internal boundary-value problem (i.b.v.p.) *is defined by replacing $\mathscr{I}$ with $\mathscr{E}$.*

Any of the three problems already introduced are said to satisfy existence if they possess at least one solution for every admissible data; uniqueness when the only solution of the problem with vanishing data is the zero solution; almost uniqueness when any solution of the latter problem belongs to the null set $\mathcal{N}$. For some purposes, it is too restrictive to require uniqueness; owing to this fact, a problem will be said to be well posed if it satisfies existence and almost uniqueness.

It can be proved easily [10], that *the problem of diffraction is well posed if and only if every $u \in D$ can be written as $u = u_1 + u_2$ with $u_1 \in \mathcal{I}$, $u_2 \in \mathcal{E}$ and this representation is almost unique.* In addition, *if for some B satisfying* (38) *the e.b.v.p. satisfies existence, then $\mathcal{I}$ is a complete connectivity condition* [10]. Finally, *if the diffraction problem is well posed, then* $\exists B : D \to D^*$, *which satisfies* (38), *and the e.b.v.p. satisfies existence.* This latter assertion can be proved defining $Bu = Au_1$, because such B fulfils the required conditions. The above results imply the main property advanced in the introduction of this section.

Theorem 4 *Let $\mathcal{I} \subset D$ be a connectivity condition. If the diffraction problem associated with $\mathcal{I}$ is well posed, then $\mathcal{I}$ is a complete connectivity condition.*

7 Theory of connectivity

In this section, the relation between the problem of diffraction and the problem of connecting is discussed. This latter is basic to the formulation of the finite-element method. It is shown that these two problems are closely connected; furthermore, it is shown that the problem of connecting associated with formally symmetric operators leads to complete connectivity conditions whenever it is well posed.

The problem of connecting consists in constructing solutions in a region such as $R \cup E$ of Fig. 1, by connecting those corresponding to individual subregions such as R and E. The theory to be presented is an abstract

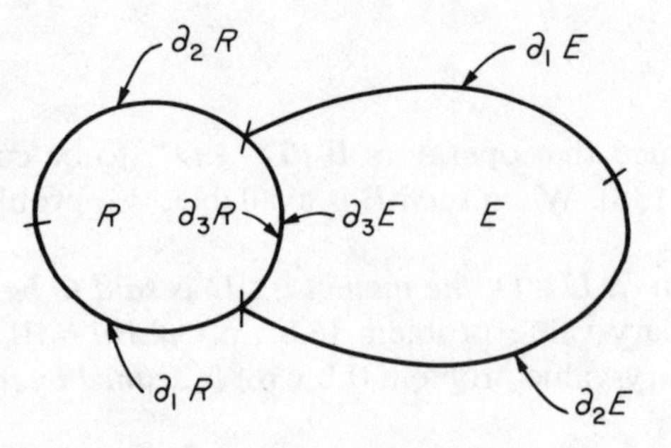

Fig. 1. Regions considered in the application to potential theory and elasticity.

one and can be applied in more general situations as long as the postulates are satisfied.

Let $\hat{P} : \hat{D} \to \hat{D}^*$ be a formally symmetric operator and D, D_E a decomposition of $\hat{D}$, so that every $\hat{u} \in \hat{D}$ can be written in a unique manner as $\hat{u} = u + u_E$, with $u \in D$ and $u_E \in D_E$. It will be assumed that this decomposition is such that

$$\langle \hat{P}\hat{u}, \hat{v} \rangle = \langle \hat{P}u, v \rangle + \langle \hat{P}u_E, v_E \rangle \tag{39}$$

for every $\hat{u}, \hat{v} \in \hat{D}$. This permits the definition of $P : \hat{D} \to \hat{D}^*$ and $P_E : \hat{D} \to \hat{D}^*$ by the equations

$$\langle P\hat{u}, \hat{v} \rangle = \langle \hat{P}u, v \rangle, \qquad \langle P_E\hat{u}, \hat{v} \rangle = \langle \hat{P}u_E, v_E \rangle. \tag{40}$$

They satisfy

$$\hat{P} = P + P_E. \tag{41}$$

At the same time, due to (40), they can be thought of as $P : D \to D^*$ and $P_E : D_E \to D_E^*$.

Definition *Assume $\mathcal{S} \subset \hat{D}$ has the following properties:*

(*a*) *$\mathcal{S}$ is a complete connectivity condition for $\hat{P}$;*
(*b*) *for every $u \in D$, $\exists u_E \in D_E \ni u + u_E \in \mathcal{S}$;*
(*c*) *for every $u_E \in D_E$, $\exists u \in D \ni u + u_E \in \mathcal{S}$.*

Then $\mathcal{S}$ will be said to be a smoothness relation *or* condition.

Elements $u \in D$ and $u_E \in D_E$ are said to be smooth extensions of each other when $u + u_E \in \mathcal{S}$. In this section, it is assumed that a smoothness relation $\mathcal{S}$ is given.

Definition *Let $U \in D$ and $U_E \in D_E$ be given. Then $u \in D$ is said to be a* solution of the problem of connecting,† *when*

$$Pu = PU_E \tag{42}$$

and $\hat{u} = u + u_E \in \mathcal{S}$ for some $u_E \in D_E \ni$

$$P_E u_E = P_E U_E. \tag{43}$$

Formulations of more general problems of connecting which include prescribed jumps have been given [10], as well as associated variational principles.

† Probably, it would be better to call this *problem of connecting in the restricted sense*, because preference is given to the subspace D, while, in a more general treatment, D and D_E play a symmetric role.

To reduce the problem of connecting to a problem of diffraction, it is enough to define the set $\mathscr{I} \subset D$ by

$$\mathscr{I} = \{u \in D \mid \exists \hat{u} = u + u_E \in S, P_E u_E = 0\}. \tag{44}$$

This is because *the set $\mathscr{I}$ given by Eqn* (44) *is necessarily a connectivity condition for $P : D \to D^*$, which is, in addition, complete when the diffraction problem is well posed,* as has been shown in [10].

8 Applications

Dual variational principles for the heat and wave equations have been given in [1–3] and can be easily extended to elastodynamics and other fields. Here, only applications of the general diffraction problem and of the problem of connecting will be considered.

For applications to elasticity and potential theory, consider functions $u = (u_1, \ldots, u_m)$ defined in the region $R \cup E$ of the n-dimensional Euclidean space (Fig. 1), with boundary $\partial_1 R \cup \partial_2 R \cup \partial_1 E \cup \partial_2 E$; the common boundary between R and E is $\partial_3 R = \partial_3 E$. The unit normal vector $\mathbf{n}$ is taken pointing outwards, from $R \cup E$ on its boundary, and from E on $\partial_3 R = \partial_3 E$. The subspaces D and D_E can be taken as the set of functions which are C^2 on R and on E, respectively. The space $\hat{D}$ is made by couples of such functions. Define $\hat{P} : \hat{D} \to \hat{D}^*$ by

$$\langle \hat{P}\hat{u}, \hat{v} \rangle = \int_{R \cup E} v_\alpha \mathscr{L}_\alpha(\mathbf{u})\, d\mathbf{x} + \int_{\partial_1(R \cup E)} u_\alpha T_\alpha(\mathbf{v})\, d\mathbf{x} - \int_{\partial_2(R \cup E)} v_\alpha T_\alpha(\mathbf{u})\, d\mathbf{x}, \tag{45}$$

where $\partial_1(R \cup E) = \partial_1 R \cup \partial_1 E$, $\partial_2(R \cup E) = \partial_2 R \cup \partial_2 E$, while $\mathscr{L}$ and T are the differential operators:

$$\mathscr{L}_\alpha(\mathbf{u}) = \frac{\partial}{\partial x_j}\left(C_{\alpha j \beta q} \frac{\partial u_\beta}{\partial x_q}\right) + \rho u_\alpha, \tag{46a}$$

$$T_\alpha(\mathbf{u}) = C_{\alpha j \beta q} \frac{\partial u_\beta}{\partial x_q} n_j. \tag{46b}$$

With these definitions, Eqns (39) are obviously satisfied. Here, as in what follows, Latin indices run from 1 to n, while Greek ones run from 1 to m. The coefficients ρ and $C_{\alpha j \beta q}$ are smooth functions on R and E separately, such that

$$C_{\alpha j \beta q} = C_{\beta q \alpha j}. \tag{47}$$

By integration by parts, it is seen that

$$\langle \hat{A}\hat{u}, \hat{v}\rangle = \int_{\partial_3 R} \{\bar{v}_\alpha[T_\alpha(\mathbf{u})] - [u_\alpha]\bar{T}_\alpha(\mathbf{v})\, d\mathbf{x} + \int_{\partial_3 R} \{\bar{T}_\alpha(\mathbf{u})[v_\alpha] - \bar{u}_\alpha[T_\alpha(\mathbf{v})]\}\, d\mathbf{x}, \quad (48)$$

where the square bracket stands for the jumps on $\partial_3 R$ of the function, while the bar over a symbol, e.g., $\bar{v}$, indicates its corresponding average. In view of this equation, a sufficient condition for $\hat{u} = u + u_E \in \hat{D}$ to be in the null set $\hat{\mathcal{N}}$ of $\hat{A}$ is that

$$u_\alpha = u_{E\alpha} = 0 \quad \text{and} \quad T_\alpha(\mathbf{u}) = T_\alpha(\mathbf{u}_E) = 0 \quad \text{on} \quad \partial_3 R. \quad (49)$$

From this fact, it can be seen that (27) is satisfied, and, therefore, $\hat{P}$ is formally symmetric.

It will be assumed in what follows that the coefficients $C_{\alpha j\beta q}$ are such that $\hat{u} = u + u_E \in \hat{\mathcal{N}}$ if and only if (49) are satisfied. When $m = 1$, $n = 3$ and $C_{1j1q} = \delta_{jq}$, $\mathcal{L}_1$ is Laplace's operator if $\rho \equiv 0$ and the reduced wave operator if $\rho \equiv 1$. If $m = n = 3$, $\mathcal{L}_i$ is the operator of static elasticity if $\rho = 0$ and the reduced operator of elastodynamics if $\rho \equiv 1$. In applications to potential theory, Eqn (49) is always satisfied, while the strong ellipticity of C_{ijpq} grants the same condition in applications to elasticity [9, 10].

Let the set $\mathcal{S} \subset \hat{D}$ be defined by the condition that $\hat{u} = u + u_E \in \mathcal{S}$ if and only if

$$[u_\alpha] = [T_\alpha(\mathbf{u})] = 0 \quad \text{on} \quad \partial_3 R. \quad (50)$$

Then, $\hat{\mathcal{N}} \subset \mathcal{S}$. In addition, by inspection of (48), it is seen that $\langle \hat{A}\hat{u}, \hat{v}\rangle$ vanishes whenever $\hat{u}, \hat{v} \in \mathcal{S}$; thus, $\mathcal{S}$ is a connectivity condition. Finally, it can be seen that this connectivity condition is always complete in applications to potential theory and the reduced wave equation, while, again, the same property is enjoyed in applications to elasticity when C_{ijpq} is strongly elliptic. This shows that $\mathcal{S}$ is a smoothness condition, because properties (b) and (c) are obvious, at least if $\partial_3 R$ is assumed to be sufficiently smooth.

References

1 Herrera, I., A general formulation of variational principles, Instituto de Ingenieria, UNAM, E-10, 1974.

2 Herrera, I. and Bielak, J., Dual variational principles for diffusion equations, *Quart. Appl. Math.* **34,** 85–102, 1976.

3 Herrera, I. and Sewell, M. J., Dual extremum principles for non-negative unsymmetric operators, *J. Inst. Math. Applics,* **21,** 95–115, 1978; also available as Technical Summary Report No. 1743, Mathematics Research Center, University of Wisconsin-Madison, 1977.

4 Noble, B. and Sewell, M. J., On dual extremum principles in applied mathematics, *J. Inst. Math. Applics.*, **9,** 123–193, 1972.
5 Arthurs, A. M., *Complementary Variational Principles*, Oxford University Press, Oxford, 1970.
6 Brezis, H. and Ekeland, I., Un principle variationnel associé a certaines équations paraboliques. Les cas indépendant du temps, *C. R. Acad. Sci. Paris, Sér. A*, **282,** 971–974, 1976.
7 Sewell, M. J. and Noble, B., Estimation of linear functionals in nonlinear problems, *Proc. R. Soc. Lond. A*, **361,** 293–324 (1978); also Technical Summary Report No. 1703, Mathematics Research Center, University of Wisconsin-Madison, 1976.
8 Herrera, I. and Bielak, J., A simplified version of Gurtin's variational principles, *Arch. Rat. Mech. Anal.*, **53,** 131–149, 1974.
9 Herrera, I., General variational principles applicable to the hybrid element method, *Proc. Nat. Acad. Sci.*, **74,** 2595–2597, 1977.
10 Herrera, I., Theory of connectivity for formally symmetric operators, *Proc. Nat. Acad. Sci.*, **74,** 4722–4725, 1977.
11 Nashed, M. Z., *Differentiability and Related Properties of Nonlinear Operators*; *Some Aspects of the Role of Differentials in Nonlinear Functional Analysis and Applications*, Academic Press, New York, 1971, pp. 102–309.
12 Vainberg, M. M., *Variational Methods in the Study of Non-linear Operator Equations*, Holden-Day, San Francisco, 1964.

Professor Ishmael Herrera
Istituto de Investigaciones en Matemáticas Aplicandas y en Sistemas, and Istituto de Inginiería,
National University of Mexico,
Apartado Postal 20–726,
México,
D. F. Mexico

D. D. Joseph

A new separation of variables theory for problems of Stokes flow and elasticity

Summary

Some classes of fourth-order boundary-value problems arising in the theory of Stokes flow and elasticity are solved by the method of biorthogonal series. The eigenfunctions are formed from separable solutions when the separation constants (eigenvalues) are chosen to make the solution and its normal derivative vanish at the side walls. A general and unified algorithm is presented for solving such problems on strips, in wedges, between disks, in cylinders and between cylinders and in cones. The method applies to many different equations and it always leads to the same general algorithms, the same type of biorthogonal expansion, the same type of reduction to ordinary differential equations, the same biorthogonality condition and the same type of formulas for the biorthogonal coefficients. The method leads to a new theory of biorthogonal 'Fourier' series of two-component vector-valued functions. Convergence of the series is proved for sufficiently smooth but otherwise arbitrary data. Completeness of the representations is established in a smaller, but still large, class of functions. Questions of summability by Féjers method, Gibbs phenomenon, representation of functions in weak classes and other points of analysis in the theory of trigonometric series are important open questions in this new theory.

1. Introduction

The theory described here can be regarded as the extension to fourth-order problems of the method of generalized 'Fourier series' used in the study of second-order problems. This type of extension was first introduced by R. C. T. Smith [1] in his study of stresses in a semi-infinite strip clamped at its side and loaded at its top edge. Smith's ideas were used by Joseph and Fosdick [2] to study a narrow-gap approximation for secondary motions generated by the Weissenberg effect. A more complete analysis, including numerical computations, was given by Joseph and Sturges [3] in their study of the free surface on a liquid filling a rectangular trench heated from its side. In that paper, it is shown that Smith's biorthogonal series are formally analogous to complex Fourier series and, though the biorthogonal eigenfunctions are much more complicated than trigonometric functions, the biorthogonal 'Fourier' coefficients may be computed from simple algorithms. Joseph and Sturges [3] also showed how the eigenfunction expansions should be used to compute solutions for other boundary conditions and in strips of finite height. The same type of biorthogonal expansions were used by Joseph [4] in a study of the free surface on the round edge of a flowing liquid filling a torsion flow viscometer. This is the first case where this type of eigenfunction expansion arises for a problem which is not biharmonic. Similar eigenfunction expansions are required for the axisymmetric problems of Stokes flow between concentric cylinders studies by Yoo and Joseph [5] and for the problem of axisymmetric flow in a cone studied by Liu and Joseph [6]. Yoo's thorough study [7] of secondary motions indiced by the Weissenberg effect is a notable achievement of this method of analysis. Liu and Joseph [8] showed how the corner eigenfunctions of Dean and Montagnon [9] and Moffat [10] may be used to generate biorthogonal series solution of Stokes-flow problems in wedge-shaped circular sectors. In their example, a Stokes flow is generated by buoyancy which is induced by density differences associated with heating one side wall. The secondary motion associated with the motion of a free surface on a viscoelastic fluid between oscillatory planes (Sturges and Joseph [11]) falls within the domain of application of the method of biorthogonal series. This problem may be reduced to the study of $\nabla^4\psi+\lambda^2\nabla^2\psi=0$ (λ^2 is complex) where ψ and its normal derivative vanish on the side wall. Such problems are also common in the linear theory of dynamic elasticity and in the linearized theory of buckling.

The list of problems given in the last paragraph are a small sample of those which can be solved by biorthogonal eigenfunction expansions. The eigenfunctions required in these different problems depend on the given data and the shape of the boundary; though these differ from problem to problem, the expansions for different problems share common properties which appear to be intrinsic to Stokes flow in cavities and to problems of elasticity with built-in side walls.

The method of solution in biorthogonal series requires the expansion of two-component vector-valued functions into a series of vector-valued biorthogonal eigenfunctions. The representation of arbitrary vectors with biorthogonal series is an independent problem of pure analysis with only weak connections to boundary-value problems. Smith [1] established conditions on the data sufficient to gaurantee the convergence of the biorthogonal series. But Smith's conditions eliminate most applications.

Joseph [12] and Joseph and Sturges [13] showed that the biorthogonal series will converge for arbitrary, sufficiently smooth, data. The rate of convergence depends on the data to be expanded. For 'bad' data like step functions and 'ramp' functions, convergence is conditional; better data gives absolute and even uniform convergence, as in the elementary theory of trigonometric series. And, as in the elementary theory, Féjer's method of computing Cesaro sums seems to greatly improve the rate of convergence.

The aim of this paper is to bring the recent results, just reviewed, into one place so as to emphasize their common features and to create propaganda for the method. I believe the method is very important, it bears the same relation to fourth-order problems as generalized Fourier series do to second-order problems. There are also some new results in this paper; in particular, the results stated in section 5 on the completeness of the eigenfunction expansions appear not to have noticed before.

In section 2, we consider the canonical problem first posed and solved formally by Smith [1]. In section 3, we review various extensions and generalizations of the biorthogonal series expansion of boundary-value problems. In section 4, we prove convergence of the series for arbitrary smooth data and establish rates of convergence. In section 5, we consider the problem of completeness in the sense of justifications for the expansions.

2 The canonical edge problem in a semi-infinite strip

The canonical problem is defined as follows. We seek a bounded biharmonic function $\hat{\Psi}(t, y)$ in the semi-infinite strip $\mathscr{V}=[t, y: -1\leqslant t\leqslant 1, y\leqslant 0]$ satisfying

$$(\tilde{\Psi}(1, y), \tilde{\Psi}(-1, y), \tilde{\Psi}_{,t}(1, y), \tilde{\Psi}_{,t}(-1, y)]=[c_1, c_2, c_3, c_4], \tag{2.1}$$

where c_1, c_2, c_3 and c_4 are constants and $\tilde{\Psi}_{,tt}(t, 0)$ are $\tilde{\Psi}_{,yy}(t, 0)$ are arbitrary, prescribed, sufficiently-smooth functions.

The function

$$\Psi(t, y)=\tilde{\Psi}(t, y)-\frac{c_3-c_4}{4}(t^2-1)+\frac{c_1-c_2-c_3-c_4}{4}(3t-t^3)+\frac{c_3+c_4}{2}t+\frac{c_1+c_2}{2} \tag{2.2}$$

satisfies

$$\nabla^4\Psi=0 \quad \text{in} \quad \mathscr{V}. \tag{2.3a}$$

$$\Psi=\Psi_{,t}=0 \quad \text{when} \quad t=\pm 1, \tag{2.3b}$$

$$\{\Psi_{,yy}(t, 0), \Psi_{,tt}(t, 0)\}=\{f(t), g(t)\}, \tag{2.3c}$$

where $f(t)$ and $g(t)$ are arbitrary, prescribed, sufficiently-smooth functions and Ψ is bounded as $y \to -\infty$. The prescribed edge function $g(t)$ is compatible with the side-wall boundary condition (2.3b) if and only if $g(t)$ satisfies the compatibility

$$\langle g \rangle = \langle tg \rangle = 0, \tag{2.4}$$

where

$$\langle \cdot \rangle = \int_{-1}^{1} \cdot \, dt.$$

R. C. T. Smith [1] gave a formal solution of (2.3) and justified the solution under the conditions that $f(\pm 1) = f'(\pm 1) = g(\pm 1) = g'(\pm 1) = 0$. These restrictions on the data rule out most of the applications. They also make the theory uninteresting from the point of view of pure 'Fourier' analysis. Fortunately, Smith's restrictions are not intrinsic and they may be dropped (*see* section 4 and 5). For the moment, we pursue the formal theory.

Separable solutions

$$\Psi^{(n)} = \phi_1^{(n)}(t) \exp(S_n y)$$

satisfy (2.3a) and (2.3b) if

$$\phi_{1,tttt}^{(n)} + 2S_n^2 \phi_{1,tt}^{(n)} + S_n^4 \phi_1^{(n)} = 0, \tag{2.5a}$$

$$\phi_1^{(n)}(\pm 1) = \phi_{,t}^{(n)}(\pm 1) = 0. \tag{2.5b}$$

The complex constants S_n are eigenvalues which arise as follows. We may always decompose the solution of (2.5) into even and odd sets. The even solutions may be written as

$$\phi_1^{(n)}(t) = S_n \sin S_n \cos S_n t - S_n t \cos S_n \sin S_n t. \tag{2.6}$$

Evidently, $\phi_1^{(n)}(\pm 1) = 0$ and we may verify that $\phi_{1,t}^{(n)}(\pm 1) = 0$ if and only if

$$\sin 2S_n + 2S_n = 0. \tag{2.7}$$

The odd solutions may be written as

$$\phi_1^{(n)}(t) = S_n \cos S_n \sin S_n t - S_n t \sin S_n \cos S_n t, \tag{2.8}$$

where

$$\sin 2S_n = 2S_n. \tag{2.9}$$

The eigenfunctions (2.6) and (2.8) are sometimes called the Papkovich–Fadle [14, 15] functions in honor of the two gentlemen who first introduced them in the study of problems of elasticity.

We note that there are no real-valued solutions of the eigenvalue equations (2.7) and (2.9) other than $S_0 = 0$. There is no eigenfunction belonging to $S_0=0$ satisfying (2.5a) and the *four* conditions (2.5b). It follows that the eigenvalues are all complex constants. *If S_n is an eigenvalue, then so is $-S_n$ and so is $\bar{S}_n$, the complex conjugate of S_n.* All of the eigenvalues are known if all be known in the first quadrant of the complex S plane. Following Joseph and Sturges [3], we shall index the eigenvalues to accentuate the analogy between the biorthogonal series (2.13) and (2.14), below, and the complex form of Fourier's trigonometric series. We first identify the roots of, say, (2.7) in the first quadrant and order them in a sequence according to the size of their real parts; that is, $0<\text{Re } S_1<\text{Re } S_2<\text{Re } S_3$, etc. We then identify the roots of (2.7) in the fourth quadrant as

$$S_{-n}=\bar{S}_n. \tag{2.10}$$

Roots in the second quadrant are given by $-S_n$ and in the third quadrant by $-S_{-n}$. Identical conventions are adopted for the roots of (2.9). It follows from these conventions that

$$\phi_1^{(-n)}(t, S_{-n})=\phi_1^{(n)}(t, \bar{S}_n)=\bar{\phi}_1^{(n)}(t, S_n) \tag{2.11}$$

and

$$\phi_1^{(n)}(t, -S_n)=\phi_1^{(n)}(t, S_n). \tag{2.12}$$

We now seek the solution of (2.3) as a series of separable solutions in the form $(n\neq 0)$

$$\Psi(t, y)=\sum_{-\infty}^{\infty} C_n\phi_1^{(n)}(t)\exp(S_ny)/S_n^2. \tag{2.13}$$

This series is a formal solution of (2.3) if the constants C_n can be selected so that

$$\begin{bmatrix} f(t) \\ g(t) \end{bmatrix}\equiv\mathbf{f}(t)=\sum_{-\infty}^{\infty} C_n\boldsymbol{\phi}^{(n)}(t)=\sum_{-\infty}^{\infty} C_n\begin{bmatrix} \phi_1^{(n)}(t) \\ \phi_2^{(n)}(t) \end{bmatrix}, \tag{2.14}$$

where

$$\phi_2^{(n)}(t)=\phi_{1,tt}^{(n)}/S_n^2. \tag{2.15}$$

One of Smith's basic contributions is an algorithm for computing the constants C_n. Combining (2.15) and (2.4) we find that

$$\phi_{2,tt}^{(n)}+S_2^n(2\phi_2^{(n)}+\phi_1^{(n)})=0. \tag{2.16}$$

Equations (2.15) and (2.16) may be written as

$$\boldsymbol{\phi}_{,tt}^{(n)}+S_n^2\mathbf{A}\cdot\boldsymbol{\phi}=0, \tag{2.17}$$

where

$$\mathbf{A}=\begin{bmatrix} 0 & -1 \\ 1 & 2 \end{bmatrix}. \tag{2.18}$$

The matrix $\mathbf{A}$ has an important place in the theory. We call it the biorthogonality matrix.

An adjoint (row) vector

$$\mathbf{\Psi}^{(r)} = [\Psi_1, \Psi_2]$$

belonging to S_n satisfies the differential system

$$\mathbf{\Psi}^{(n)}_{,tt} + S_n^2 \mathbf{\Psi} \cdot \mathbf{A} = 0 \tag{2.19}$$

and

$$\Psi_2^{(n)}(\pm 1) = \Psi_{2,t}^{(n)}(\pm 1) = 0. \tag{2.20}$$

Equations (2.19) and (2.20) are derived in the usual way. We introduce a scalar product

$$\mathbf{\Psi} \cdot \mathbf{\phi} = [\Psi_1, \Psi_2] \begin{bmatrix} \phi_1 \\ \phi_2 \end{bmatrix} = \Psi_1 \phi_1 + \Psi_2 \phi_2$$

and an inner product identity

$$\langle \mathbf{\Psi} \cdot (\mathbf{\phi}_{,tt} + S_n^2 \mathbf{A} \cdot \mathbf{\phi}) \rangle = [\mathbf{\Psi} \cdot \mathbf{\phi}_{,t} - \mathbf{\Psi}_{,t} \cdot \mathbf{\phi}]_{-1}^{1} + \langle (\mathbf{\Psi}_{,n} + S_n^2 \mathbf{A}) \cdot \mathbf{\phi} \rangle.$$

which holds for all $\mathbf{\Psi}$, $\mathbf{\phi} \in C_2\ [1, -1]$. If $\mathbf{\phi} = \mathbf{\phi}^{(n)}$, then

$$[\Psi_2 \phi_{2,t} - \Psi_{2,t} \phi_2]_{-1}^{1} + \langle (\mathbf{\Psi}_{,tt} + S_n^2 \mathbf{\Psi} \cdot \mathbf{A}) \cdot \mathbf{\phi} \rangle = 0. \tag{2.21}$$

The adjoint problem is the subset of $\Psi \in C_2[-1, 1]$, for which (2.21) holds when $\mathbf{\phi}$ is allowed to range over $C_2[-1, 1]$.

The vectors $\mathbf{\phi}^{(n)}$ and $\mathbf{\Psi}^{(n)}$ form a biorthogonal set. Using (2.14), (2.15), (2.19) and (2.20) we find that

$$\langle \mathbf{\Psi}^{(m)} \cdot \mathbf{A} \cdot \mathbf{\phi}^{(n)} \rangle = \delta_{nm} k_m, \tag{2.22}$$

where

$$\langle \mathbf{\Psi}^{(m)} \cdot \mathbf{A} \cdot \mathbf{\phi}^{(m)} \rangle = k_m, \tag{2.23}$$

The 'Fourier' coefficients C_n in the biorthogonal series (2.14) may be computed using (2.23):

$$C_n = \frac{1}{k_n} \langle \mathbf{\Psi}^{(n)} \cdot \mathbf{A} \cdot \mathbf{f} \rangle. \tag{2.24}$$

If the data vector $\mathbf{f}(t)$ is real-valued, then

$$C_{-n} = \bar{C}_n \tag{2.25}$$

and the series (2.13) giving $\Psi(t, y)$ is real-valued.

It is useful to maintain a distinction between the eigenfunction $\phi_1^{(n)}$ of (2.5) and the first component of the eigenvector $\mathbf{\phi}^{(n)}$ satisfying (2.17) and (2.15b). These sets are identical except that $S_0 = 0$ is not an eigenvalue of

(2.5). When $S_0=0$, we find a nontrivial eigenvector of (2.17) and (2.5*b*) and an adjoint eigenvector:

$$\begin{bmatrix}\phi_1^{(0)}\\ \phi_2^{(0)}\end{bmatrix}=\begin{bmatrix}0\\ 1\end{bmatrix}, \qquad [\Psi_1^{(0)}; \Psi_2^{(0)}]=[1, 0], \qquad k_0=-2 \tag{2.26}$$

and

$$\begin{bmatrix}\phi_1^{(0)}\\ \phi_2^{(0)}\end{bmatrix}=\begin{bmatrix}0\\ t\end{bmatrix}, \qquad [\Psi_1^{(0)}, \Psi_2^{(0)}]=[t, 0], \qquad k_0=-2. \tag{2.27}$$

These eigenvectors do not enter into the solution of (2.3), because (2.4) follows from (2.3) and rules them out.

Explicit formulae for the eigenvectors are listed below. For $n=0, \pm 1, \pm 2, \ldots$

$$\phi_1^{(n)}=\Psi_2^{(n)}, \qquad \Psi_1^{(n)}-\phi_2^{(n)}=2\phi_1^{(n)}. \tag{2.28}$$

When $n=\pm 1, \pm 2, \ldots$ and S_n are the roots of $\sin 2S+2S=0$, the eigenfunctions are even functions of t; $\phi_1^{(n)}(t)$ is given by (2.6) and

$$\phi_2^{(n)}=-\phi_1^{(n)}-2\cos S_n \cos S_n t, \qquad k_n=-4\cos^4 S_n. \tag{2.29}$$

When $n=\pm 1, \pm 2, \ldots$ and S_n are the roots of $\sin 2s-2s=0$, the eigenfunctions are odd functions of t; $\phi_1^{(n)}(t)$ is given by (2.8) and

$$\phi_2^{(n)}=-\phi_1^{(n)}+2\sin S_n \sin S_n t, \qquad k_n=-4\sin^4 S_n.$$

Whenever the formal solution of (2.3) is justified, we have an immediate and precise mathematical realization of St Venant's principle. Stating this principle in an informal way, we note that the solutions of (2.3) in the strip decay very rapidly to

$$\Psi(t, y) \sim C_1\phi_1^{(1)}(t) \exp S_1 y. \tag{2.30}$$

That the rapidity of this decay is exponential with a large decay constant (~ 3.2) can be inferred from the numerical values of the first quadrant roots of (2.7) and (2.9). The first three roots of (2.7) are given by Robbins and Smith [16] as

$$S_1=2.106\,196+i\,1.125\,365,$$
$$S_2=5.356\,269+i\,1.551\,575,$$
$$S_3=8.536\,683+i\,1.775\,544.$$

The first three roots of (2.9) are given by Hillman and Salzer [17] as

$$S_1=3.748\,838+i\,1.384\,339,$$
$$S_2=6.949\,980+i\,1.676\,105,$$
$$S_3=10.119\,259+i\,1.858\,384.$$

It follows that the interior form of the solution is independent of the precise details of the edge loads when these loads are self-equilibrated; that is, when (2.4) holds. The edge loads enter only through the constant C_1 and this constant is determined by a projection $\langle \mathbf{\Psi}^{(1)} \cdot \mathbf{A} \cdot \mathbf{f} \rangle$ of the data vector. The interior solution is a decaying system of closed eddies with a fixed spatial period (*see* Fig. 1).

3 Extensions of the theory

In this section we will describe the extensions of the theory presented in Section 2 to other problems. These extensions show that the method of biorthogonal series should have a distinguished place in the education of mathematically-minded scientists. The method of biorthogonal series is the method of 'Fourier series', which is appropriate to many problems of fourth order. Such problems cannot be avoided, in any event, in the study of mechanics.

3.1 Other boundary conditions (Joseph and Sturges [13]).

Suppose the data vector is given by

$$\begin{bmatrix} \Psi_{,y}(t,0) \\ \Psi_{,tt}(t,0) \end{bmatrix} = \begin{bmatrix} f(t) \\ g(t) \end{bmatrix} \equiv \mathbf{f}(t). \tag{3.1}$$

This is the form which the data takes in fluid mechanics, when the velocity is given by a stream function in $\mathscr{V}$ and is prescribed on $y=0$. Then, $\Psi(t, y)$ is given by (2.13) if we can find the C_n for which

$$\mathbf{f}(t) = \begin{bmatrix} f(t) \\ g(t) \end{bmatrix} = \lim_{N\to\infty} \sum_{-N}^{N} C_n \left\{ \begin{bmatrix} \phi_1^{(n)}(t) \\ \phi_2^{(n)}(t) \end{bmatrix} + \left(\frac{1}{S_n} - 1\right) \begin{bmatrix} \phi_1^{(n)} \\ 0 \end{bmatrix} \right\}, \tag{3.2}$$

where $C_0 = 0$. Using the biorthogonality condition (2.22), we get an infinite number ($N \to \infty$) of inhomogeneous equations for the coefficient C_n:

$$\langle \mathbf{\Psi}^{(n)} \cdot \mathbf{A} \cdot \mathbf{f} \rangle = C_n k_n + \sum_{-N}^{N} C_m B_{mn}, \tag{3.3}$$

where

$$B_{nm} = \left(\frac{1}{S_m} - 1\right) \langle \phi_1^{(m)} \cdot \phi_1^{(n)} \rangle.$$

We solve the infinite system in the usual way–by truncation. There are $2N$ equations for the real and imaginary parts of C_n. Proofs of convergence

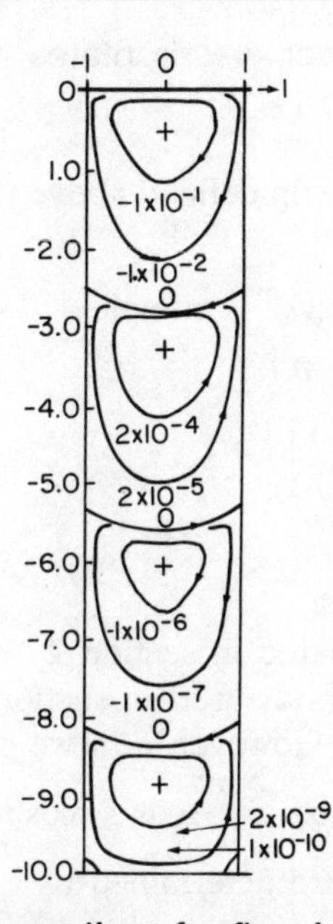

Fig. 1. Level lines of the streamlines for flow in a cavity with depth/width ratio of 5. $\Psi(t, y)$ is biharmonic of Ψ and Ψ_t vanish at the $t=\pm 1$. The normal component of velocity vanishes on the top and bottom, the tangential component vanishes at the bottom and is equal to one on the top: i.e.,

$$\Psi_y(t,0)-1=\Psi(t,0)=\Psi_y(t,-H)=\Psi(t,-H)=0.$$

This problem is of the type solved in section 3.1. The data is 'bad' and convergence of the series for $\Psi_y(t,0)$ and $\Psi(t,0)$ is conditional (*see* Joseph and Sturges [13]). Away from the edges, the flow is given by (2.30) with $S_1=2.106\,196+i\,1.253\,65$.

are presently unknown, but the numerical tests of convergence always work out (*see* Fig. 1).

If the trench has a bottom at $y=-d$, then a data vector $\mathbf{f}(t)$ is prescribed at the top ($y=0$) and a data vector $\mathbf{g}(t)$ is also prescribed at the bottom. It is necessary to retain the second and third quadrant eigenvalues, which now correspond to *bounded* eigensolutions. We get a solution in the form

$$\Psi(t, y)=\sum_{-\infty}^{\infty}(C_n \exp(S_n y)+D_n \exp(-S_n y))\phi_1^{(n)}(t). \qquad (3.5)$$

where $C_0=D_0=0$. The coefficients C_n and D_n may be computed by applying the biorthogonality condition (2.22) to the data vector at the bottom and top of the strip (*see* Fig. 1).

3.2 Vibration problems for elastic plates and viscoelastic fluids (Sturges and Joseph [11]).

Let $\mathscr{V}$ be the semi-infinite strip defined above (2.1). We want to solve the following problem:

$$\left.\begin{aligned}
&\nabla^4\Psi-\lambda^2\nabla^2\Psi=0 \text{ in } \mathscr{V} \ (\lambda^2 \text{ is complex}),\\
&\Psi(\pm 1, y)=\Psi_{,t}(\pm 1, y)=0,\\
&\mathbf{f}(t)=\begin{bmatrix} f(t)\\ g(t)\end{bmatrix}=\begin{bmatrix}\Psi_{,yy}(t,0)\\ \Psi_{,tt}(t,0)\end{bmatrix},\\
&\langle g\rangle=\langle tg\rangle=0,\\
&\Psi(t,y) \text{ is bounded in } \mathscr{V}.
\end{aligned}\right\} \tag{3.6}$$

This problem, like that treated in section 2, is in canonical form and it may be solved by the methods used in section 2. We get $\Psi(t, y)$ in the series form given by (2.13). However, now we get the even eigenfunctions

$$\phi_1^{(n)}(t)=\cos\sqrt{(S_n^2-\lambda^2)}\cos S_n t-\cos S_n\cos t\sqrt{(S_n^2-\lambda^2)}, \tag{3.7}$$

where the eigenvalues S_n are determined by

$$S_n\tan S_n=\sqrt{(S_n^2-\lambda^2)}\tan\sqrt{(S_n^2-\lambda^2)}. \tag{3.8}$$

Since λ is a prescribed complex constant, it is no longer true that $S_{-n}=\bar{S}_n$. Now we order the fourth quadrant eigenvalues in a sequence $\operatorname{Re} S_{-1}<\operatorname{Re} S_{-2}<\operatorname{Re} S_{-3}\ldots$. If S_n is a eigenvalue, so is $-S_n$.

The computation of C_n follows along the lines laid out in section 2. We get

$$\mathbf{f}(t)=\sum_{-\infty}^{\infty} C_n\boldsymbol{\phi}^{(n)}(t),\qquad \boldsymbol{\phi}^{(n)}=\begin{bmatrix}\phi_1^{(n)}(t)\\ \phi_2^{(n)}(t)\end{bmatrix}, \tag{3.9}$$

where $C_0=0$.

$$\begin{aligned}
&\boldsymbol{\phi}_{,tt}+S_n^2\mathbf{B}_n\cdot\boldsymbol{\phi}=0,\ \phi_1^{(n)}(\pm 1)=\phi_{1,t}^{(n)}(\pm 1)=0,\\
&\mathbf{B}_n\begin{bmatrix} 0 & -1\\ 1-\dfrac{\lambda^2}{S_n^2} & 2-\dfrac{\lambda^2}{S_n^2}\end{bmatrix},
\end{aligned} \tag{3.10}$$

$$C_n=\frac{1}{k_n}\langle\boldsymbol{\Psi}^{(n)}\cdot\mathbf{A}\cdot\mathbf{f}\rangle \tag{3.11}$$

($\mathbf{A}$ = biorthogonality matrix),

$$\boldsymbol{\Psi}_{,tt}+S_n^2\boldsymbol{\Psi}^{(n)}\cdot\mathbf{B}_n=0,\qquad \boldsymbol{\Psi}^{(n)}=[\Psi_1^{(n)},\phi_1^{(n)}]. \tag{3.12}$$

The expression (3.11) for C_n and the definition of k_n arise from the biorthogonality condition

$$\langle\boldsymbol{\Psi}^{(m)}\cdot\mathbf{A}\cdot\boldsymbol{\phi}^{(n)}\rangle=k_n\delta_{mn}. \tag{3.13}$$

We can solve (3.6) for other boundary conditions, as in section 3.1 above, or in finite domains. The same basic method of solution will work for many variations in the problem (3.6).

3.3 Biharmonic problems in circular sectors (Liu and Joseph [81])

The pie-shaped planar region $\mathcal{V}$ is specified in polar coordinates (r, θ) as

$$\mathcal{V}=[r, \theta:\ 0 \leqslant \eta \leqslant r < 1,\ -\beta \leqslant \theta \leqslant \beta].$$

We consider the following biharmonic problem:

$$\begin{aligned} &\nabla^4 \Psi(r, \theta)=0 \text{ in } \mathcal{V}. \\ &\Psi(r, \pm\beta)=\Psi_{,\theta}(r, \pm\beta)=0, \\ &\mathbf{f}(\theta)=\begin{bmatrix} r\left(\dfrac{1}{r}\Psi_{,r}\right)_{,r} \\ \dfrac{1}{r^2}\Psi_{,\theta\theta} \end{bmatrix}=\begin{bmatrix} f(\theta) \\ g(\theta) \end{bmatrix} \text{ is prescribed on } r=\eta, 1, \end{aligned} \tag{3.14}$$

and

$$\langle g \rangle = \langle \theta g(\theta) \rangle = 0,$$

where

$$\langle \cdot \rangle \equiv \int_{-\beta}^{\beta} \cdot \, d\theta.$$

This problem comes up, for example, in the study of the motion under a free surface on a liquid in a pie-shaped trench heated from its side (*see* Fig. 2 and Table 1) or in the study of the normal displacements of a thin elastic strip clamped at $\theta=\pm\beta$ with displacements and couples prescribed on the radial boundaries.

The solution of this problem can be found by the method of section 2. We get

$$\Psi=\sum_{-\infty}^{\infty}[C_n t^{\lambda_n}+D_n t^{-\lambda_n+2}]\phi_1^{(n)}(\theta)/\lambda_n(\lambda_n-2), \tag{3.15}$$

where $C_0=D_0=0$. The solutions split into even and odd sets:

$$\left.\begin{aligned} &\phi_1^{(n)}(\theta)=\cos(\lambda_n-2)\beta \cos\lambda_n\theta-\cos\lambda_n\beta \cos(\lambda_n-2)\theta, \\ &\sin[2\beta(\lambda_n-1)]+(\lambda_n-1)\sin 2\beta=0; \end{aligned}\right\} \tag{3.16}$$

$$\left.\begin{aligned} &\phi_1^{(n)}(\theta)=\sin(\lambda_n-2)\beta \sin\lambda_n\theta-\sin\lambda_n\beta \sin(\lambda_n-2)\theta, \\ &\sin[2\beta(\lambda_n-1)]-(\lambda_n-1)\sin 2\beta=0. \end{aligned}\right\} \tag{3.17}$$

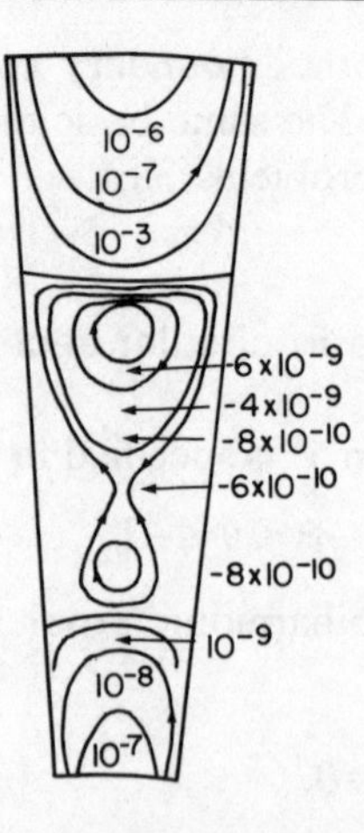

Fig. 2. (After Liu and Joseph [8].) Stokes flow in a wedge. $2\beta = 10°$, $a/b = 0.5$. Level lines of

$$\Psi(r, \theta) = \sum_{-\infty}^{\infty} C_n t^{\lambda_n - 3} + D_n t^{-\lambda_n - 1} \frac{\phi_1(n)}{\lambda_n(\lambda_n - 2)}.$$

Table 1 *Convergence of the top edge series and the bottom edge series when* $2\beta = 10°$ *and* $a/b = 0.5$ *in Fig.* 2

$$\frac{\Psi}{t^3} \times 10^6 = \frac{10^6 \times f(\theta, \beta)}{16} = 10^6 \times \sum_{-N}^{N} (C_n t^{\lambda_n - 3} + D_n t^{-\lambda_n - 1}) \frac{\phi_1^{(n)}}{\lambda_n(\lambda_n - 2)}$$

θ (degrees)	$\frac{10^6 \times f(\theta, \beta)}{16}$	$N=1$	$N=3$	$N=5$	$N=9$	$N=10$
On the top: $t=1$						
0	2.433 72	2.447 76	2.433 52	2.433 63	2.433 68	2.433 69
1	2.242 67	2.253 77	2.242 92	2 242 68	2.242 63	2.242 64
2	1.716 47	1.718 89	1.716 09	1.716 42	1.716 42	1.716 44
3	0.995 85	0.987 91	0.995 74	0.995 79	0.995 81	0.995 83
4	0.314 85	0.306 05	0.315 50	0.314 92	0.314 81	0.314 82
5	0	0	0	0	0	0
On the bottom: $t=0.5$						
0	2.433 72	2.324 07	2.445 25	2.433 88	2.433 40	2.433 69
1	2.242 67	2.180 42	2.242 55	2.242 50	2.242 35	2.242 63
2	1.716 47	1.753 39	1.718 76	1.716 39	1.716 17	1.716 44
3	0.995 85	1.090 52	0.994 26	0.996 20	0.995 60	0.995 83
4	0.314 85	0.371 74	0.312 85	0.314 07	0.314 68	0.314 82
5	0	0	0	0	0	0

The eigenvalues $\mu_n = \lambda_n - 1$ are symmetrically distributed in the complex μ-plane; if λ_n is an eigenvalue, then $\bar{\lambda}_n$, $2-\lambda_n$ and $2-\bar{\lambda}_n$ are also eigenvalues. Moreover, $\phi_1^{(n)}(\theta, \lambda_n) = -\phi_1^{(n)}(\theta, 2-\lambda_n)$.

The coefficients C_n and D_n are determined from the edge data using the biorthogonality condition (3.20). This condition may be obtained after the eigenvector problem is reduced to ordinary differential equations by methods like those discussed in section 2. We get

$$\mathbf{f}(\theta) = \sum_{-\infty}^{\infty} C_n \boldsymbol{\phi}^{(n)}(\theta), \qquad \boldsymbol{\phi}^{(n)} = \begin{bmatrix} \phi_1^{(n)}(\theta) \\ \phi_2^{(n)}(\theta) \end{bmatrix}, \tag{3.18}$$

$$\boldsymbol{\phi}^{(n)}_{,\theta\theta} + \mathbf{A}_n \cdot \boldsymbol{\phi}^{(n)} = 0, \qquad \phi_1^{(n)}(\pm\beta) = \phi_{1,\theta}^{(n)}(\pm\beta) = 0$$

$$\mathbf{A}_n = \begin{bmatrix} 0 & -\lambda_n(\lambda_n - 2) \\ \lambda_n(\lambda_n - 2) & (\lambda_n - 2)^2 + \lambda_n^2 \end{bmatrix}, \tag{3.19}$$

$$C_n = \frac{1}{k_n} \langle \boldsymbol{\Psi}^{(n)} \cdot \mathbf{A} \cdot \mathbf{f} \rangle \tag{3.20}$$

($\mathbf{A}$ = biorthogonality matrix),

$$\boldsymbol{\Psi}^{(n)}_{,\theta\theta} + \boldsymbol{\Psi}^{(n)} \cdot \mathbf{A}_n = 0, \qquad \boldsymbol{\Psi}^{(n)} = [\Psi_1^{(n)}(\theta), \phi_1^{(n)}(\theta)]. \tag{3.21}$$

The expression (3.11) for C_n and the definition of k_n arise from the biorthogonality condition

$$\langle \boldsymbol{\Psi}^{(m)} \cdot \mathbf{A} \cdot \boldsymbol{\phi}^{(n)} \rangle = k_n \delta_{mn}. \tag{3.22}$$

Many variations of the canonical problem (3.14) are tractable to analysis by the method of separation of variables and biorthogonal series.

3.4 Stokes-flow problems between parallel disks (Joseph [4])

We now consider an axisymmetric problem of Stokes flow between parallel disks. The problem statement is given in the caption to Fig. 3.

The solution of this Stokes-flow problem is given by

$$\Psi(r, t) = \sum_{-\infty}^{\infty} C_n \phi_1^{(n)}(t) F(S_n, r)/S_n^2,$$

where

$$F(S_n, r) = \frac{r}{R} \frac{I_1(S_n r)}{I_1(S_n R)}$$

and $I_1(S_n r)$ is a modified Bessel function of the first kind. We find that

$$\mathbf{f}(t) = \sum_{-\infty}^{\infty} C_n \boldsymbol{\phi}^{(n)}(t),$$

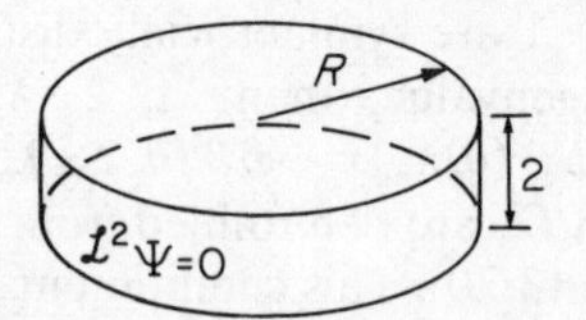

Fig. 3. Stokes flow between parallel disks. The stream function $\Psi(r, t)$ satisfies the Stokes–Beltrami equation $\mathscr{L}\Psi = 0$, where

$$\mathscr{L} = \frac{\partial^2}{\partial r^2} - \frac{1}{r}\frac{\partial}{\partial r} + \frac{\partial^2}{\partial t^2}.$$

On the disks $\Psi(r, \pm 1) = \Psi_{,t}(r \pm 1) = 0$. The data vector

$$\mathbf{f}(t) = \begin{bmatrix} f(t) \\ g(t) \end{bmatrix} \equiv \begin{bmatrix} R\left(\frac{1}{r}\Psi_{,r}\right)_{,r} \\ \Psi_{,tt} \end{bmatrix}$$

is prescribed on the round edge. $g(t)$ satisfies the compatibility condition (2.4).

$$\mathscr{V} = [r, t\colon 0 \leq r \leq R, -1 \leq t \leq 1].$$

where $\boldsymbol{\phi}^{(n)}(t)$, S_n and C_n are exactly the eigenvectors, eigenvalues and biorthogonal coefficients given in section 2.

3.5 Stokes-flow problems between coaxial cylinders (Yoo and Joseph [5])

This problem arises, for example, in the study of the secondary motions associated with thermally induced convective currents or in the study of secondary motions induced by the Weissenberg effect in viscoelastic fluids (Yoo [7]).

The problem statement is as follows. A stream function $\Psi(r, y)$ for axisymmetric flow in cylindrical coordinates is defined in the region $\mathscr{V}$ between coaxial cylinders,

$$\mathscr{V} = [r, y\colon 0 < a \leq r \leq b,\ b - a = 2,\ y \leq 0].$$

The streamfunction satisfies

$$\mathscr{L}^2\Psi = 0 \text{ in } \mathscr{V}, \tag{3.23}$$

where

$$\mathscr{L} = \left(\frac{\partial^2}{\partial r^2} - \frac{1}{r}\frac{\partial}{\partial r} + \frac{\partial^2}{\partial y^2}\right).$$

$$\Psi(a, y) = \Psi(b, y) = \Psi_{,r}(a, y) = \Psi_{,r}(b, y) = 0, \tag{3.24}$$

and

$$\Psi(r, y) \text{ is bounded in } \mathscr{V}. \tag{3.25}$$

At the edge $y=0$, the values of the normal component of velocity and the shear stress are prescribed in terms of the data vector

$$\mathbf{f}(r)=\begin{bmatrix} f(r) \\ g(r) \end{bmatrix}=\begin{bmatrix} \Psi_{,yy} \\ r\left(\frac{1}{r}\Psi_{,r}\right)_{,r} \end{bmatrix}. \tag{3.26}$$

The prescription of $g(r)$ on $y=0$ is compatible with the wall boundary conditions (3.24) if and only if

$$\langle g(r)\rangle=\langle r^2 g(r)\rangle=0, \tag{3.27}$$

where

$$\langle\cdot\rangle=\int_a^b \frac{1}{r}\cdot \mathrm{d}r.$$

The solution of the Stokes-flow problem (3.23)–(3.27) is given by

$$\Psi(r, y)=\sum_{-\infty}^{\infty} C_n \phi_1^{(n)}(r)/p_n^2,$$

where $C_0=0$ and

$$\phi_1^{(n)}(r)=A_1^{(n)} rJ_1(p_n r)+A_2^{(n)} rY_1(p_n r) + A_3^{(n)} r^2 J_0(p_n r)+A_4^{(n)} r^2 Y_0(p_n r),$$

where $J_l(p_n r)$ and $Y_l(p_n r)$ are Bessel functions. The constants $A_l^{(n)}$ and the eigenvalues p_n are selected to satisfy the side-wall boundary conditions (3.24) and will not be given here (*see* Yoo and Joseph [5]). The eigenvalues are symmetrically located in the four quadrants of the complex p_n plane and the numbering convention used for S_n in Section 2 applies also to p_n.

The coefficients C_n may be obtained by methods like those used in section 2. We find that

$$r\left(\frac{1}{r}\boldsymbol{\phi}_{,r}^{(n)}\right)_{,r}+p_n^2 \mathbf{A}\cdot\boldsymbol{\phi}^{(n)}=0, \qquad \boldsymbol{\phi}^{(n)}=\begin{bmatrix}\phi_1^{(n)} \\ \phi_2^{(n)}\end{bmatrix},$$

where $\phi_1^{(n)}(r)$ and $\phi_{1,r}^{(n)}(r)$ vanish at $r=a$ and $r=b$,

$$r\left(\frac{1}{r}\boldsymbol{\Psi}_{,r}^{(n)}\right)_{,r}+p_n^2 \boldsymbol{\Psi}^{(n)}\cdot\mathbf{A}=0, \qquad \boldsymbol{\Psi}^{(n)}=[\Psi_1^{(n)}, \phi_1^{(n)}]$$

and $\mathbf{A}$ is the biorthogonality matrix. From these equations, we find the biorthogonality condition;

$$\langle\boldsymbol{\Psi}^{(m)}\cdot\mathbf{A}\cdot\boldsymbol{\phi}^{(n)}\rangle=k_n\delta_{mn}.$$

The C_n are then given by

$$C_n = \frac{1}{k_n} \langle \mathbf{\Psi}^{(n)} \cdot \mathbf{A} \cdot \mathbf{f} \rangle.$$

Stokes-flow problems between coaxial cylinders can be worked when the cylinder is bounded and not semi-infinite and for other boundary conditions on the top and bottom. The same type of analysis can be carried out for Stokes-flow edge problems in circular cylinders ($a = 0$).

3.6 Stokes flow in cones (Liu and Joseph [6])

Let $\mathscr{V}$ be the right circular cone of polar radius $r = 1$ and polar angle $\theta = \theta_0$ shown in Fig. 4. The motion inside $\mathscr{V}$ is governed by the Stokes–Beltrami equation

$$\left(\frac{\partial^2}{\partial r^2} + \frac{1-\xi^2}{r^2}\frac{\partial^2}{\partial \xi^2}\right)^2 \Psi(r, \xi) = 0, \tag{3.28}$$

where $\xi = \cos\theta$. On the conical side walls, $\xi = \xi_0$,

$$\Psi(r, \xi_0) = \Psi_{,\xi}(r, \xi_0) = 0. \tag{3.29}$$

The data vector is prescribed at the spherical edge $r = 1$:

$$\mathbf{f}(\xi) \equiv \begin{bmatrix} f(\xi) \\ g(\xi) \end{bmatrix} = \begin{bmatrix} r^4\left(\frac{1}{r}\Psi_{,r}\right)_{,r} \\ (1-\xi^2)\Psi_{,\xi\xi} \end{bmatrix}. \tag{3.30}$$

This data prescription is equivalent to the statement that the normal component of velocity and the shear stress $S_{r\theta}$ are prescribed on the spherical cap at $r = 1$. The prescription (3.30) of $g(\xi)$ is compatible with

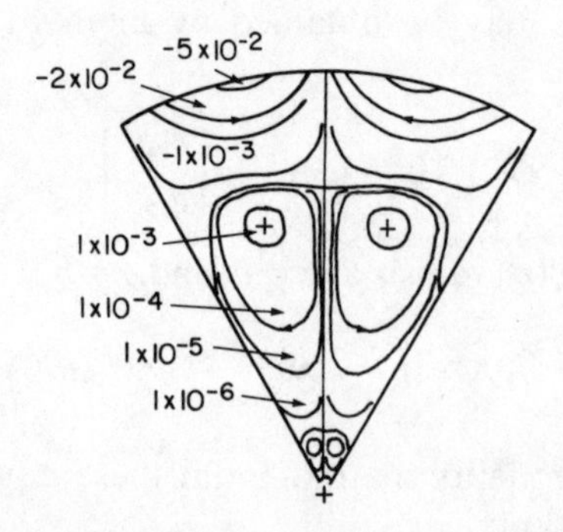

Fig. 4. (After Liu and Joseph [6].) Stokes flow in a cone. Level lines of $r^{3/2+\mu_1}\phi_1(\xi)$ for $2\theta_0 = 60°$, $\mu_1 = 5.888\,24 + i\,2.030\,12$.

the side-wall boundary condition if and only if

$$\int_{\xi_0}^{1} \frac{g(\xi)}{1-\xi^2}\,d\xi = \int_{\xi_0}^{1} \frac{\xi g(\xi)}{1-\xi^2}\,d\xi = 0. \tag{3.31}$$

The solution of the Stokes-flow problem (3.28)–(3.31) is given by

$$\Psi(r, \xi) = \sum_{-\infty}^{\infty} \frac{C_n}{\mu_n^2 - \frac{9}{4}} r^{3/2+\mu_n} \phi_1^{(n)}(\xi) \tag{3.32}$$

where $C_0 = 0$ and

$$\phi_1^{(n)}(\xi) = (1-\xi^2)^{1/2}\{P^1_{\mu_n-3/2}(\xi_0)P^1_{\mu_n-1/2}(\xi) - P^1_{\mu_n-1/2}(\xi_0)P^1_{\mu_n-3/2}(\xi)\}, \tag{3.33}$$

where P^1_μ (cos θ) is an associated Legendre function. It is obvious that $\phi_1^{(n)}(\xi_0) = 0$. $\phi_{1,\xi}^{(n)}(\xi_0)$ also vanishes if $\mu = \mu_n$ are selected as eigenvalues; that is, as roots $\mu = \mu_n$ of the characteristic equation

$$\begin{aligned}\{(\mu+3/2)P^1_{\mu+3/2}(\xi_0) + (\mu-3/2)P^1_{\mu+1/2}(\xi_0)\}& \\ \times\{(\mu-1/2)P^1_{\mu+1/2}(\xi_0) + (\mu+1/2)P^1_{\mu-3/2}(\xi_0)\}& \\ = 4\mu^2\xi_0^2 P^1_{\mu+1/2}(\xi_0)P^1_{\mu-3/2}(\kappa_0).&\end{aligned} \tag{3.34}$$

In deriving (3.34) we used recursion relations.

The eigenvalues μ_n are all complex-valued. It is easy to verify, using the fourth-order differential equation satisfied by $\phi_1^{(n)}(\xi)$, that if μ_n is an eigenvalue, so is $\bar{\mu}_n$ and $-\mu_n$. The μ_n in the first quadrant of the complex μ-plane are arranged in an increasing sequence according to the size of their real parts.

The coefficients C_n may be obtained by methods like those used in section 2. We find that

$$(1-\xi^2)\boldsymbol{\phi}^{(n)}_{,\xi\xi} + \mathbf{A}_n \cdot \boldsymbol{\phi}^{(n)} = 0, \qquad \boldsymbol{\phi}^{(n)} = \begin{bmatrix}\phi_1^{(n)} \\ \phi_2^{(n)}\end{bmatrix},$$

where

$$\phi_1^{(n)}(\xi_0) = \phi_{1,\xi}^{(n)}(\xi_0) = 0,$$
$$(1-\xi^2)\boldsymbol{\Psi}^{(n)}_{,\xi\xi} + \boldsymbol{\Psi}^{(n)} \cdot \mathbf{A}_n = 0 \qquad \boldsymbol{\Psi}^{(n)} = [\boldsymbol{\Psi}_1^{(n)}, \phi_1^{(n)}]$$

and

$$\mathbf{A}_n = \begin{bmatrix} 0 & -(\mu_n^2 - 9/4) \\ \mu_n^2 - 1/4 & 2(\mu_n^2 - 5/4) \end{bmatrix}.$$

From these equations, we find the biorthogonality condition

$$\langle \boldsymbol{\Psi}^{(m)} \cdot \mathbf{A} \cdot \boldsymbol{\phi}^{(n)} \rangle = k_n \delta_{mn},$$

where $\mathbf{A}$ is the biorthogonality matrix and

$$\langle\cdot\rangle = \int_{\xi_0}^{1} \left(\frac{1}{1-\xi^2}\right) \mathrm{d}\xi.$$

The C_n are then given by

$$C_n = \frac{1}{k_n} \langle \mathbf{\Psi}^{(n)} \cdot \mathbf{A} \cdot \mathbf{f} \rangle.$$

In dealing with fourth-order problems in cones, it is necessary to compute associated Legendre functions of arbitrary complex order. Though representations of these functions in terms of integrals and hypergeometric series have been given by mathematicians of antiquity, the computational value of such representations is unknown. Liu and Joseph [6] worked successfully with

$$P_\nu^1(\cos\theta) = \frac{i(\nu+1)}{2\pi} \int_0^{2\pi} (\cos\theta + i\sin\theta\cos t)^\nu \cos t \,\mathrm{d}t$$

for complex and unrestricted values of ν.

The methods used here can be applied to problems in truncated cones and problems with other boundary conditions. It can also be applied to problems of Stokes flow between cones. The analysis of flow between cones is important in understanding the flow in cone and plate rheometers used for rheological measurements. In the analysis of flow between cones it is necessary to introduce the other kind of Legendre functions $Q_\mu^1(\xi)$ which are singular at $\xi = 1$. There are some ancient representations of $Q_\mu^1(\xi)$ which are supposed to be good for complex, unrestricted values of μ.

4 Mathematical theory of biorthogonal expansions of vector-valued functions

In sections 2 and 3, we studied different boundary-value problems governed by partial differential equations of the fourth order. These different problems were solved formally. The formal solutions can be justified provided only that in each and every problem a biorthogonal series representation

$$\mathbf{f}(t) = \sum_{-\infty}^{\infty} C_n \mathbf{\phi}^{(n)}(t) = \sum_{-\infty}^{\infty} \frac{C_n}{k_n} \langle \mathbf{\psi}^{(n)} \cdot \mathbf{A} \cdot \mathbf{f} \rangle \mathbf{\phi}^{(n)}(t) \tag{4.1}$$

can be justified.

The representation (4.1) asserts that a certain large class of functions $f(t)$ and $g(t)$ may be expanded as a two-component vector

$$\mathbf{f}(t) = \begin{bmatrix} f(t) \\ g(t) \end{bmatrix} \tag{4.2}$$

in a series of characteristic vectors

$$\boldsymbol{\phi}^{(n)}(t) = \begin{bmatrix} \phi_1^{(n)}(t) \\ \phi_2^{(n)}(t) \end{bmatrix}$$

generated as eigensolutions of a second-order ordinary differential equation. The biorthogonal coefficients C_n are expressed through a scalar product

$$C_n = \frac{1}{k_n} \langle \boldsymbol{\Psi}^{(n)} \cdot \mathbf{A} \cdot \mathbf{f} \rangle \tag{4.3}$$

involving the data vector $\mathbf{f}$, the biorthogonal matrix $\mathbf{A} = \begin{bmatrix} 0 & 1 \\ -1 & 2 \end{bmatrix}$, and adjoint characteristic vectors

$$\boldsymbol{\Psi}^{(n)} = [\Psi_1^{(n)}, \phi_1^{(n)}]$$

generated as eigensolutions of the second-order problem adjoint to the one for $\boldsymbol{\phi}^{(n)}$. The formula (4.3) always arises from a biorthogonality condition of the form

$$\langle \boldsymbol{\Psi}^{(m)} \cdot \mathbf{A} \cdot \boldsymbol{\phi}^{(n)} \rangle = k_m \delta_{nm}. \tag{4.4}$$

The representation (4.1) is notable because of the following.

1. The form of the expansion, the form of the biorthogonality condition (4.4) and the value of the biorthogonality matrix are always the same, though the eigenvector bases $\boldsymbol{\phi}^{(n)}$, $\boldsymbol{\Psi}^{(n)}$ and the eigenvalues differ from problem to problem.
2. Though two functions $f(t)$ and $g(t)$ are expanded, only one set of biorthogonal coefficients are needed. In general, we expect to have nontrivial representations of the function $g(t) = 0$ or $f(t) = 0$ with $C_n = 0 \forall n$ if and only if $f(t) = g(t) = 0$.
3. Since each different boundary-value problem generates a different basis $\boldsymbol{\phi}^{(n)}(t)$, we may have many different representations of the vector $\mathbf{f}(t)$ in terms of biorthogonal series.

The problem of representing $\mathbf{f}(t)$ in biorthogonal series can be studied as an independent mathematical problem with only incidental connections to the separation of variables theory for boundary-value problems satisfying partial differential equations of order four. This independent problem is not more strongly tied to differential equations than trigonometric series are tied to theory of differential equations for harmonic oscillators.

Now we are going to consider results which establish the conditions on $f(t)$ which justify the expansion (4.1). There are two main types of results which are required.

1. The conditions under which the series (4.1) converges.
2. The conditions under which the series (4.1) converges to $\mathbf{f}(t)$: i.e., the conditions under which the bases $\boldsymbol{\phi}^{(n)}(t)$ are complete.

From now on we will confine our attention to the basis $\boldsymbol{\phi}^{(n)}(t)$ and eigenvalues S_n generated by the canonical edge problem in the semi-infinite strip of width two. This is the problem considered by Smith [1] and the convergence proofs of Joseph [12] and Joseph and Sturges [13] apply to it. The restriction of the analysis to the basis $\boldsymbol{\phi}^{(n)}(t)$ of section 2 has greater generality than might first be supposed. To prove convergence, we need to establish the asymptotic distribution of the eigenvalues and the asymptotic forms of the eigenfunctions. It turns out that these asymptotic forms are practically the same for the $\boldsymbol{\phi}^{(n)}$ of section 2 and for the many different $\boldsymbol{\phi}^{(n)}(t)$ of section 3. Hence, all of our convergence proofs may be expected to carry over with only slight and fairly obvious modifications. The method we use in section 5 to establish completeness for the basis $\boldsymbol{\phi}^{(n)}$ of section 2 is Smith's extension of the residue method of Titchmarsh [18]. This method carries over immediately to the analysis of completeness of the bases generated by the problems treated in section 3.

In the study of the separation of variables solutions of fourth-order problems, we required that the edge data and side-wall boundary values should be compatible. This compatibility condition is expressed by (2.4) and it eliminates the eigenvalue $S_0=0$ and the eigenfunctions (2.26) and (2.27). The restriction (2.4) on the functions $g(t)$ which may be expanded in biorthogonal series is apparent and not real. For if (2.4) does not hold for $g(t)=g_e(t)+g_0(t)$, where $g_e=\frac{1}{2}[g(t)+g(-t)]$ and $g_0=\frac{1}{2}[g(t)-g(-t)]$, then (2.4) does hold for $\hat{g}(t)=f(t)-\langle g_e\rangle-t\langle tg_0\rangle$. Hence, *in the sequel, we* shall *assume*, without loss of generality, *that* $\langle g\rangle=\langle tg\rangle=0$.

We can obtain an expansion formula for arbitrary $\mathbf{f}(t)$ by superposition from the expansion formula

$$\mathbf{f}(t)=\mathbf{f}(-t)=\sum_{-\infty}^{\infty}C_n\boldsymbol{\phi}^{(n)}(t), \qquad C_0=0, \tag{4.5}$$

for even data ($\boldsymbol{\phi}^{(n)}$ belong to S_n satisfying $\sin 2S_n+2S_n=0$) and the expansion formula

$$\mathbf{f}(t)=-\mathbf{f}(-t)=\sum_{-\infty}^{\infty}C_n\boldsymbol{\phi}^{(n)}(t), \qquad C_0=0 \tag{4.6}$$

for odd data ($\boldsymbol{\phi}^{(n)}$ belonging to S_n satisfying $\sin 2S_n-2S_n=0$). It is sufficient to consider, say, the even data. The results and proofs for odd data are essentially the same.

Smith [1] has justified (4.5) and (4.6) under the restrictions that

$$f(\pm1)=g(\pm1)=f'(\pm1)=g'(\pm1)=0 \tag{4.7}$$

and

$$(f''(t),\ g''(t)) \text{ are of bounded variation.} \tag{4.8}$$

He notes that if $f(t)$ and $g(t)$ are merely of bounded variation, then the series on the left of (4.5) and (4.6) may diverge. For example, we have the following asymptotic results for the eigenvalues S_n satisfying $\sin 2S_n + 2S_n$ and the associated (even) eigenvectors (2.6), (2.28) and (2.29)

$$2S_n \to (2n-\tfrac{1}{2})\pi + i\log(4n-1)\pi, \tag{4.9}$$

$$\begin{bmatrix}\sin S_n t\\ \cos S_n t\end{bmatrix} = \frac{1}{2}\begin{bmatrix} i\\ 1\end{bmatrix}[(4n-1)\pi]^{|t|/2} e^{-i(n-1/4)\pi|t|} + O(n^{-|t|/2}) \tag{4.10}$$

and

$$k_n = -4\cos^4 S_n = -\frac{(4n-1)^2\pi^2}{4} + O(n). \tag{4.11}$$

When n is very large,

$$S_n = O(n), \quad k_n = O(n^2) \tag{4.12}$$

and

$$\boldsymbol{\phi}^{(n)}, \boldsymbol{\Psi}^{(n)} = O(n^{(3+|t|)/2}). \tag{4.13}$$

Smith notes that, if it is assumed that $f(t)$ and $g(t)$ are of bounded variation, then

$$C_n = O(n^{-1}), \qquad C_n\boldsymbol{\phi}^{(n)} = O(n)$$

and the series (4.5) will diverge. Smith argued that the divergence of (4.5) and (4.6) need not necessarily effect the practical value of the solution since (2.13) converges rapidly for any $y > 0$, however small (*see* Smith [1], p. 23).

Smith's restrictions (4.7) are too severe. They rule out the applications in which the values on $y = 0$ are important. And they rule out the possibility of a mathematical theory of biorthogonal expansions of the form (4.5) and (4.6) except in the very restricted class satisfying (4.7) and (4.8). He says (p. 237) that the 'Details of the calculation make it unlikely that the conditions imposed on $(f(t), g(t))$ can be much relaxed.' Fortunately, Smith's statements are incorrect; we get uniform absolute convergence to $\mathbf{f}(t)$ even when $g(\pm 1) \neq 0$ and $g'(\pm 1) \neq 0$ and we get conditional and not uniform convergence with no conditions on the boundary values of $f(t)$ or $g(t)$.

The most important type of hypothesis on $f(t)$ and $g(t)$ are, like (4.7), associated with the boundary values at $t = \pm 1$. We consider three cases. In the first case,

$$f(\pm 1) = f'(\pm 1) = 0. \tag{*}$$

Uniform convergence of (4.5) and (4.6) is possible under the hypothesis

(*) because

$$f(t)=\sum_{-\infty}^{\infty} C_n\phi_1^{(n)}(t) \tag{4.14}$$

is compatible with $\phi_1^{(n)}(\pm 1)=\phi_{1,t}^{(n)}(\pm 1)=0$. From the point of view of differential equations, (*) is normal for nice problems because $f=\Psi_{,yy}(t,0)$ and $\Psi(y,t)=\Psi_{,t}(y,t)$ vanish on $t\pm 1$. In the second case,

$$f(\pm 1)=0. \tag{**}$$

Uniform convergence to $f'(t)$ is impossible because the left and right side of the first derivative of (4.14) do not match when $t=\pm 1$. In the third case,

$$f(\pm 1),\ f'(\pm 1),\ g(\pm 1),\ g'(\pm 1) \text{ are unrestricted.} \tag{***}$$

Uniform convergence to $f(t)$ is impossible. In all three cases, we get interior, point-wise convergence but the convergence is uniform only under the hypothesis (*).

The proof of convergence under hypotheses (*) and (**) and certain smoothness assumptions follow from easy estimates of C_n and majorization by numerical series (Joseph [12]). In the case (***), the convergence is conditional; the proof is delicate and will not be given here (*see* Joseph and Sturges [13]).

To prove convergence under (*), we assume that $f(t)$ and $g(t)$ are in the class $C_2[-1,1]\cap C_4^p[-1,1]$ with a finite number of jumps, at most, in the third and fourth derivative. Then, integrating by parts, we find that

$$\begin{aligned}
k_nC_n &= \langle \mathbf{\Psi}^{(n)}\cdot\mathbf{A}\cdot\mathbf{f}\rangle=\langle(2\phi_1^{(n)}(t)-\Psi_1^{(n)}(t)g(t)+\phi_1^{(n)}(t)f(t)\rangle\\
&= -\frac{1}{S_n^2}\langle\phi_{1,tt}^{(n)}(t)g(t)+\Psi_{1,tt}^{(n)}(t)f(t)\rangle\\
&= -\frac{1}{S_n^2}[\phi_{1,t}^{(n)}(t)g(t)+\Psi_{1,t}^{(n)}(t)f(t)]_{-1}^{1}\\
&\quad +\frac{1}{S_n^2}[\phi_1^{(n)}(t)g'(t)+\Psi_1^{(n)}(t)f'(t)]_{-1}^{1}\\
&\quad -\frac{1}{S_n^2}\langle\phi_1(t)g''(t)+\Psi_1(t)f''(t)\rangle.
\end{aligned} \tag{4.15}$$

Since $f(\pm 1)=f'(\pm 1)=\phi_1^{(n)}(\pm 1)=\phi_{1,t}^{(n)}(\pm 1)=0$ and (from (2.28) and (2.29)) $\Psi_1^{(n)}=\phi_1^{(n)}-2\cos S_n\cos S_nt$, we have

$$-k_nS_n^2C_n=\langle(g+f)''\phi_1^{(n)}(t)-2\cos S_n\cos S_ntf''\rangle. \tag{4.16}$$

The largest term in the integrand is $\phi_1^{(n)}(t)=O(n^{(3+|t|)/2})$ and

$$\begin{aligned}
\langle(g+f)''\phi_1^{(n)}(t)\rangle &= S_n[\sin S_n\langle(g+f)''\cos S_nt\rangle-\cos S_n\langle t(g+f)''\sin S_nt\rangle]\\
&= 2(g+f)''-\sin S_n\langle(g+f)''\sin S_nt\rangle-\cos S_n\langle(t(g+f)'')'\cos S_nt\rangle,
\end{aligned} \tag{4.17}$$

where we have used the fact that f and g are even functions of t. A similar reduction holds when f and g are odd. The last term of (4.16) and the last two terms of (4.17) can be integrated once more by parts over the intervals where the integrands are continuous. The integration by parts introduces S_n into the denominator so that the remaining integrals are of $O(1)$ when n is large. We therefore have proved that

If $f(\pm1)=f'(\pm1)=0$ *and* $f(t)$ *and* $g(t)\in C_2[-1,1]\cap C_4^p[-1,1]$, *then, when n is large,*

$$C_n\to\frac{1}{k_nS_n^2}O(1)\to O\left(\frac{1}{n^4}\right) \tag{4.18}$$

and, for each t, $-1\leqslant t\leqslant 1$, the series (4.5) may be majorized by a convergent numerical series

$$C\sum_{n=1} 1/n^{(5-|t|)/2},\ -1\leqslant t\leqslant 1. \tag{4.19}$$

So, in case (*), we get uniform absolute convergence without conditions on the boundary values of $g(t)$.

In case (**), we find (Joseph [12]) that $C_n=O(1/n^3)$ and the series (4.5) may be majorized by the convergent numerical series

$$C\sum_{n=1} 1/n^{(3-|t|)/2},\qquad 1<t<1.$$

The convergence is absolute but it need not be uniform. As we have already noted, the differentiated series (4.5) *cannot converge to* f' (±1), but it may converge conditionally.

If all conditions on the boundary values of $f(t)$ are discarded, we get $C_n=O(1/n^2)$. The series converges, but not absolutely and not uniformly (*see* Joseph and Sturges [13]).

It is necessary to maintain a distinction between convergence and convergence to $\mathbf{f}(t)$. Smith proved convergence to $\mathbf{f}(t)$ under the hypotheses (4.7) and (4.8). If one checks through Smith's computation in the appendix to his paper, one finds that the conditions $g(\pm1)=g'(\pm1)=0$ which he required are not used in his demonstration (*see* sections 4 and 5 of this paper). We therefore have convergence to $\mathbf{f}(t)$ under hypothesis (*) when $\mathbf{f}''(t)$ is of bounded variation or for $\mathbf{f}(t)$ in the class $C_2[-1,1]\cap C_4^p[-1,1]$. We shall call these conditions normal for nice problems, and for such problems we get uniform pointwise convergence and the formal solution of the boundary-value problem is justified.

There is, at present, no theorem of completeness when $\mathbf{f}(t)$ is in a bigger class. Indeed, we have already noted that if no boundary conditions are prescribed for $\mathbf{f}(t)$ at $t=\pm1$, or if (**) is prescribed, we cannot have uniform convergence since, at least, $f'(\pm1)\neq0$ cannot be represented by the series (4.5) or (4.6). The interior convergence of (4.5) and (4.6) can, however, be guaranteed in the cases (**) and (***) provided that $\mathbf{f}(t) \in C_1[-1, 1] \cap C_3^P[-1, 1]$ in case (**) and to $C_0[-1, 1] \cap C_2^P[-1, 1]$ in case (**). There is as yet no theorem guaranteeing interior pointwise convergence to $\mathbf{f}(t)$ for case (**) or case (***), but our numerical work shows that we do get such convergence.

The situation which prevails in case (***), when $f'(\pm 1) \neq 0$ and $f(\pm 1) \neq 0$, is analogous to Fourier series, where, for example, the representation

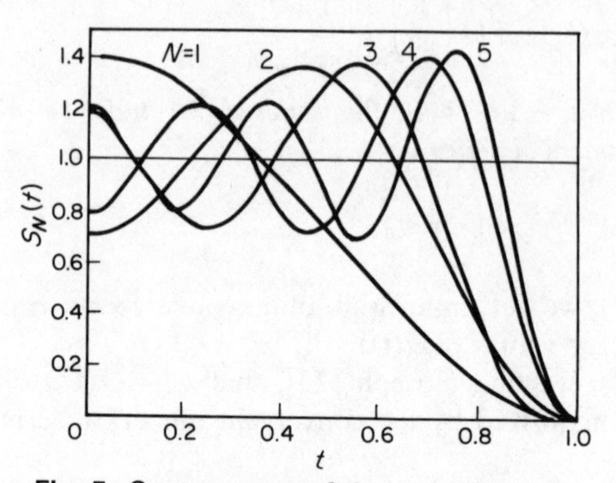

Fig. 5. Convergence of the partial sums

$$S_N(t) = \sum_{-N}^{N} \frac{-1}{\cos^4 S_n} \phi_1^{(n)}(t)$$

of the biorthogonal series $\lim_{N\to\infty} S_N(t)$ representing the unit step function $f(t) = 1, -1 < t < 1$.

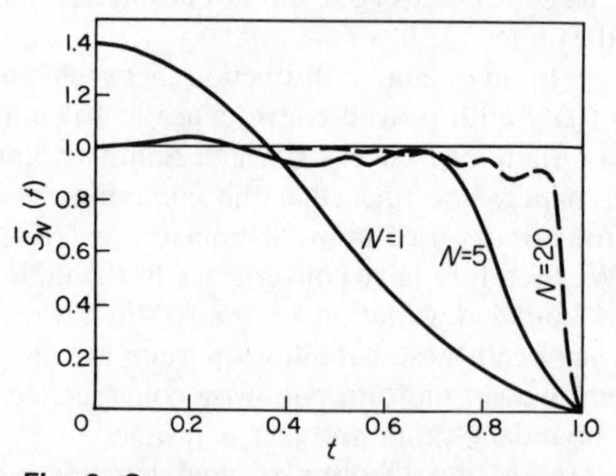

Fig. 6. Convergence of the Cesaro sums

$$\bar{S}_N(t) = \frac{1}{N} \sum_{M=1}^{N} S_M(t) \text{ to the unit step function } f(t) = 1, -1 < t < 1.$$

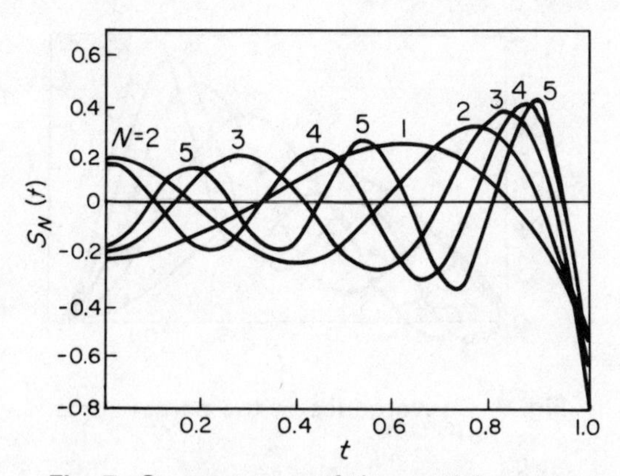

Fig. 7. Convergence of the partial sums

$$S_N(t)=\sum_{-N}^{N}\frac{-1}{\cos^4 S_n}\,\phi_2^{(n)}(t)$$

of the biorthogonal series $\lim_{N\to\infty} S_N(t)$ representing the zero function $g(t)=0$, $-1<t<1$.

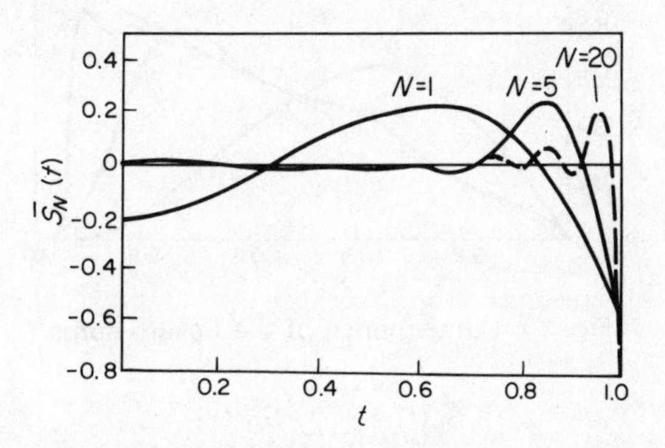

Fig. 8. Convergence of the Cesaro sums

$\bar{S}_N(t)=\dfrac{1}{N}\sum_{M=1}^{N} S_M(t)$ to the zero function $g(t)=0$, $-1<t<1$.

$$1=\sum_{n=1}^{\infty}\frac{2-2\cos n\pi}{n\pi}\sin n\pi t, \qquad 0<t<1, \tag{4.20}$$

must fail when $t=\pm 1$. We get conditional interior convergence to 1 at each interior point and the convergence is not uniform.

The step function may also be represented by a biorthogonal series of even eigenvectors

$$\mathbf{f}(t)=\begin{bmatrix}1\\0\end{bmatrix}=\sum_{-\infty}^{\infty} C_n\begin{bmatrix}\phi_1^{(n)}(t)\\ \phi_2^{(n)}(t)\end{bmatrix}, \qquad C_0=0, \tag{4.21}$$

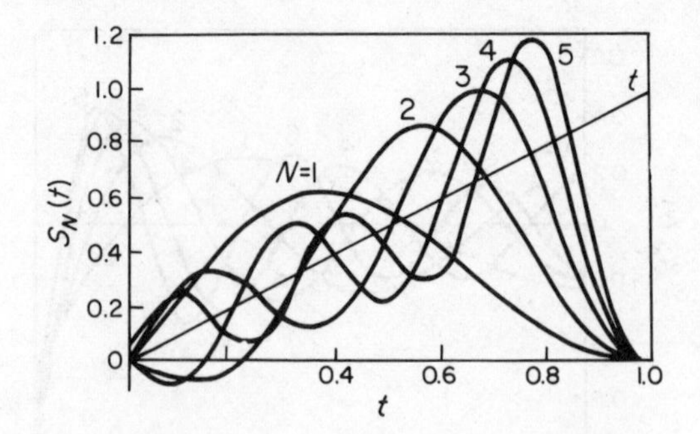

Fig. 9. Convergence of the partial sums

$$S_N(t)=\sum_{-N}^{N}-\frac{1}{S_n^2}\phi_1^{(n)}(t)$$

of the biorthogonal series $\lim_{N\to\infty} S_N(t)$ representing the unit ramp function $f(t)=t, -1<t<1$.

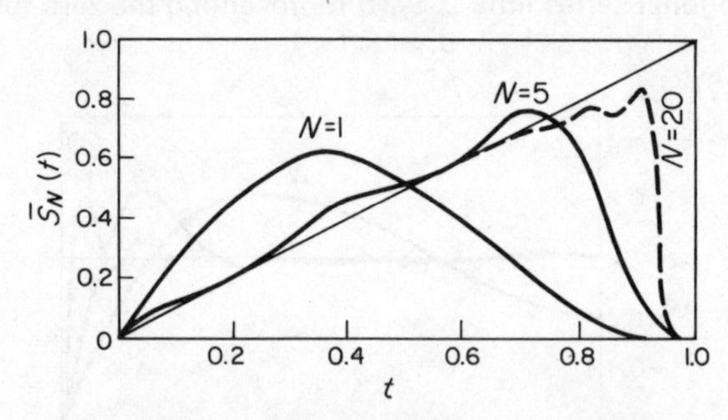

Fig. 10. Convergence of the Cesaro sums

$$\bar{S}_N(t)=\frac{1}{N}\sum_{M=1}^{N} S_M(t)$$

to the unit ramp function $f(t)=t, -1,<t<1$.

where $C_n=-1/\cos^4 S_n$. Similarly, we can expand the ramp function

$$\mathbf{f}(t)=\begin{bmatrix} t \\ 0 \end{bmatrix}=\sum_{-\infty}^{\infty}\begin{bmatrix} \phi_1^{(n)}(t) \\ \phi_2^{(n)}(t) \end{bmatrix}, \qquad C_0=0, \tag{4.22}$$

in the odd set of eigenfunctions. We get different nontrivial representations of $g(t)=0$ from (4.21) and (4.22).

In Figs 5 through 8 (taken from Joseph and Sturges [13]), we have exhibited the convergence of the partial sums

$$\mathbf{S}_N(t)=\sum_{-N}^{N}\frac{1}{\cos^4 S_n}\begin{bmatrix} \phi_1^{(n)}(t) \\ \phi_2^{(n)}(t) \end{bmatrix}\to\begin{bmatrix} 1 \\ 0 \end{bmatrix}$$

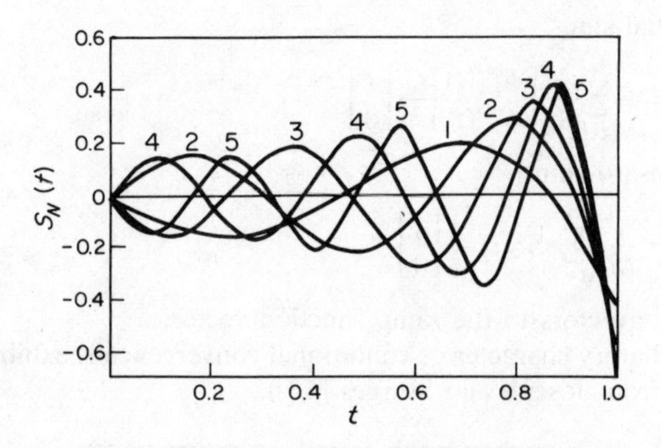

Fig. 11. Convergence of the partial sums

$$S_N(t)=\sum_{-N}^{N}-\frac{1}{S_n^2}\phi_2^{(n)}(t)$$

of the biorthogonal series $\lim_{N\to\infty} S_N(t)$ representing the zero function $g(t)=0$, $-1<t<1$.

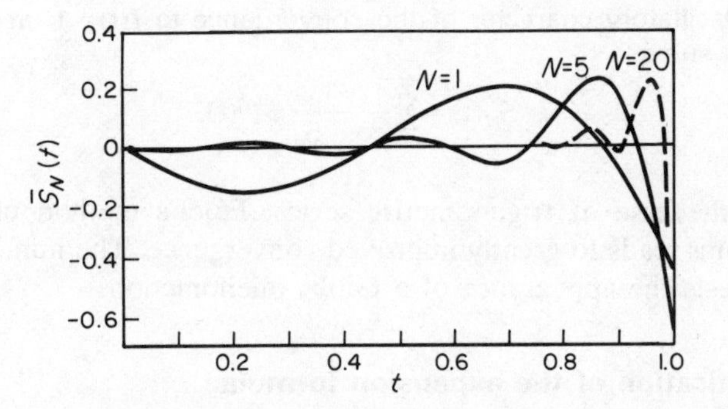

Fig. 12. Convergence of the Cesaro sums

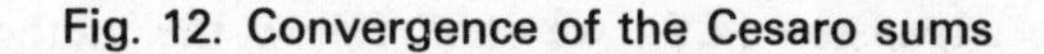

$$\bar{S}_N(t)=\frac{1}{N}\sum_{N=1}^{N} S_M(t)$$

to the zero function $g(t)=0$, $-1<t<1$.

and the Cesaro sum

$$\mathbf{S}_M(t)=\frac{1}{M}\sum_{N=1}^{M}\mathbf{S}_N(t)\to\begin{bmatrix}1\\0\end{bmatrix}$$

of even eigenfunction to the step-function vector. In Figs 9 through 12 (taken from Joseph and Sturges [13]), we have exhibited the convergence of the partial sums

$$\mathbf{S}_N(t) = -\sum_{-N}^{N} \frac{1}{S_n^2}\begin{bmatrix}\phi_1^{(n)}(t)\\ \phi_2^{(n)}(t)\end{bmatrix} \to \begin{bmatrix} t \\ 0\end{bmatrix}$$

and the Cesaro sum

$$\mathbf{S}_M(t) = \frac{1}{M}\sum_{M=1}^{M} \mathbf{S}_N(t) \to \begin{bmatrix} t \\ 0\end{bmatrix}$$

of odd eigenvectors to the ramp-function vector.

The oscillatory character of conditional convergence is exhibited in Fig. 13 (taken from Joseph and Sturges [13]).

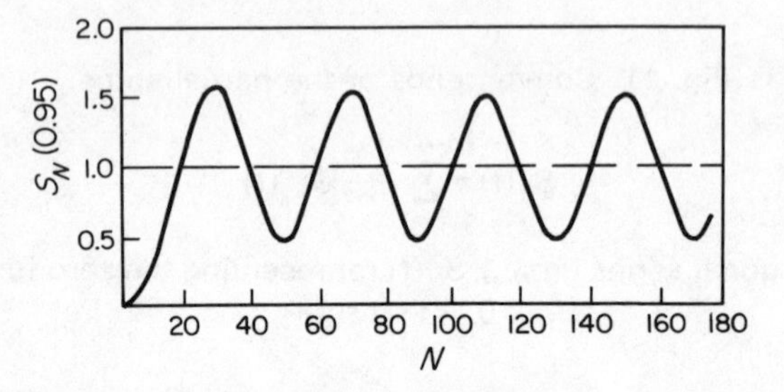

Fig. 13. Oscillatory character of the convergence to $f(t) = 1$ (at $t = 0.95$) of the partial sums

$$S_N(t) = \sum_{-N}^{N} \frac{-1}{\cos^4 S_n}\phi_1^{(n)}(t).$$

As in the case of trigonometric series, Féjer's method of summing Cesaro sums leads to greatly improved convergence. The numerical work also suggests the appearance of a Gibbs phenomenon.

5. Justification of the expansion formula

To justify the expansion (4.5), we shall follow the method of residues used by Smith [1]. The method is a generalization of one used by Titchmarsh [18] to study eigenfunction expansions associated with second-order differential equations. In fact, both authors make perfectly explicit the calculations used in the general spectral theory for linear operators. Of course, explicit computation leads to explicit results without a lot of hard-to-verify hypotheses. In the general theory, one considers an operator $\mathbf{T}(=\mathbf{A}^{-1}\,\mathrm{d}^2/\mathrm{d}t^2)$ with eigenvalues S^2 and domain $\mathcal{D}$, mapping $\mathcal{D} \to \mathcal{R} \supset \mathcal{D}$. The resolvent $\mathbf{R}(S)$ of $\mathbf{T} - S^2\mathbf{1}$ is the operator from $\mathcal{R} \to \mathcal{D}$ which inverts the problem $(\mathbf{T} - S^2\mathbf{1})\mathbf{x} = \mathbf{f}$, $\forall \mathbf{f} \in \mathcal{R}$ when S^2 is in the resolvent set; that is when S^2 is not an eigenvalue of $\mathbf{T}$. The expansion theorem can be obtained computing residues in the complex S plane at the singularities of the resolvent, provided that (a) there is a sequence of closed contours C_N not passing through eigenvalues of $\mathbf{T}$, (b) the minimum distance from the origin of the S-plane to C_N tends to infinity with N and (c)

$$\lim_{N\to\infty} \{\sup_{S\in C_N} \|\mathbf{R}(S)\mathbf{f}\|\} = 0, \qquad \forall \mathbf{f} \in \mathcal{R}.$$

In Smith's analysis, the class $\mathcal{R}$ of functions are the vector-valued fields satisfying (4.7) and (4.8). We will now show that the conditions $q(\pm 1) = g'(\pm 1) = 0$, which Smith assumed, are not required in his proof. It follows that we may take $\mathcal{R}$ as the larger class of vector fields satisfying (4.8) and (*); that is,

$$\mathbf{f}(t) = \begin{bmatrix} f(t) \\ g(t) \end{bmatrix}, \qquad f(\pm 1) = f'(\pm 1) = \langle g \rangle = \langle tg \rangle = 0.$$

$$\mathbf{f}(t) \in C_1[-1, 1], \qquad f''(t) \text{ of bounded variation.}$$

Smith solves the nonhomogeneous equation

$$\mathbf{M}''(t) + S^2\mathbf{A} \cdot \mathbf{M}(t) = \mathbf{A} \cdot \mathbf{f}(t), \tag{5.1}$$

where $\mathbf{A}$ is the biorthogonality matrix; $\mathbf{M}(t)$ is a two-component column vector such that

$$\mathbf{E} \cdot \mathbf{M}(\pm 1) + \hat{\mathbf{E}} \cdot \mathbf{M}'(\pm 1) = 0, \qquad \mathbf{E} = \begin{bmatrix} 1 & 0 \\ 0 & 0 \end{bmatrix}, \qquad \hat{\mathbf{E}} = \begin{bmatrix} 0 & 0 \\ 1 & 0 \end{bmatrix}. \tag{5.2}$$

A Green-function solution of this problem may be obtained by the method of variation of parameters. We seek $\mathbf{M}$ in the form

$$\mathbf{M}(t) = \mathbf{X}_1(t) \cdot \mathbf{F}_1(t) + \mathbf{X}_2(t) \cdot \mathbf{F}_2(t), \tag{5.3}$$

where $\mathbf{X}_1(t) = \mathbf{X}(t-1)$, $\mathbf{X}_2(t) = \mathbf{X}(t+1)$ and the 2×2 matrix $\mathbf{X}(t)$ satisfies

$$\mathbf{X}''(t) + S^2\mathbf{A} \cdot \mathbf{X} = 0, \qquad \mathbf{E} \cdot \mathbf{X}(0) + \hat{\mathbf{E}} \cdot \mathbf{X}'(0) = 0 \tag{5.4}$$

and $\mathbf{F}_1(t)$ and $\mathbf{F}_2(t)$ are such that

$$\mathbf{X}_1 \cdot \mathbf{F}_1' + \mathbf{X}_2 \cdot \mathbf{F}_2' = 0 \tag{5.5}$$

with $\mathbf{F}_1(-1) = \mathbf{F}_2(1) = 0$. Equations (5.1), (5.3) and (5.5) imply that

$$\mathbf{X}_1' \cdot \mathbf{F}_1' + \mathbf{X}_2' \cdot \mathbf{F}_2' = \mathbf{A} \cdot \mathbf{f}. \tag{5.6}$$

We next introduce the 2×2 matrices $\mathbf{Y}_1(t) = \mathbf{Y}(t-1)$ and $\mathbf{Y}_2(t) = \mathbf{Y}(t+1)$ satisfying

$$\mathbf{Y}''(t) + S^2\mathbf{Y} \cdot \mathbf{A} = 0, \qquad \mathbf{Y}[0] \begin{bmatrix} 0 & 0 \\ 1 & 0 \end{bmatrix} + \mathbf{Y}'(0) \begin{bmatrix} 0 & 0 \\ 0 & 1 \end{bmatrix} = 0 \tag{5.7}$$

and define the Wronskians

$$\mathbf{W}_{ij}(t) = \mathbf{Y}_j \cdot \mathbf{X}_i' - \mathbf{Y}_j' \cdot \mathbf{X}_i \qquad (i, j = 1, 2). \tag{5.8}$$

Clearly, $\mathbf{W}'_{ij}=0$ and $\mathbf{W}_{ij}$ is a constant matrix. The boundary conditions imply that $\mathbf{W}_{11}=\mathbf{W}_{22}=0$. Premultiplying (5.6) by $\mathbf{Y}_2$ and (5.5) by Y_2, we find that

$$\mathbf{W}_{12}\mathbf{F}'_1=\mathbf{Y}_2\cdot\mathbf{A}\cdot\mathbf{f}$$

and similarly,

$$\mathbf{W}_{21}\mathbf{F}'_2=\mathbf{Y}_1\cdot\mathbf{A}\cdot\mathbf{f}.$$

Hence, after solving for $\mathbf{F}'_1$ and $\mathbf{F}'_2$ and integrating, we get

$$\begin{aligned}\mathbf{F}_1&=\mathbf{W}_{12}^{-1}\int_{-1}^{t}\mathbf{Y}_2(u)\cdot\mathbf{A}\cdot\mathbf{f}(u)\,du,\\ \mathbf{F}_2&=-\mathbf{W}_{21}^{-1}\int_{t}^{1}\mathbf{Y}_1(u)\cdot\mathbf{A}\cdot\mathbf{f}(u)\,du.\end{aligned}\tag{5.9}$$

Equations (5.3) and (5.9) solve (5.1) and (5.2). Moreover,

$$\mathbf{X}(t)=\begin{bmatrix}\sin St-St\cos St & St\sin St\\ \sin St+St\cos St & 2\cos St-St\sin St\end{bmatrix},$$

$$\mathbf{Y}(t)=\begin{bmatrix}2\cos St+St\sin St & St\sin St\\ 3\sin St-St\cos St & \text{Sin } St-St\cos St\end{bmatrix}.$$

$$\mathbf{W}_{12}^{-1}=\frac{1}{2S(\sin^2 2S-4S^2)}\begin{bmatrix}-2S\sin 2s & \sin 2S+2S\cos 2S\\ -\sin 2S+2S\cos 2S & 2S\sin 2S\end{bmatrix}.\tag{5.10}$$

and

$$\mathbf{W}_{21}^{-1}=\frac{1}{2S(\sin^2 2S-4S^2)}\begin{bmatrix}-2S\sin 2s & -\sin 2S-2S\cos 2S\\ \sin 2S-2S\cos 2S & 2S\sin 2S\end{bmatrix}.\tag{5.11}$$

Smith notes that $\mathbf{SM}(t,S)$, $-1\leqslant t\leqslant 1$, is a meromorphic function of S with poles at the zeros S_n of

$$\sin^2 2S-4S^2=(\sin 2S+2S)(\sin 2S-2S)=0.$$

The zeros of $\sin 2S+2S$ and $\sin 2S-2S$ are symmetric in the four quadrants of the S-plane. Smith shows that the residue at a zero of sin $2S+2S$ is

$$\tfrac{1}{2}C_n\begin{bmatrix}\phi_1^{(n)}(t)\\ \phi_2^{(n)}(t)\end{bmatrix},\qquad n=\pm1,\pm2,\ldots,$$

where $\phi_1^{(n)}(t)$, $\phi_2^{(n)}(t)$ are the even eigenfunctions (2.6) and (2.29) and C_n is computed from (2.24) on the even adjoint eigenfunctions $\Psi_1^{(n)}(t)$ and $\Psi_2^{(n)}(t)$. Similarly, the residue of $\mathbf{MS}$ at a zero of sin $2S-2S$ is given by (5.12) computed on the odd eigenfunction and adjoint eigenfunctions. $\mathbf{SM}(S,t)$ has no residue at $S=0$.

To verify (5.12), for example at $S=S_n$, where $\sin 2S_n = -2S_n$, note that

$$(S-S_n)\mathbf{W}_{12}^{-1} = -\frac{1}{8S\cos^2 S_n}\begin{bmatrix} S_n & -\sin^2 S_n \\ \cos^2 S_n & -S_n \end{bmatrix},$$

$$(S-S_n)\mathbf{W}_{21}^{-1} = -\frac{1}{8S_n\cos^2 S_n}\begin{bmatrix} S_n & \sin^2 S_n \\ -\cos^2 S_n & -S_n \end{bmatrix}$$

and

$$(S-S_n)S_n\mathbf{M}(S_n) = -\frac{1}{8\cos^2 S_n}\left\{\begin{bmatrix} -\phi_1^{(n)}(t) & \dfrac{\sin^2 S}{S}\phi_1^{(n)}(t) \\ -\phi_2^{(n)}(t) & \dfrac{\sin^2 S}{S}\phi_2^{(n)}(t) \end{bmatrix} \times \int_{-1}^{t} \mathbf{Y}_2(u)\cdot\mathbf{A}\cdot\mathbf{f}(u)\,\mathrm{d}u - \begin{bmatrix} \phi_1^{(n)}(t) & \dfrac{\sin^2 S}{S}\phi_1^{(n)}(t) \\ \phi_2^{(n)}(t) & \dfrac{\sin^2 S}{S}\phi_2^{(n)}(t) \end{bmatrix}\int_t^1 \mathbf{Y}_1(u)\cdot\mathbf{A}\cdot\mathbf{f}(u)\,\mathrm{d}u\right\}$$

$$= \tfrac{1}{2}C_n\begin{bmatrix} \phi_1^{(n)}(t) \\ \phi_2^{(n)}(t) \end{bmatrix}.$$

Let us assume that

$$\mathbf{M} = \frac{1}{S_2}\begin{bmatrix} f(t) \\ g(t) \end{bmatrix} - \frac{1}{S^2}\mathbf{X}_1(t)\cdot\mathbf{W}_{12}^{-1}\cdot\int_{-1}^{t}\mathbf{Y}_2(u)\cdot\mathbf{f}''(u)\,\mathrm{d}u + \frac{1}{S^2}\mathbf{X}_2(t)\cdot\mathbf{W}_{21}^{-1}\cdot\int_t^1 \mathbf{Y}_1(u)\cdot\mathbf{f}''(u)\,\mathrm{d}u$$

$$= \frac{1}{S^2}\begin{bmatrix} f(t) \\ g(t) \end{bmatrix} + \mathbf{G}(s,t) \tag{5.13}$$

and

$$\lim_{N\to\infty}\frac{1}{2\pi i}\int_{C_N} S\mathbf{G}(S,t)\,\mathrm{d}s = 0, \qquad 1\leqslant t\leqslant 1. \tag{5.14}$$

C_N is the square with vertices $2N\pi\ (\pm 1 \pm i)$ and the contour is described in the anticlockwise sense. Smith proved that (5.14) holds provided only that $\mathbf{f}''(t)$ is of bounded variation. Taking account of the numbering convention established in section 2, $(S_n, \boldsymbol{\phi}^{(n)}(S_n), C_n(S_n)) = (-S_n, \boldsymbol{\phi}^{(n)}(-S_n), C_n(-S_n))$, we find, using (5.3), (5.9) and (5.12), that

$$\lim_{N\to\infty}\frac{1}{2\pi i}\int_{C_N} S\mathbf{M}(t,S)\,\mathrm{d}s = \sum_{-\infty}^{\infty} C_n\begin{bmatrix} \phi_1^{(n)}(t) \\ \phi_2^{(n)}(t) \end{bmatrix} \tag{5.15}$$

and, using (5.13) and (5.14), that

$$\lim_{N\to\infty} \frac{1}{2\pi i} \int_{C_N} S\mathbf{M}(t, S) = \begin{bmatrix} f(t) \\ g(t) \end{bmatrix}. \tag{5.16}$$

It follows that, if (5.13) holds, then (4.5) and (4.6) are valid and the series on the right converges pointwise, absolutely and uniformly to the data vector on the left.

It was evidently the derivation of (5.13) which induced Smith to introduce the restrictions $g(\pm 1) = g'(\pm 1) = 0$. But his demonstration does not require these restrictions because his P satisfies the equation under his (X) without requiring $g(\pm 1) = g'(\pm 1)$. More directly, we shall now show that (5.3) and (5.9) imply (5.13) provided only that $f(\pm 1) = f'(\pm 1) = 0$. Noting first that, by (5.7), $\mathbf{Y}_i(u) \cdot \mathbf{A} = -\mathbf{Y}_i''/S^2$ $(i = 1, 2)$, we integrate by parts:

$$\begin{aligned} S^2\mathbf{M} &= \mathbf{X}_1(t) \cdot \mathbf{W}_{12}^{-1} \cdot \int_{-1}^{t} \mathbf{Y}_2(u) \cdot \mathbf{A} \cdot \mathbf{f}(u)\, du \\ &\quad -\mathbf{X}_2(t) \cdot \mathbf{W}_{21}^{-1} \cdot \int_{t}^{1} \mathbf{Y}_1(u) \cdot \mathbf{A} \cdot \mathbf{f}(u)\, du \\ &= -\mathbf{X}_1(t) \cdot \mathbf{W}_{21}^{-1} \cdot \{[\mathbf{Y}_2' \cdot \mathbf{f}]_{-1}^{t} - [\mathbf{Y}_2 \cdot \mathbf{f}']_{-1}^{t}\} \\ &\quad + \mathbf{X}_2(t) \cdot \mathbf{W}_{21}^{-1} \cdot \{[\mathbf{Y}_i' \cdot \mathbf{f}]_t^1 - [\mathbf{Y}_i \cdot \mathbf{f}]_t^1\} \\ &\quad + S^2\mathbf{G}(S, t). \end{aligned} \tag{5.17}$$

We next observe that

$$\mathbf{Y}_2(-1) = \mathbf{Y}_1(1) = \mathbf{Y}(0) = \begin{bmatrix} 2 & 0 \\ 0 & 0 \end{bmatrix}$$

and

$$\mathbf{Y}_2'(-1) = \mathbf{Y}'(1) = \mathbf{Y}'(0) = S \begin{bmatrix} 0 & 0 \\ 2 & 0 \end{bmatrix}.$$

Since, by hypothesis,

$$\mathbf{f}(\pm 1) = \begin{bmatrix} 0 \\ g(\pm 1) \end{bmatrix}, \qquad f'(\pm 1) = \begin{bmatrix} 0 \\ g'(\pm 1) \end{bmatrix},$$

we calculate

$$\begin{aligned} S^2\mathbf{M} &= -\{\mathbf{X}_1(t) \cdot \mathbf{W}_{21}^{-1} \cdot \mathbf{Y}_2'(t) + \mathbf{X}_2(t) \cdot \mathbf{W}_{21}^{-1} \cdot \mathbf{Y}_1(t)\} \cdot \mathbf{f}(t) \\ &\quad + (\mathbf{X}_1(t) \cdot \mathbf{W}_{21}^{-1} \cdot \mathbf{Y}_2(t) + \mathbf{X}_2(t) \cdot \mathbf{W}_{21}^{-1} \cdot \mathbf{Y}(t)\} \cdot \mathbf{f}(t) \\ &\quad + S^2\mathbf{G}(S, t). \end{aligned} \tag{5.18}$$

But the second bracket in (5.18) vanishes by (5.5) and, differentiating the second bracket, we get (5.18) in the form

$$S^2\mathbf{M}=\{\mathbf{X}_1'(t)\cdot\mathbf{W}_{12}^{-1}\cdot\mathbf{Y}_2(t)+\mathbf{X}_2'(t)\cdot\mathbf{W}_{21}^{-1}\cdot\mathbf{Y}_1(t)\}\cdot\mathbf{f}(t)+S^2\mathbf{G}(S,t) \tag{5.19}$$

To prove that the bracket in (5.19) is $\mathbf{1}$, we note that $\mathbf{W}_{22}=0$, $\mathbf{Y}_2(t)$ is not singular and $\mathbf{X}_2\cdot\mathbf{W}_{21}^{-1}\cdot\mathbf{Y}_1=-\mathbf{X}_1\cdot\mathbf{W}_{21}^{-1}\cdot\mathbf{Y}_2$. Then

$$\begin{aligned}0&=\mathbf{Y}_2^{-1}\cdot(\mathbf{Y}_2\cdot\mathbf{X}_2'-\mathbf{Y}_2'\cdot\mathbf{X}_2)\cdot\mathbf{W}_{21}^{-1}\cdot\mathbf{Y}_1\\&=\mathbf{X}_2'\cdot\mathbf{W}_{21}^{-1}\cdot\mathbf{Y}_1+\mathbf{Y}_2^{-1}\cdot(\mathbf{Y}_2'\cdot\mathbf{X}_1)\cdot\mathbf{W}_{21}^{-1}\cdot\mathbf{Y}_2\\&=\mathbf{X}_2'\cdot\mathbf{W}_{21}^{-1}\cdot\mathbf{Y}_1+\mathbf{Y}_2^{-1}(-\mathbf{W}_{21}+\mathbf{Y}_2\cdot\mathbf{X}_1')\cdot\mathbf{W}_{12}^{-1}\cdot\mathbf{Y}_2\\&=\mathbf{X}_2'\cdot\mathbf{W}_{21}^{-1}\cdot\mathbf{Y}_1'+\mathbf{X}_1\cdot\mathbf{W}_{12}^{-1}\cdot\mathbf{Y}_2-\mathbf{1}.\end{aligned}$$

Hence,

$$S^2\mathbf{M}=\mathbf{f}(t)+S^2\mathbf{G}(S,t);$$

i.e., (5.13) holds.

Acknowledgement I am grateful to Simon Rosenblat and George Habetler for very helpful discussions on the question of completeness treated in section 5. The work was supported by the NSF (Grant no. 19047) and by the U.S. Army Research Office.

References

1 Smith, R. C. T., The bending of a semi-infinite strip. *Aust. J. Sci. Res.*, **5,** 277–237, 1952.

2 Joseph, D. D. and Fosdick, R., The free surface on a liquid between cylinders rotating at different speeds. Part I, *Arch. Rat. Mech. Anal.*, **49,** 321–401, 1973.

3 Joseph, D. D. and Sturges, L., The free surface on a liquid filling a trench heated from its side, *J. Fluid Mech.* **69,** 565, 1975.

4 Joseph, D. D., Slow motion and viscometric motion. Stability and bifurcation of the rest state of a simple fluid, *Arch. Rat. Mech. Anal.*, **56,** 99, 1974.

5 Yoo, J. Y. and Joseph, D. D., Stokes flow in a trench between concentric cylinders, *SIAM J. Appl. Math.*, **34,** 247–285, 1978.

6 Liu, C. H. and Joseph, D. D., Stokes flow in conical cavities, *SIAM J. Appl. Math.*, **34,** 286–296, 1978.

7 Yoo, J. Y., PhD Thesis, University of Minnesota, 1977.

8 Liu, C. H. and Joseph, D. D., Stokes flow in wedge-shaped trenches, *J. Fluid Mech.*, **80,** 443, 1977.

9 Dean, W. R. and Montagnon, P. E., On the steady motion of viscous liquid in a corner, *Proc. Camb. Phil. Soc.*, **45,** 389, 1949.

10 Moffatt, H. K., Viscous and resistive eddies near a sharp corner, *J. Fluid Mech.*, **18,** 1, 1964.

11 Sturges, L. and Joseph, D. D., The free surface on a simple fluid between cylinders undergoing torsional oscillations. Part III. Oscillating planes, *Arch. Rat. Mech. Anal.*, **69,** 245, 1977.

12 Joseph, D. D., The convergence of biorthogonal series for biharmonic and Stokes flow edge problems. Part 1, *SIAM J. Appl. Math.*, **33**, 337–347, 1977

13 Joseph, D. D. and Sturges, L. The convergence of biorthogonal series for biharmonic and Stokes flow edge problems. Part 2, *SIAM J. Appl. Math.*, **34,** 7–26, 1978.

14 Papkowich, P. R., Uber eine Form der Lösung des biharmonischen Problems für das Rechteck, *Dokl. Acad. Sci. USSR*, **27,** 337, 1940.

15 Fadle, J., Die Selbstspannungs-Eigenwert funktionen der quadratischen Scheibe, *Ingenieur Archiv*, **11,** 125, 1941.

16 Robbins, C. I. and Smith, R. C. T., A table of roots of sin $z = z$, *Phil. Mag.*, **84,** 1004–1005, 1948.

17 Hillman, A. P. and Salzer, H. E., Roots of sin $z = z$, *Phil. Mag.*, **39,** 575–577, 1943.

18 Titchmarsh, E. C., *Eigenfunction Expansions Associated with Second Order Differential Equations*, Oxford University Press, London, 1946.

Professor Daniel D. Joseph,
Department of Aerospace Engineering
and Mechanics,
The University of Minnesota,
Minneapolis,
Minn. 55112, USA

M. Kac

Semiclassical quantum mechanics and Morse's theory

1

What I propose to say in this paper is almost purely expository and most of it is explicitly or implicitly contained in the work of people other than myself. Connections between the semiclassical approach to quantum mechanics and certain aspects of Morse's theory have been emphasized by Martin Gutzwiller [1] and L. C. Schulman [2]. More recently, my attention was drawn to a paper [3] and a number of preprints by S. Levit and U. Smilansky, which pursue the same line and contain a number of interesting new observations and results.

2

I shall be mostly concerned with systems of one degree of freedom ($n = 1$) with the Lagrangian

$$L = \frac{1}{2}\left(\frac{dx}{dt}\right)^2 - V(x). \tag{2.1}$$

The extension to systems with more degrees of freedom is not entirely trivial and will be discussed briefly at the end of the paper.

The fundamental concept of quantum mechanics is the 'propagator' $K(x_0, t_0 \mid x, t)$, which allows one to calculate the probability that a system in state ϕ_0 at t_0 will be found in the state ϕ at tx by the formula

$$\left|\int_{-\infty}^{\infty}\int_{-\infty}^{\infty} \phi_0(x_0)K(x_0, t_0 \mid x, t)\phi(x)\,dx_0\,dx\right|^2. \tag{2.2}$$

(It should be recalled that in quantum mechanics the 'state' of a system is a (wave) function.)

According to Feynman, the propagator K is given by the formula

$$K(x_0, t_0 \mid x, t) = \int \mathrm{d(path)} \exp\left\{\frac{i}{\hbar} S[x(\tau)]\right\} \tag{2.3}$$

where

$$S[x(\tau)] = \int_0^t \left[\frac{1}{2}\left(\frac{\mathrm{d}x}{\mathrm{d}t}\right)^2 - V(x(\tau))\right] \mathrm{d}\tau. \tag{2.4}$$

$\hbar$ the Planck constant divided by 2π and d(path) signifies integration over the space of paths subject to the condition

$$x(0) = x_0, \qquad x(t) = x. \tag{2.5}$$

The 'integral over paths' is defined (*à la* Riemann) as the limit of

$$\left(\frac{1}{\sqrt{(2\pi i \Delta \hbar)}}\right)^N \int_{-\infty}^{\infty}\!\!\int \exp\left\{\frac{i}{h}\left[\frac{(x_1 - x_0)^2}{2\Delta} + \frac{(x_2 - x_1)^2}{2\Delta} + \ldots + \frac{(x - x_{N-1})^2}{2\Delta} - \Delta \sum_1^{N-1} V(x_K)\right]\right\} \mathrm{d}x_1 \ldots \mathrm{d}x_{N-1} \tag{2.6}$$

as $N \to \infty$, $\Delta \to 0$ and $N\Delta = t$.

For the sake of definiteness, we set

$$\sqrt{i} = \exp\left\{\frac{\pi}{4} i\right\}.$$

3

The semiclassical approximation to the propagator is obtained as follows. Let $x_{\mathrm{cl}}(\tau)$, $x_{\mathrm{cl}}(0) = x_0$, $x_{\mathrm{cl}}(t) = x$ be the classical path which satisfies the Hamilton–Maupertuis principle

$$\delta S = 0. \tag{3.1}$$

Set

$$x(\tau) = x_{\mathrm{cl}}(\tau) + y(\tau), \qquad y(0) = y(t) = 0, \tag{3.2}$$

and expand the action $S[x(\tau)]$ about $x_{\mathrm{cl}}(\tau)$. Thus,

$$S[x(\tau)] = S[x_{\mathrm{cl}}(\tau)] \to \int_0^t \left[\frac{1}{2}\left(\frac{\mathrm{d}y}{\mathrm{d}\tau}\right)^2 + \tfrac{1}{2} V''(x_{\mathrm{cl}}(\tau)) y^2(\tau)\right] \mathrm{d}\tau + \ldots, \tag{3.3}$$

where the term linear in y vanishes because of (3.1).

The semiclassical approximation corresponds to replacing $S[x(\tau)]$ by the two terms on the right-hand side of (3.3). In other words, the semiclassical propagator is given by the formula

$$\tilde{K}(x_0, t_0 \mid x, t) = \exp\left\{\frac{i}{\hbar} S[x_a]\right\} \int \mathrm{d(path)} \exp\left\{\frac{i}{\hbar} \int_0^t \left[\frac{1}{2}\left(\frac{\mathrm{d}y}{\mathrm{d}\tau}\right)^2 \tfrac{1}{2} V''(x_a(\tau) y^2(\tau)\right]\right\} \mathrm{d}\tau. \tag{3.4}$$

It should be understood that

$$S[x_{\mathrm{cl}}(\tau)] = \int_0^t \left[\frac{1}{2}\left(\frac{\mathrm{d}x_{\mathrm{cl}}}{\mathrm{d}\tau}\right)^2 V(x_{\mathrm{cl}}(\tau)\right] \mathrm{d}\tau = S(x_0, x; t) = S_{\mathrm{cl}}(x_0, x; t) \tag{3.5}$$

and the path integral in (3.4) is defined as the limit ($N\Delta = t$)

$$\lim_{N\to\infty} \left(\frac{1}{\sqrt{(2\pi i \Delta \hbar)}}\right)^N \int_{-\infty}^{\infty} \int_{-\infty}^{\infty} \exp\left\{\frac{\mathrm{i}}{\hbar}\left[\frac{y_1^2}{2\Delta} + \frac{(y_2 - y_1)^2}{2\Delta} + \dots + \frac{y_{N-1}^2}{2\Delta} - \frac{\Delta}{2}\sum_1^{N-1} r(k\Delta) y^2\right]\right\} \mathrm{d}y_1 \dots \mathrm{d}y_{N-1}, \tag{3.6}$$

where I have used the abbreviation

$$r(\tau) = V''(x_{\mathrm{cl}}(\tau)). \tag{3.7}$$

4

The calculation of the path integral

$$\int \mathrm{d(path)} \exp\left\{\frac{\mathrm{i}}{\hbar} \int \left[\frac{1}{2}\left(\frac{\mathrm{d}y}{\mathrm{d}\tau}\right)^2 - \tfrac{1}{2} V''(x_{\mathrm{cl}}(\tau) y^2(\tau))\right]\right\} \mathrm{d}\tau \tag{4.1}$$

is usually accomplished by the method of Cameron and Martin (*see*, for example, Dashen, Hasslacher and Neven [4]) or by following Pauli [5].

The result is

$$\frac{1}{\sqrt{(2\pi i \hbar)}} \sqrt{\left(\pm \frac{\partial^z S}{\partial x_0\, \partial x}\right)}, \tag{4.2}$$

with some uncertainty as to the sign.

I shall use a different method of evaluating (4.1) via (3.6), which is completely straightforward and was first used in Kac and Siegert [6] (*see*, in particular, section VI) (*see also* A. M. Mark [7]).

Setting

$$y_k = \sqrt{\Delta}\sqrt{\hbar}\, \bar{y}_k,$$

the integral (3.6) becomes

$$\left(\frac{1}{\sqrt{(2\pi i)}}\right)^N \frac{1}{\sqrt{(h\Delta)}} \int_{-\infty}^{\infty} \dots \exp\left\{\frac{i}{2}\left[\bar{y}_1^2\right.\right.$$

$$\left.\left. + (\bar{y}_2 - \bar{y}_1)^2 + \dots + \bar{y}_{N-1}^2 - \Delta^2 \sum_1^{N-1} r(K\Delta \bar{y}_k^2\right]\right\} d\bar{y}_1 \dots d\bar{y}_{N-1} \tag{4.3}$$

and the matrix of the quadratic form in the exponent is

$$\mathbf{A}_0 - \mathbf{R},$$

where

$$\mathbf{A}_0 = \begin{bmatrix} 2 & -1 & 0 & \dots & 0 \\ -1 & 2 & -1 & \dots & \\ 0 & -1 & 2 & \dots & -1 \\ 0 & \dots & -1 & -1 & 2 \end{bmatrix} \quad ((n-1)\times(n-1))$$

and

$$\mathbf{R} = \Delta^2 \begin{bmatrix} r(\Delta) & & & \\ & r(2\Delta) & & 0 \\ & & \ddots & \\ & 0 & & \\ & & & r(N-1)\Delta \end{bmatrix}$$

Let $\mu_1, \dots, \mu_{N-1}$ be the eigenvalues of the matrix $\mathbf{A}_0 - \mathbf{R}$, then, since

$$\frac{1}{\sqrt{(2\pi)}} \int_{-\infty}^{\infty} e^{i\mu x^2/2}\, dx = \begin{cases} \displaystyle\int_{-\infty}^{\infty} \frac{e^{i\pi/4}}{\sqrt{\mu}}, & \mu > 0, \\ \displaystyle\int_{-\infty}^{\infty} \frac{e^{-i\pi/4}}{\sqrt{-\mu}}, & \mu < 0, \end{cases}$$

we obtain

$$\frac{1}{\sqrt{(2\pi i)}} \frac{1}{\sqrt{(\hbar\Delta)}} \frac{e^{-(i\pi/2)\times \text{number of negative } \mu\text{'s}}}{\sqrt{|\det(\mathbf{A}_0 - \mathbf{R})|}}$$

for the integral (4.3).

Writing

$$\mathbf{A}_0 - \mathbf{R} = \mathbf{A}_0(\mathbf{I} - \mathbf{A}_0^{-1}\mathbf{R})$$

and using the easily verifiable facts that

$$\det \mathbf{A}_0 = (N-1) + 1 = N$$

and

$$(\mathbf{A}_0^{-1})_{ij} = N\left\{\min\left(\frac{i}{N}, \frac{j}{N}\right) - \frac{ij}{N^2}\right\},$$

we see that

$$\Delta \det(\mathbf{A}_0 - \mathbf{R}) = t \det(\mathbf{I} - \mathbf{A}_0^{-1}\mathbf{R})$$

and the (i, j)th element of $\mathbf{I} - \mathbf{A}_0^{-1}\mathbf{R}$ is

$$\delta_{ij} - t\Delta\left\{\min\left(\frac{i\Delta}{t}, \frac{j\Delta}{t}\right) - \left(\frac{i\Delta}{t}\right)\left(\frac{j\Delta}{t}\right)\right\} r(j\Delta).$$

It thus follows by a classical theorem of Hilbert that the limit of $\det(\mathbf{I} - \mathbf{A}_0^{-1}\mathbf{R})$ as $\Delta \to 0$ is the Fredholm determinant of the kernel

$$K(x, y) = t\left\{\min\left(\frac{x}{t}, \frac{y}{t}\right) - \frac{xy}{t^2}\right\} r(y), \qquad 0 \leq x, y \leq t. \tag{4.4}$$

The Fredholm determinant of (4.4) is easily seen to be the product

$$\prod_{j=1}^{\infty} (1 - \lambda_j(t)),$$

where the λ's are the eigenvalues of the Sturm–Liouville equation

$$\lambda\phi''(\tau) + r(\tau)\phi(\tau) = 0, \qquad \phi(0) = \phi(\tau) = 0, \tag{4.5}$$

where, of course,

$$r(\tau) = V''(x_{\text{cl}}(\tau)).$$

Combining these results with that of Pauli, we obtain the identity

$$t\prod_{j=1}^{\infty} (1 - \lambda_j(t)) = \pm\left(\frac{\partial^2 S}{\partial x_0\, \partial x}\right)^{-1}$$

or, more precisely,

$$t\prod_{j=1}^{\infty} (1 - \lambda_j(t)) = (-1)^{(\text{number of } \lambda\text{'s which are} > 1)} \left|\frac{\partial^2 S}{\partial x_0\, \partial x}\right|^{-1}. \tag{4.6}$$

It is also clear that, in the limit $N \to \infty$, the negative μ's approach $(1 - \lambda)$'s, which are negative.

5

As t increases, so do the λ_j's. If for some t the number of λ_j's which are greater than 1 is, let us say, l, i.e.,

$$\lambda_1(t) > \lambda_2(t) > \ldots > \lambda_l(t) > 1,$$

then it follows that at certain times (assuming non-degeneracy) $t_1 < t_2 < \ldots < t_l$ the eigenvalues crossed the value 1:

$$\lambda_1(t_1) = 1, \qquad \lambda_2(t_2) = 1, \ldots, \lambda_l(t_l) = 1.$$

Each time such a crossing occurred, $(\partial^2 S/\partial x_0\, \partial x)^{-1}$ is equal to 0 and, hence, it follows that (keeping x_0 and x fixed) the number of zeros (up to t) of $(\partial^2 S/\partial x_0\, \partial x)^{-1}$ is equal to the number of eigenvalues $\lambda_j(t)$ which are greater than 1.

On the other hand, each time

$$\left(\frac{\partial^2 S}{\partial x_0\, \partial x}\right)^{-1} = -\frac{\partial x(x_0, p_0; t)}{\partial p_0}$$

is equal to zero, the point (t, x) is conjugate to (t_0, x_0), so that the number of t_i's, $t_0 < t_i < t$, such that (t_i, x) is conjugate to (t_0, x_0) and is equal to the number of eigenvalues $\lambda_j(t)$ which are greater than 1.

Finally, if t_l is the *last* time before t such that (x, t_l) is conjugate to (x_0, t_0), then 1 is the lth eigenvalue of

$$\lambda\phi'' + r(\tau)\phi = 0, \qquad \phi(t_0) = \phi(t_l) = 0$$

or, in other words,

$$\phi'' + r(\tau)\phi = 0, \qquad \phi(t_0) = \phi(t_l) = 0$$

and, by Sturman theory, ϕ has l nodes (counting t_l but *not* counting t_0).

Thus, the number of nodes of a nontrivial ϕ,

$$\phi'' + V''(x_{\mathrm{cl}}(\tau))\phi = 0, \qquad \phi(t_0) = \phi(t_l) = 0,$$

(assumed to exist), is equal to the number of zeros of $(\partial^2 S/\partial x_0\, \partial x)^{-1}$ in the interval $t_0 < \tau \leq t_l$.

6

The extension to the case of more degrees of freedom follows similar lines.

First of all, Pauli's result is valid if one replaces $\partial^2 S/\partial x_0\, \partial x$ by the Hessian

$$\left\| \frac{\partial^2 S}{\partial x_0^i\, \partial x^j} \right\| \tag{6.1}$$

(although the derivation now requires somewhat nontrivial manipulations in the Hamilton–Jacobi theory).

The analog of the matrix $\mathbf{A}_0 - \mathbf{R}$ is now (I consider, for the sake of simplicity, the case of two degrees of freedom)

$$\left(\begin{array}{c|c} \mathbf{A}_0 - \mathbf{R}_{11} & \mathbf{R}_{12} \\ \mathbf{R}_{21} & \mathbf{A}_0 - \mathbf{R}_{22} \end{array}\right), \tag{6.2}$$

where

$$R_{ij} = \Delta^2 \begin{bmatrix} r_{ij}(\Delta) & & & 0 \\ & r_{ij}(2\Delta) & & \\ & & \ddots & \\ 0 & & & r_{ij}((N-1)\Delta) \end{bmatrix}, \quad i, j = 1, 2, \tag{6.3}$$

and

$$r_{ij}(t) = \frac{\partial^2 V}{\partial x^i \, \partial x^j} \tag{6.4}$$

(evaluated for the classical path).

Instead of the Fredholm determinant of an ordinary kernel, we have a matrix of kernels

$$\begin{pmatrix} K_{11}(x, y) & K_{12}(x, y) \\ K_{21}(x, y) & K_{22}(x, y) \end{pmatrix} \tag{6.5}$$

with

$$K_{ij}(x, y) = t\left\{\min\left(\frac{x}{t}, \frac{y}{t}\right) - \frac{xy}{t^2}\right\} r_{ij}(y), \tag{6.6}$$

yielding the eigenvalue problem

$$\int_0^t (K_{11}(x, y)\phi_1(y) + K_{12}(x, y)\phi_2(y)) \, \mathrm{d}y = \lambda\phi_1(x),$$

$$\int_0^t (K_{21}(x, y)\phi_1(y) + K_{22}(x, y)\phi_2(y)) \, \mathrm{d}y = \lambda\phi_2(x),$$

which is equivalent to the system

$$\lambda\phi_1''(x) + r_{11}\phi_1 + r_{12}\phi_2 = 0. \tag{6.7a}$$

$$\lambda\phi_2''(x) + r_{21}\phi_1 + r_{22}\phi_2 = 0, \tag{6.7b}$$

with the boundary conditions

$$\phi_1(0) = \phi_1(t) = 0, \qquad \phi_2(0) = \phi_2(t) = 0. \tag{6.8}$$

As before, one can associate the number of eigenvalues of (6.7) (with boundary conditions (6.8)) which are greater than 1 with a number of appropriate conjugate points and arrive at the conclusion usually associated with Morse's theory.

Acknowledgement My thanks are due to M. C. Gutzwiller and L. C. Schulman for clarifying discussions and to L. C. Schulman for letting me see a copy of his lecture on quantum mechanics in which aspects of Morse's theory are elegantly discussed.

References

1 Gutzwiller, M. C., Periodic orbits and classical quantization conditions, *J. Math. Phys.*, **12,** 343–354, 1971 and earlier papers by the author quoted therein.

2 Schulman, L. C., Caustics and multi-valuedness, in *Functional Integration and its Applications* (A. M. Arthurs, ed.), Clarendon Press, Oxford, 1974, pp. 144–156.

3 Levit, S. and Smilansky, U., A new approach to Gaussian path integrals and the evaluation of the semiclassical propagator, *Ann. Phys.*, **103,** 198–207, 1977.

4 Dashen, R. F., Hasslacher, B. and Neven, A., Nonperturbative methods and extended-hadron models in field theory I. Semiclassical functional methods, *Phys. Rev. D*, **10,** 4114–4129, 1974.

5 Pauli, W., *Lectures on Physics*, (C. Enz, ed.), MIT Press, Cambridge, Mass., 1973.

6 Kac, M. and Siegert, A. F. J., On the theory of noise in radio receivers with square law detectors, *J. Appl. Phys.*, **18,** 383–397, 1947.

7 Mark, A. M., Some probability limit theorems, *Bull. Am. Math. Soc.* **55,** 885–900, 1949.

Professor M. Kac
The Rockefeller University,
1230 New York Avenue,
New York City,
NY 10021,
USA

I. A. Kunin

Group deformations and quantizations

We often meet a situation in which a theory contains parameters in such a way that there exists a non-trivial limit when parameters tend to zero (or to infinity). The best-known and most important cases are transitions from relativity and quantum mechanics to the corresponding classical theories. Opposite operations are group deformation (from Galilean to Lorentz group) and quantization. It seems at first sight that these two operations have nothing in common.

The goal of this paper is to show that such operations may be considered from the same group-theoretic point of view. We shall confine ourselves in this paper to a brief outline of the connections between group deformations and quantizations.

We begin with some mathematical preliminaries.

1. Lie groups with an involution†

In what follows, for simplicity, we shall denote Lie groups and the corresponding Lie algebras by the same letters, since it will always be clear from the context which we are really dealing with.

Let a Lie algebra G admit a decomposition $G = L \oplus X$, satisfying the commutation relations

$$[L, L] \subset L, \qquad [L, X] \subset X, \qquad [X, X] \subset L, \tag{1.1}$$

where L is a semisimple subalgebra of G. Then, there exists an automorphism

$$s(z) = \begin{Bmatrix} z & \text{for} & z \in L \\ -z & \text{for} & z \in X \end{Bmatrix} \tag{1.2}$$

and s is an involution, i.e., $s^2 = 1$.

† For more detailed information, *see*, for example, the excellent book by Hermann [1].

Conversely, if s is an involution of G, then L and X can be defined by (1.2) and it is readily seen that (1.1) follows.

In terms of the corresponding Lie groups, the coset space G/L is a homogeneous symmetric space and G is the group of motions (or structural group) of this space. G acts transitively on G/L, i.e., $Gp = G/L$ for every point p of G/L.

As an example that will be of use in what follows, let us consider the rotation group of a four-dimensional space $G = SO(4)$. In this case, $L = SO(3)$ and we have the well-known commutation relations

$$\begin{aligned} &[l_\alpha, l_\beta] = i\varepsilon_{\alpha\beta\lambda} l_\lambda, \qquad [x_\alpha, l_\beta] = i\varepsilon_{\alpha\beta\lambda} x_\lambda, \\ &[x_\alpha, x_\beta] = i\varepsilon_{\alpha\beta\lambda} l_\lambda \qquad (1 \leqslant \alpha, \beta, \lambda \leqslant 3), \end{aligned} \tag{1.3}$$

where $l_\alpha \in L$ are 'angular momenta' and $x_\alpha \in X$ are 'displacements'. Note that

$$\dim G = \dim L + \dim X = 3 + 3 = 6$$

and G/L is the 3-sphere.

In an analogous way, the Lorentz group $SO(3, 1)$ may be decomposed.

2 Contraction of Lie groups

Let us consider an operation that permits one to obtain a new and more degenerate group from a given one. This means that the commutators of the new Lie algebra are the same, except some of them which reduce to zero. This operation of 'contraction' of one Lie group to another is not always possible because of the Jacobi identities. The theory of contractions was developed by Inonu and Wigner [2] and others (*see*, for example, [1]). We shall confine ourselves here to a simple example.

Let $G = SO(4)$ and introduce in (1.3) the new basis of the Lie algebra:

$$l_\alpha \to l_\alpha, \qquad x_\alpha \to a^{-1} x_\alpha. \tag{2.1}$$

Then, instead of (1.3), we have

$$[l_\alpha, l_\beta] = i\varepsilon_{\alpha\beta\lambda} l_\lambda, \qquad [x_\alpha, l_\beta] = i\varepsilon_{\alpha\beta\lambda} x_\lambda, \qquad [x_\alpha, x_\beta] = a^2 i\varepsilon_{\alpha\beta\lambda} l_\lambda. \tag{2.2}$$

Thus, for $a \neq 0$ we have a sequence of isomorphic Lie groups. But when $a \to 0$ the group G contracts to the Euclidean group of rigid motions G_0 in three-dimensional vector space.

In an analogous way, we may consider the contraction of the Lorentz group $SO(3, 1)$ to the (proper) Galilean group isomorphic to G_0.

The groups $SO(4)$, $SO(3, 1)$ and G_0 form a typical triplet in which the contracted group G_0 lies between compact and non-compact groups.

The theory of contractions permits one to know all contractions of a given Lie algebra, and, conversely, in how many ways a given Lie algebra can be exhibited as a contraction.

We would like, however, to go in the opposite direction. Starting from a given Lie group and its representations, we want to construct a new deformed Lie group and its representations in an explicit way. We shall call this operation the group deformation or quantization (a justification of the last term will be given below). Unfortunately, the theory of contractions cannot help us in the group deformation, therefore we have to develop a quite different approach.

3 Initial mathematical model

Let $G = L \oplus X$ be a Lie algebra with a semisimple subalgebra L and the commutation relations

$$[L, L] \subset L, \qquad [L, X] \subset X, \qquad [X, X] \subset (0),$$
$$\dim G = \dim L + \dim X = m + n. \tag{3.1}$$

Thus, in this case, G/L is an n-dimensional vector space and the group G is the semidirect product $G = L \otimes X$.

In a more explicit form,

$$[l_i, l_j] = c_{ijk} l_k, \qquad [l_i, x_\alpha] = c_{i\alpha\beta} x_\beta,$$
$$[x_\alpha, x_\beta] = 0 \qquad (1 \leqslant i, j, k \leqslant m, \qquad 1 \leqslant \alpha, \beta \leqslant n), \tag{3.2}$$

where $l_i \in L$ and $x_\alpha \in X$ are angular momenta and displacements.

Let $\Phi(X)$ be a proper functional space. It is natural to consider operators

$$x_\alpha : \varphi(x) \to x_\alpha \varphi(x),$$
$$\partial_\alpha : \varphi(x) \to \frac{\partial \varphi(x)}{\partial x_\alpha}, \qquad \varphi \in \Phi, \tag{3.3}$$

and pseudodifferential operators $f = f(x_\alpha, \partial_\beta)$.

In particular, the angular momenta l_i admit the representation

$$l_i = c_{i\alpha\beta}\, \partial_\alpha x_\beta, \tag{3.4}$$

which satisfies the commutation relations (3.2). This follows from the obvious identity

$$[\partial_\alpha, x_\beta] = \delta_{\alpha\beta} \tag{3.5}$$

and the Jacobi identities on c_{ijk} and $c_{i\alpha\beta}$.

To complete the mathematical structure of the corresponding physical theory, we should introduce the basic physical notions: observables (pseudodifferential operators), states (linear functionals on observables), etc. We shall not proceed in this direction; instead, we shall describe the situation in more abstract algebraic terms.

Let $X \ni x_\alpha$ and $K \ni k_\alpha$ be two dual n-dimensional vector spaces. Consider a $(2n+1)$-dimensional Lie algebra H with the basis

$$x_1, \ldots, x_n, \qquad k_1, \ldots, k_n, \qquad 1 \tag{3.6}$$

and the only non-trivial commutation relations (cf., (3.5))

$$[k_\alpha, x_\beta] = \delta_{\alpha\beta} \cdot 1. \tag{3.7}$$

This is the well-known Heisenberg algebra, and the corresponding nilpotent group H is the Heisenberg group.

The Weyl algebra $W(H)$ is the universal enveloping algebra of H, i.e., the free algebra with generators (3.6) and relations (3.7). It is useful to imagine $W(H)$ as an infinite number of storeys:

$$\begin{array}{ccccc} \cdots & \cdots & \cdots & \cdots & \cdots \\ x_\alpha x_\beta & & x_\alpha k_\beta & & k_\alpha k_\beta \\ & x_\alpha & & k_\alpha & \\ & & 1 & & \end{array} \tag{3.8}$$

the zeroth, first and second floors constituting the subalgebra of $W(H)$.

It is important that the representation of the group G is defined naturally on $W(H)$ and observables are tensor functions of the generators (3.6), i.e., elements of $W(H)$. In particular, the angular momenta

$$l_i = c_{i\alpha\beta} k_\alpha x_\beta \tag{3.9}$$

(cf., (3.4)) are on the second floor.

In what follows, all events will occur within the bounds of the Weyl algebra $W(H)$.

4 The deformation of *G* (quantization).

We want to realize the deformation

$$G = L \oplus X \to G' = L' \oplus X', \tag{4.1}$$

where L' is isomorphic to L, i.e., G' satisfies the commutation relations (cf., (3.2))

$$\left.\begin{array}{ll} [l'_i, l'_j] = c_{ijk} l_k, & [l'_i, x'_\alpha] = c_{i\alpha\beta} x'_\beta, \\ [x'_\alpha, x'_\beta] = a^2 c_{\alpha\beta\lambda} l'_\lambda & (a \neq 0). \end{array}\right\} \tag{4.2}$$

We shall now formulate two basic problems connected with the group deformation.

Problem I *To find l'_i and x'_α as elements of $W(H)$,*

$$l'_i = l'_i(x, k), \qquad x'_\alpha = x'_\alpha(x, k), \tag{4.3}$$

covariant with respect to G (or its subgroup).

Note that the covariance of the functions (4.3) is a mathematical equivalent to the physical condition of the isotropy (with respect to L) and homogeneity (with respect to X) of the deformation $G \to G'$.

Problem II *To find the transformation law for observables and states induced by the deformation $G \to G'$.*

Note that when Problem I has been solved, the solution of Problem II can be obtained in the way analogous to the case of elementary quantum mechanics (Weyl's formula). For this reason, in what follows, we shall consider only Problem I.

5 The space quantization

To introduce the elementary length a in the quantum field theory, Snyder [3] (*see also* [4, 5]) postulated that the momentum space (or k-space in our notation) is the 3-sphere with radius a^{-1}. As a consequence, he obtained the following expressions for the angular momentum and the coordinate operators:

$$l'_\alpha = l_\alpha = \varepsilon_{\alpha\beta\lambda} x_\beta k_\lambda, \qquad x'_\alpha = x_\alpha + a^2 k_\alpha k_\lambda x_\lambda. \tag{5.1}$$

These operators satisfy the commutation relations (2.2) for

$$G' = SO(4).$$

Let us introduce the basic assumption that the space quantization is equivalent to the group deformation

$$G = SO(3) \otimes X \to G' = SO(4). \tag{5.2}$$

We shall consider the solution of Problem I briefly for this case.

Let us assume, for simplicity, that $l'_\alpha = l_\alpha$ and formulate explicitly the requirements on $x'_\alpha = x'_\alpha(x, k)$:

1. The covariance with respect to G

$$\Rightarrow x'_\alpha = R_{\alpha\beta}(k) x_\beta \tag{5.3}$$

where

$$R_{\alpha\beta}(k) = f\delta_{\alpha\beta} + g k_\alpha k_\beta + h \varepsilon_{\alpha\beta\lambda} k_\lambda \tag{5.4}$$

and f, g, h are functions of $k^2 = k_\lambda k_\lambda$.

2. The correspondence principle (the existence of the limit $x'_\alpha \to x_\alpha$)

$$\Rightarrow f(0)=1. \tag{5.5}$$

Note that the last requirement means that as $k \to 0$ (long wave approximation in physical terms) the group G' contracts to the Euclidean group of motions G and $x' \to x$.

From the commutation relations for $G' = SO(4)$ and (5.5), we can obtain conditions on functions f, g, h:

$$h=0, \; 2ff' + (2k^2 f' - f)g + 1 = 0 \tag{5.6}$$

with the 'initial' condition (5.5).

The substitution $k_\alpha \to ak_\alpha$ leads to the following property of the dependence of x'_α on the elementary length a:

$$a \to 0 \Rightarrow x'_\alpha \to x_\alpha. \tag{5.7}$$

This is a full equivalent to the correspondence principle.

Let us consider two particular solutions of (5.6).

1. $f=1, \qquad g=1 \Rightarrow$ Snyder's model. (5.8)
2. $f=\sqrt{(1-a^2k^2)} \equiv \rho^2, \qquad g=0 \Rightarrow x'_\alpha = \rho x_\alpha \rho$ (5.9)

The latter case admits a simple interpretation: the initial vector k-space contracts into the ball $k^2 \leqslant a^{-2}$ with an induced new measure proportional to $\rho(k)$. Note that in this case the solution of Problem II is particularly easy.

Thus, we have many possible realizations for space quantization and it seems that, from the physical point of view, all of them must be equivalent.

Let us mention some distinguishing features of the quantized space. The spectrum of the coordinate operator x'_α (as well as of l'_α) is discrete. For the corresponding eigenvalues, we have

$$\lambda_n = na, \tag{5.10}$$

i.e., the interval between two neighbouring eigenvalues is equal to the elementary length.

Because x'_1 and x'_2 do not commute, it is impossible for them to take exact values in the same state. We have the following analogy of the uncertainty principle:

$$\Delta x'_1 \cdot \Delta x'_2 \geqslant a^2 \langle l'_3 \rangle. \tag{5.11}$$

These features justify the application of the term quantization to the group deformation.

Now a remark about the physical validity of this scheme of the space quantization. The metric of the classical Newtonian world is not Euclidean but rather conformally Euclidean. As a consequence, the space quantization has to be closely connected with the scale group

$$\varphi(x) \to \varphi(e^{\alpha} x) \qquad (-\infty < \alpha < \infty). \tag{5.12}$$

The corresponding generator $x_\lambda \, \partial_\lambda$ should be included in G and G', which is not the case here. But the development of a more correct space quantization exceeds the scope of this paper.

Note that the deformation of the Galilean group to the Lorentz group $SO(3, 1)$ may be performed, *mutatis mutandis*, in an analogous way.

6 The transition from classical to quantum mechanics as a group deformation

Let $X = Q \oplus P$ be the classical phase-space with $2n$ coordinates

$$x_\alpha = \begin{cases} q_\alpha & \text{for} \quad 1 \leqslant \alpha \leqslant n, \\ p_\alpha & \text{for} \quad n+1 \leqslant \alpha \leqslant 2n. \end{cases} \tag{6.1}$$

The geometry of X is defined by the symplectic group $Sp(n)$ that has dimension $n(2n+1)$.

For the same reason as above, we shall add to X and $Sp(n)$ the scale group

$$\varphi(x) \to \varphi(e^{\alpha} x) \tag{6.2}$$

Then, for the initial group G, we have

$$\dim G = 2n + n(2n+1) + 1 = 2n^2 + 3n + 1. \tag{6.3}$$

To find the deformed group G' of the same dimension and possessing a subgroup isomorphic to $Sp(n)$, let us consider the group $Sp(n+1)$ and its subgroup $St(n)$, preserving a $(2n+2)$-dimensional vector. We have

$$\dim St(n) = \dim Sp(n+1) - (2n+2) = 2n^2 + 3n + 1. \tag{6.4}$$

The comparison with (6.3) shows that we may assume $G' = St(n)$.

It may be shown [6] that the corresponding Lie algebra admits the decomposition

$$G' = L \oplus H, \tag{6.5}$$

where L is the Lie algebra of the symplectic group $Sp(n)$ and H is the Heisenberg algebra. It is well known [6], however, that this is exactly the algebra that governs quantum mechanics. Thus, the usual quantization may be described in terms of the group deformation $G \to G'$.

Explicit formulae for generators of $G' = St(n)$ can be obtained in the same way as was done above. In this case, however, for good physical reasons, we require the covariance with respect to a subgroup of G only.

7 Concluding remarks

1. The quantizations may be considered from the group-theoretic point of view as a special kind of induced representation of groups.
2. It is interesting to consider some one-parameter quantizations that have not been investigated before, e.g., classical electrodynamics with charge quantization.
3. The most non-trivial results may be expected when investigating multi-parameter quantizations with parameters of different dimensions such as length, velocity, action, charge, etc.

References

1 Hermann, R., *Lie Groups for Physicists*, Benjamin, New York, 1966.
2 Inonu, E. and Wigner, E. P., *Proc. Natl Acad. Sci.*, **39,** 510, 1953.
3 Snyder, H., *Phys. Rev.*, **71,** 38, 1974; **72,** 68, 1947.
4 Kadishevski, V., *J. Exp. and Theor. Phys.*, **41,** 1885, 1961.
5 Golfand, Y., *J. Exp. and Theor. Phys.*, **43,** 256, 1962.
6 Kirillov, A., *Elements of Representation Theory*, Nauka, Moscow, 1972 (in Russian).

Professor I. A. Kunin,
Chair of Mathematics
Electrotechnical Institute,
ul. K Marksa 20,
630087 Novosibirsk,
USSR

V. D. Kupradze

On dynamic and related problems of elasticity

In recent years, interest in three-dimensional problems of elasticity has notably increased.

Primarily, this increase of interest has been concerned with dynamic problems and especially those which, besides hyperbolic equations of elasticity, contain other types of differential equation, mostly parabolic ones.

Problems of thermoelasticity and elastothermodiffusion are typical examples of such problems.

At the first of these Symposia (Lecce, Italy, 1975), a lecture by Professor W. Nowacki was devoted to these and other related problems of mechanics [7].

The methods for solving such complex problems in the general case have until recently been unknown.

Lions and Raviart [6], Dafermos [1] and Fichera [2] have obtained their solutions in the last decade. These authors use functional methods, prove the existence of weak solutions and derive the existence of the classical solution from the familiar embedding theorems on the basis of the Hilbert-space technique.

In 1971, Kupradze and Burchuladze [3] solved the dynamical problems of elasticity by the methods of the potential theory and singular integral equations; later, it was shown that the same method also gives the solutions for thermoelasticity and elastothermodiffusion.

In some cases, the new approach allows us to prove the existence of classical solutions under more general conditions than any other method and, what is more important, leads to an explicit form of solutions suitable for numerical computations.

The solution technique in all cases is the same as in the purely elastic case; that is why we shall consider one of the initial-boundary problems of elasticity in detail; the results may be easily extended to other problems.

1 First initial boundary-value problem of elasticity [3, 5]

Notations $\mathscr{D}$ is a finite three-dimensional domain, occupied by a homogeneous, isotropic, elastic body with the Lamé constants λ and μ, S is the boundary of $\mathscr{D}$, Q_T is the four-dimensional cylinder $\{(x, t): x \in \mathscr{D}, t \in [0, T)\}$, $S_T \equiv \{(x, t): x \in S, t \in [0, T)\}$ is a lateral surface of Q_T.

We seek a vector $u(x, t)$, regular in Q_∞, satisfying the conditions

$$\left.\begin{aligned}
&\text{(a)}\quad L(\partial_x)u(x, t) - \frac{\partial^2 u(x, t)}{\partial t^2} = -h(x, t),\\
&\qquad (L(\partial_x) - \text{Lamé operator}\\
&\text{(b)}\quad \lim_{t\to+0} u(x, t) = \varphi^{(0)}(x), \quad \lim_{t\to+0} \frac{\partial u(x, t)}{\partial t} = \varphi^{(1)}(x);\\
&\text{(c)}\quad \lim_{x\to y\in S} u(x, t) = f(y, t);\\
&\text{(d)}\quad \left|\frac{\partial^i u(x, t)}{\partial x_1^{i_1} \partial x_2^{i_2} \partial x_3^{i_3} \partial t^{i_4}}\right| < Ce^{\sigma_0 t},\ i = 0, 1, 2;\ \sigma_0 = \text{const.} \geqslant 0 \text{ for large } t.
\end{aligned}\right\} \quad (1)$$

We assume that the 'data' satisfy the conditions of two types: *smoothness* and *compatibility*.

Smoothness conditions

1°. There exist derivatives

$$\left.\begin{aligned}
&\frac{\partial^{p+q} h(x, t)}{\partial x_1^{p_1} \partial x_2^{p_2} \partial x_3^{p_3} \partial t^q}, \qquad p_1 + p_2 + p_3 = p,\ 0 \leqslant p + q \leqslant 4;\\
&2°.\ \varphi^{(0)}(x) \in C^7(\bar{\mathscr{D}}),\ \varphi^{(1)}(x) \in C^6(\bar{\mathscr{D}});\\
&3°.\ f(\cdot, t) \in C^{2,\alpha}(S),\ \alpha > 0; \qquad f(y, \cdot) \in C^7[0, \infty);\\
&4°.\ S \subset \Lambda_2(\beta),\ \beta > 0.
\end{aligned}\right\} \quad (2)$$

Compatibility conditions

$$\left.\begin{aligned}
&1°.\ \left(\frac{\partial^m f(y, t)}{\partial t^m}\right)_{t=0} = \varphi^{(m)}(y), \qquad m = 0, 1, \ldots 5,\\
&\text{where}\\
&\varphi^{(m)}(x) = L(\partial_x)\varphi^{(m-2)}(x) - \left(\frac{\partial^{m-2} h(x, t)}{\partial t^{m-2}}\right)_{t=0}, \qquad m = 2, \ldots 5.
\end{aligned}\right\} \quad (3)$$

Basic Theorem *Under the conditions* (2) *and* (3), *the problem has a regular solution* $u \in C^1(\bar{\mathscr{D}} \cap C^2(\mathscr{D}))$; *it is unique and represented by the Laplace integral*

$$u(x, t) = \frac{1}{2\pi i} \int_{\sigma - i\infty}^{\sigma + i\infty} e^{\tau t} v(x, \tau)\, d\tau,$$

where $\tau=\sigma+i\omega$ *and* $\sigma\geqslant\sigma_0'>\sigma_0$; $v(x,\tau)$ *is a uniformly convergent series and its approximate value is found by solving a finite system of linear algebraic equations.*

The proof will be outlined below. First of all, the existence of the vector function $s(x, t)$ satisfying the conditions

1°. $\left(\dfrac{\partial^m s(x,t)}{\partial t^m}\right)_{t=0}=\varphi^{(m)}(x), \qquad m=0,1,\ldots 5,$

2°. $s(x, t)=0$ for large values of t,

is proved.

Further, we introduce the difference

$$u_0(x,t)=u(x,t)-s(x,t),$$

for which we obtain a similar initial boundary-value problem with the distinction that now the initial conditions are trivial (equal to zero), and the right-hand side of the equation (we denote it by $h_0(x, t)$) and the boundary value (we denote it by $f_0(y, t)$) satisfy the conditions

$$\left.\begin{aligned}&\left(\frac{\partial^m h_0(x,t)}{\partial t^m}\right)_{t=0}=0, \qquad m=0,\ldots 3;\ x\in\bar{\mathscr{D}},\\ &\left(\frac{\partial^m f_0(y,t)}{\partial t^m}\right)_{t=0}=0, \qquad m=0,\ldots 5;\ y\in S.\end{aligned}\right\} \tag{4}$$

The homogeneity of the initial conditions and Eqns (4) caused by the introduction of the vector function $s(x, t)$ play an important role. These very properties, together with the smoothness and compatibility conditions, allow us to show that the boundary-value problem

$$L(\partial_x)\tilde{u}_0(x,\tau)-\tau^2\tilde{u}_0(x,\tau)=\tilde{h}_0(x,\tau), \quad \lim_{x\to y\in S}\tilde{u}_0(x,\tau)=\tilde{f}_0(y,\tau),$$

where

$$\tilde{h}_0(x,\tau)=\int_0^\infty e^{-\tau t}h_0(x,t)\,dt, \qquad \tilde{f}_0(y,\tau)=\int_0^\infty e^{-\tau t}f_0(y,t)\,dt$$

for $\sigma\geqslant\sigma_0'>\sigma_0$, is soluble; the solution is regular, unique and an analytic function of τ which, for large values of $|\tau|$, satisfies, uniformly with respect to x, the inequalities

$$|\tilde{u}_0(x,\tau)|\leqslant c\,|\tau|^{-4}, \qquad \left|\frac{\partial\tilde{u}_0(x,\tau)}{\partial x_i}\right|\leqslant c\,|\tau|^{-8/3}, \qquad i=1,2,3;\ x\in\bar{\mathscr{D}},$$

$$\left|\frac{\partial^2\tilde{u}_0(x,\tau)}{\partial x_i\,\partial x_j}\right|\leqslant c\,|\tau|^{-10/9}, \qquad i,j=1,2,3;\ x\in\bar{\mathscr{D}}_*\subset\mathscr{D}. \tag{5}$$

Now, we define the vector $u_0(x, t)$ by the integral equation

$$\int_0^\infty e^{-\tau t} u_0(x, t)\, dt = \tilde{u}_0(x, \tau). \tag{6}$$

In view of the properties of $\tilde{u}_0(x, \tau)$, it is possible to invert Eqn (6),

$$u_0(x, t) = \frac{1}{2\pi i} \int_{\sigma - i\infty}^{\sigma + i\infty} e^{\tau t} \tilde{u}_0(x, \tau)\, dt,$$

and the direct verification, by means of inequalities (5), shows that the required solution of problem (1) is

$$u(x, t) = \frac{1}{2\pi i} \int_{\sigma - i\infty}^{\sigma + i\infty} e^{\tau t} \tilde{u}_0(x, \tau)\, d\tau + s(x, t);$$

thus, the assertion about the existence and uniqueness is proved.

To prove the assertion about the approximate calculation of the solution, we introduce the vector

$$v(x, \tau) = \int_0^\infty e^{-\tau t} u(x, t)\, dt. \tag{7}$$

Due to (5), the integral is invertible and

$$u(x, t) = \frac{1}{2\pi i} \int_{\sigma - i\infty}^{\sigma + i\infty} e^{\tau t} v(x, \tau)\, d\tau. \tag{8}$$

Moreover, the properties of $u(x, t)$ allow us to show that integral (7) is the solution of the boundary problem

$$L(\partial_x) v(x, \tau) - \tau^2 v(x, \tau) = \mathscr{F}(x, \tau), \qquad \lim_{x \to y \in S} v(x, \tau) = F(y, \tau), \tag{9}$$

where

$$\mathscr{F}(x, \tau) = -\varphi^{(1)}(x) - \tau \varphi^{(0)}(x) - \int_0^\infty e^{-\tau t} h(x, t)\, dt,$$

$$F(y, \tau) = \int_0^\infty e^{-\tau t} f(y, t)\, dt.$$

To solve this problem (approximately), it is sufficient to solve a finite system of linear algebraic equations.

In fact, for any regular solution of (9), we have

$$\forall x \in \mathscr{D}: \qquad 2v(x, \tau) = \int_S \Gamma(x - y, i\tau) g(y, \tau)\, d_y S - \Phi(x, \tau), \tag{10}$$

$$\forall x \bar{\in} \mathscr{D}: \qquad 0 = \int_S \Gamma(x - y, i\tau) g(y, \tau)\, d_y S - \Phi(x, \tau), \tag{11}$$

where $\Gamma(x-y, i\tau)$ is the fundamental matrix for the operator $L(\partial_x)-\tau^2E$; it is written out explicitly; moreover, we have designated

$$g(y, \tau)=T(\partial_y)v(y, \tau) \qquad (T(\partial_y) \text{ is the stress operator}),$$

$$\Phi(x, \tau)=\int_S (T(\partial_y)\Gamma(x-y, i\tau))'F(y, \tau)\,d_yS-\int_{\mathscr{D}} \Gamma(x-y, i\tau)\mathscr{F}(y, \tau)\,dy.$$

Let S_1 be an arbitrary closed surface enclosing S and let $\{x^{(k)}\}_{k=1}^{\infty}$ be a countable set of points on it. The set $\{\Gamma^p(x^{(k)}-y, i\tau)\}_{k=1}^{\infty}$, $p=1, 2, 3$, is linearly independent and complete in $\mathscr{L}_2(S)$. We number its elements, for example, as

$$\psi^{(k)}(y, \tau)=\Gamma^{l_k}(x^{[(k+2)/3]}-y, i\tau), \qquad k=1, 2, \ldots, \tag{12}$$

where $l_k=k-3[(k-1)/3]$ and $[k]$ is the greatest integral part of k.

Define $a_n(\tau)$, $n=1, 2, \ldots N$, from the minimum of the $L_2(S)$ norm:

$$\left\|g(y, \tau)-\sum_{n=1}^{N} a_n(\tau)\psi^{(n)}(y\,\tau)\right\|_{\mathscr{L}_2(S)}=\min;$$

then

$$\sum_{n=1}^{N} a_n(\tau)(\psi^{(n)}, \psi^{(k)})=(\psi^{(k)}, g), \qquad k=1, \ldots N,$$

where $(\psi^{(n)}, \psi^{(k)})$ is the scalar product in $\mathscr{L}_2(S)$, or, in view of (12),

$$\sum_{n=1}^{N} a_n(\tau)(\psi^{(n)}, \psi^{(k)})=\int_S \Gamma^{l_k}(x^{[(k+2)/3]}-y, i\tau)g(y, \tau)\,d_yS.$$

But, from (11), we obtain

$$\int_S \Gamma^l(x^{(k)}-y, i\tau)g(y, \tau)\,d_yS=\Phi_l(x^{(k)}, \tau), \qquad l=1, 2, 3;\ k=1, 2, \ldots;$$

hence, for $a_n(\tau)$, we obtain the system of linear equations

$$\sum_{n=1}^{N} a_n(\tau)(\psi^{(n)}, \psi^{(k)})=\Phi_{l_k}(x^{[(k+2)/3]}, \tau) \qquad k=1, \ldots N.$$

Since the determinant of the system is Gram's determinant, it is uniquely soluble.

Moreover, it is known that, in $\mathscr{L}_2(S)$,

$$g(y, \tau)=\lim_{N\to\infty} \sum_{n=1}^{N} a_n(\tau)\psi^{(n)}(y, \tau).$$

Substitute into (10) the sum $\sum_{n=1}^{N} a_n(\tau)\psi^{(n)}(y, \tau)$ instead of $g(y, \tau)$ and denote the result by $v_N(x, \tau)$. It is proved that for an arbitrary positive ε there exists a positive integer $N_0(\varepsilon)$, such that for $N>N_0(\varepsilon)$, at any point

$x \in \mathscr{D}$, we have $|v_N(x, \tau) - v(x, \tau)| < \varepsilon$, where $v(x, \tau)$ is the exact solution. Substituting $v(x, \tau) \doteq v_N(x, \tau)$ in (8), we get all assertions of the theorem.

2 The initial boundary-value problem of thermoelasticity [5]

In this case, the unknowns are the displacement vector $u(x, t)$ and the temperature $u_4(x, t)$. Thus, the four-component vector $\mathscr{U} \equiv (u, u_4)$ is sought, which is regular in Q_∞ and satisfies the conditions

$$\text{(a)}\quad L(\partial_x)u(x, t) - \gamma \operatorname{grad} u_4(x, t) - \frac{\partial^2 u(x, t)}{\partial t^2} = -h(x, t),$$

$$\forall (x, t) \in Q_\infty:\quad \Delta u_4(x, t) - \kappa \frac{\partial u_4(x, t)}{\partial t} - \eta \frac{\partial}{\partial t} \operatorname{div} u(x, t) = -h_4(x, t);$$

$$\text{(b)}\quad \forall x \in \mathscr{D}: \lim_{t \to +0} u(x, t) = \varphi^{(0)}(x), \qquad \lim_{t \to +0} \frac{\partial u(x, t)}{\partial t} = \varphi^{(1)}(x),$$

$$\lim_{t \to +0} u_4(x, t) \equiv \varphi_4^{(0)}(x);$$

$$\text{(c)}\quad \forall y \in S: \lim_{x \to y} u(x, t) = f(y, t), \qquad \lim_{x \to y} u_4(x, t) = f_4(y, t);$$

(d) the behaviour in t is the same as in the previous problem.

In the proof of the basic theorem, we did not use the fact that the operator $L(\partial_x)$ is self-adjoint; this enables us to extend the method to problems which are not self-adjoint. The differential operator of thermoelasticity is not self-adjoint. Applying the arguments of the previous section appropriately, we get an analogous theorem for the initial boundary value problem of thermoelasticity.

The smoothness and compatibility conditions coincide, essentially, with (2) and (3) [5]; the dynamic problems of couple-stress thermoelasticity are investigated similarly.

The problem of thermoelasticity was studied in [1]. To obtain the classical solution, this paper adopts more rigid assumptions than (2), (3).

3 The initial boundary-value problem of thermodiffusion in an elastic medium

The equations of thermodiffusion in an elastic medium were obtained in [8]. In this case, the unknowns are: elastic displacement, temperature and a parameter characterizing the diffusing substance, called the chemical potential.

It is required that we find a five-component vector $\mathscr{U}(x, t) \equiv (u, u_4, u_5)$, regular in Q_∞, satisfying the conditions

(a) $L(\partial_x)u(x,t)-\gamma \operatorname{grad} u_4(x,t)-\beta \operatorname{grad} u_5(x,t)-\dfrac{\partial^2 u(x,t)}{\partial t^2}=-h(x,t),$

$$\forall (x,t)\in Q_\infty:\quad k_1\Delta u_4(x,t)-c_1\frac{\partial u_4(x,t)}{\partial t}-e\frac{\partial u_5(x,t)}{\partial t}-\gamma\frac{\partial}{\partial t}\operatorname{div} u =-h_4(x,t),$$

$$k_2\Delta u_5(x,t)-c_2\frac{\partial u_5(x,t)}{\partial t}-e\frac{\partial u_4(x,t)}{\partial t}-\beta\frac{\partial}{\partial t}\operatorname{div} u=-h_5(x,t);$$

(b) $\forall x\in \mathscr{D};\ t\to +0$: $\lim u(x,t)=\varphi^{(0)}(x)$, $\quad \lim \dfrac{\partial u(x,t)}{\partial t}=\varphi^{(1)}(x)$,

$\lim u_4(x,t)=\varphi_4^{(0)}(x)$, $\quad \lim u_5(x,t)=\varphi_5^{(0)}(x)$;

(c) $\forall y\in S;\ x\to y$: $\lim u(x,t)=f(y,t)$, $\quad \lim u_4(x,t)$

$\lim u_5(x,t)=f_5(y,t)$;

(d) the behaviour in t is the same as in the previous problems.

The smoothness and compatibility conditions are similar to (2), (3).

The differential operator of the problem is not self-adjoint; using the procedure described earlier appropriately, we obtain a theorem similar to the basic one (T. Burchuladze).

This problem was studied by G. Fichera [2] for anisotropic media, using the Laplace transforms together with the technique of generalized solutions, Fichera proved the existence and uniqueness of the solution and obtained important integral estimates.

The existence of the classical solution is proved in [2] under more rigid conditions than (2), (3).

References

1 Dafermos, C. M., On the existence and the asymptotic stability of solutions to the equations of linear thermoelasticity, *Arch. Rat. Mech. Anal.*, **29,** 241–271, 1968.

2 Fichera, G., Uniqueness, existence and estimate of the solution in the dynamical problem of thermodiffusion in an elastic solid. *Arch. Mech. Stosowanej*, **26,** 903–920, 1974.

3 Kupradze, V. and Burchuladze, T., On dynamic problems of the theory of elasticity, in *Trends in Elasticity and Thermoelasticity*, W. Nowacki Anniversary Volume, Wolters-Nordhoff Publ., Amsterdam, 1971, pp. 135–149.

4 Kupradze, V. and Burchuladze, T., Dynamic problems in the theory of elasticity and thermoelasticity, *Surveys in Science and Technology. Contemporary Problems in Mathematics.* Moscow 1975, Vol. 7.

5 Kupradze, V., Gegelya, T., Basheleyshuili, M. and Burchuladze, T., *Three-dimensional Problems of the Mathematical Theory of Elasticity and Thermoelasticity*, 2nd ed. Moscow, 1976.

6 Lions, J. and Raviart, P., Remarques sur la résolution et l'approximation d'équations d'évolutions couplées, *I.C.C. Bulletin S*, 1966.

7 Nowacki, W., Coupled fields in elasticity, in *Trends in Applications of Pure Mathematics to Mechanics* (G. Fichera, ed.), Pitman, London, 1976, pp. 263–280.

8 Podstrichac, Y., Differential equations of the problem of thermodiffusion in isotropic deformable solids., *Dokl. Acad. Nauk USSR*, **2,** 169–172, 1961.

Professor V. D. Kupradze,
Institute of Mathematics,
ul. Z. Rukhadze 1,
Tbilisi-15, 380093,
USSR.

R. Leis

Exterior boundary-value problems in mathematical physics

Exterior boundary-value problems arise in acoustics, linear elasticity, electromagnetic theory, etc. In my opinion the theory of electromagnetic waves seems to be the most interesting example. Thus, in this paper, I shall try to show the problematic nature of exterior boundary-value problems by developing a theory of electromagnetic waves.

We start by introducing the general time-dependent Maxwell equations for inhomogeneous anisotropic media. Since the intention is to give a Hilbert-space theory, several notations regarding solutions in the weak sense are needed. We will then investigate the time harmonic equations. The damped case can very easily be handled; the undamped case, however, is more difficult. First, we deal with boundary and eigenvalue problems for bounded domains. Afterwards, we will solve exterior boundary-value problems using the method of limiting absorption. In order to apply this method, *a priori* estimates have to be derived, and, in order to prove uniqueness, the principle of unique continuation has to be verified.

Having made these preparations, the general time-dependent case (initial-value problems) may be handled. The spectral theorem for the Maxwell operator has to be formulated first; initial-value problems for bounded domains can be solved afterwards; and, finally, some aspects of a general theory of wave propagation will be given.

The time-dependent Maxwell equations read as follows

$$\left.\begin{aligned}&\left(\varepsilon\frac{\partial}{\partial t}+\sigma\right)\mathbf{E}-\operatorname{rot}\mathbf{H}=\mathbf{J},\\&\mu\frac{\partial}{\partial t}\mathbf{H}+\operatorname{rot}\mathbf{E}=\mathbf{K}.\end{aligned}\right\}\tag{1}$$

To this we add the boundary and initial conditions

$$\mathbf{n}\times\mathbf{E}\mid\partial G=0,\qquad \mathbf{E}(\mathbf{x},0)=\mathbf{E}^0,\qquad \mathbf{H}(\mathbf{x},0)=\mathbf{H}^0.$$

Here, G is the underlying domain; $\mathbf{E}$, $\mathbf{H}$ are electric and magnetic field forces, respectively; $\mathbf{J}$, $\mathbf{K}$ are incoming fields; ε is the dielectric constant,

μ is the permeability and σ the electric conductivity. In the general case of anisotropic inhomogeneous media, ε and μ are positive definite matrices, and $\sigma \geqslant 0$.

It is relatively easy to prove uniqueness for this initial-value problem. One has to discuss the energy of the process and to show

$$\mathscr{F} = \int_G \{\mathbf{E}'\varepsilon\mathbf{E} + \mathbf{H}'\mu\mathbf{H}\}\,\mathrm{d}\mathbf{x} = \text{const}.$$

We are not going into details here.

By comparison with other equations of mathematical physics (wave equation, linear elasticity equations), additional problems arise for Maxwell's equations. The reason is that Maxwell's equations split up in the time-independent (stationary) case, and one is no longer led to an elliptic boundary-value problem. The stationary Maxwell's equations read

$$\sigma\mathbf{E} - \operatorname{rot}\mathbf{H} = \mathbf{J}, \qquad \operatorname{rot}\mathbf{E} = \mathbf{K}.$$

The decomposition when $\sigma = 0$ is immediately observed. In this case, $\mathbf{H}$ is not 'reasonably' defined any more, and, for $\mathbf{E}$ as well, no elliptic equation exists. Therefore, existence theorems of the theory of elliptic differential equations cannot be carried over without further considerations. In particular, not all first derivatives of the solutions can be estimated; thus, a compactness criterion (Rellich's selection theorem) is missing. In contrast to elliptic equations, there are solutions for domains with non-smooth boundaries, the first derivatives of which are not square-integrable.

In what follows, we shall deal with these difficulties. First, we shall treat the time-harmonic case and afterwards the time-dependent case. Unbounded domains G are admitted as well (exterior boundary-value problems).

1 Notations

As usual, let $L_2(G)$ be the Hilbert space of square-integrable functions or fields with the scalar product

$$(\mathbf{E}, \mathbf{F}) = \int_G \mathbf{E}(\mathbf{x})\overline{\mathbf{F}(\mathbf{x})}$$

and the norm $\|\mathbf{E}\|^2 = (\mathbf{E}, \mathbf{F})$. Let $\mathring{C}(G)$ be the space of test functions or test fields of G, i.e., of infinitely differentiable functions with compact support in G. Generally, fields are denoted by capital letters and functions by small letters. The derivatives are defined in the weak sense; in particular, we call $\mathbf{F} := \operatorname{rot} \mathbf{E}$ and $f := \operatorname{div} \mathbf{E}$, if the following is valid:

$$(\mathbf{E}, \operatorname{rot} \mathbf{\Phi}) = (\mathbf{F}, \mathbf{\Phi}) \quad \text{for all} \quad \mathbf{\Phi} \in \mathring{C}(G),$$
$$(\mathbf{E}, \operatorname{grad} \phi) = -(f, \phi) \quad \text{for all} \quad \phi \in \mathring{C}(G),$$

respectively. Let R be the Hilbert space of L_2-fields $\mathbf{F}$ with rot $\mathbf{F} \in L_2$, the scalar product

$$(\mathbf{E}, \mathbf{F})_r = (\mathbf{E}, \mathbf{F}) + (\operatorname{rot} \mathbf{E}, \operatorname{rot} \mathbf{F})$$

and the norm $\|\mathbf{E}\|_r^2 = (\mathbf{E}, \mathbf{E})_r$. In order to formulate our boundary-value problems, we need another Hilbert space of fields belonging to R and satisfying the boundary condition. Therefore, we define $\mathring{R}$ as the completion of test fields under the $\|\ldots\|_r$-norm. For each $\mathbf{E} \in \mathring{R}$, a sequence of test fields $\{\mathbf{\Phi}_n\}$ exists such that

$$\|\mathbf{\Phi}_n - \mathbf{E}\| \to 0 \quad \text{and} \quad \|\operatorname{rot} \mathbf{\Phi}_n - \operatorname{rot} \mathbf{E}\| \to 0.$$

It is easily observed that in the case of a smooth boundary and continuously differentiable $\mathbf{E}$, $\mathbf{n} \times \mathbf{E} \mid \partial G = \mathbf{0}$ follows from $\mathbf{E} \in \mathring{R}$. Since $\mathbf{E} \in \mathring{R}$

$$(\mathbf{F}, \operatorname{rot} \mathbf{E}) = (\operatorname{rot} \mathbf{F}, \mathbf{E})$$

is valid for any $\mathbf{F} \in C_1(\bar{G})$. On the other hand, one gets (Gauss' theorem)

$$(\mathbf{F}, \operatorname{rot} \mathbf{E}) = (\operatorname{rot} \mathbf{F}, \mathbf{E}) + \int_{\partial G} \mathbf{F}\overline{(\mathbf{n} \times \mathbf{E})}.$$

Thus, the surface integral vanishes for all $\mathbf{F} \in C_1(\bar{G})$ and, therefore, $\mathbf{n} \times \mathbf{E} \mid \partial G = \mathbf{0}$ as well.

$\mathring{R}$ may be also defined in the following way:

$$\mathring{R} := \{\mathbf{E} \in R \mid (\mathbf{F}, \operatorname{rot} \mathbf{E}) = (\operatorname{rot} \mathbf{F}, \mathbf{E}) \quad \text{for all} \quad \mathbf{F} \in R\}.$$

Moreover, we set

$$R_0 := \{\mathbf{E} \in L_2 \mid \operatorname{rot} \mathbf{E} = 0\}$$
$$D_0 := \{\mathbf{E} \in L_2 \mid \operatorname{div} \mathbf{E} = 0\}$$
$$D_{\varepsilon 0} := \{\mathbf{E} \in L_2 \mid \operatorname{div} \varepsilon \mathbf{E} = 0\}.$$

Finally, let $\mathring{H}_1$ be the Hilbert space which results from the completion of test functions under the norm $\|f\|_1^2 = (f, f)_1$, where

$$(f, g)_1 = (f, g) + (\operatorname{grad} f, \operatorname{grad} g)$$

2 Formulation of the boundary-value problems

We now proceed to the time-harmonic case and, therefore, we assume that all functions appearing in the problem are proportional to $e^{-i\omega t}$. We then obtain the time-independent Maxwell equations

$$\left.\begin{aligned} -i\omega\eta\mathbf{E}-\operatorname{rot}\mathbf{H}&=\mathbf{J},\\ -i\omega\mu\mathbf{H}+\operatorname{rot}\mathbf{E}&=\mathbf{K},\end{aligned}\right\} \tag{2}$$

with $\eta=\varepsilon+i\sigma/\omega$ and the boundary condition $\mathbf{n}\times\mathbf{E}\,|\,\partial G=\mathbf{0}$. These equations also contain the stationary case ($\omega=0$).

The first existence theorems for boundary-value problems of the time-independent Maxwell equations for bounded and unbounded domains were originated by C. Müller in 1952. Müller deals with homogeneous isotropic media and uses the method of integral equations known from the potential theory. This procedure naturally requires a smooth boundary. Difficulties arise in comparison with the potential theory in view of the fact that, in the theory of Maxwell's equations, the analogy to the dipole potential exhibits a stronger singularity. The method of C. Müller was extended to R^n by H. Weyl.

If the general case of anisotropic inhomogeneous media and an arbitrary boundary is to be dealt with, the method of integral equations is no longer suitable, since it depends too strongly on the smoothness of the boundary and on the explicit knowledge of the fundamental solution. The application of Hilbert-space methods suggests itself; these methods have been developed from Dirichlet's principle. Such an application will be carried out in the following.

To begin with, let ω be non-zero. Then, $\mathbf{H}$ may be eliminated, and for $\mathbf{E}$, we obtain

$$\operatorname{rot}\mu^{-1}\operatorname{rot}\mathbf{E}-\omega^2\eta\mathbf{E}=\operatorname{rot}\mu^{-1}\mathbf{K}-i\omega\mathbf{J}=\mathbf{F}\quad\text{with}\quad\mathbf{n}\times\mathbf{E}\,|\,\partial G=\mathbf{0}. \tag{3}$$

For this equation, there exists a complete solution theory. We define

$$B(\mathbf{\Phi},\mathbf{E})=(\operatorname{rot}\mathbf{\Phi},\mu^{-1}\operatorname{rot}\mathbf{E})-\omega^2(\mathbf{\Phi},\eta\mathbf{E})$$

and we distinguish two cases.

Definition 1 *Let either G be bounded or G be unbounded and $\sigma>0$. We then call $\mathbf{E}\in\mathring{R}(G)$* a weak solution *of the boundary-value problem if, for all $\mathbf{\Phi}\in\mathring{C}(G)$,*

$$B(\mathbf{\Phi},\mathbf{E})=(\mathbf{\Phi},\mathbf{F}) \tag{4}$$

holds.

This proves to be the easier case. In the damped case, the boundary-value problems are clearly soluble, otherwise Fredholm's alternative is valid. More interesting are boundary-value problems for exterior domains with vanishing dampening, i.e., for domains with bounded complement and $\sigma = 0$. In order to simplify our considerations, we assume in this case that a R_0 exists such that in $G_a = \{\mathbf{x} \in G \mid |\mathbf{x}| > R_0\}$,

$$\mu_{ik} = \mu_0 \delta_{ik}, \qquad \varepsilon_{ik} = \varepsilon_0 \delta_{ik} \quad \text{and} \quad \mathbf{F} = \mathbf{0}$$

hold with positive numbers μ_0 and ε_0. In this case, we choose as the solution space

$$\mathring{R}^*(G) = \{\mathbf{E} \in L_2^{\text{loc}} \mid \exists \mathbf{\Phi}_n \in \mathring{C}(G) \quad \text{with} \quad \|\mathbf{\Phi}_n - \mathbf{E}\|_r (G_R) \to 0 \ \forall R\}$$

and use the abbreviation $G_R = \{\mathbf{x} \in \mathrm{G} \mid |\mathbf{x}| < R\}$. We then introduce

Definition 2 Let G be an exterior domain and $\sigma = 0$. Then we call $\mathbf{E} \in \mathring{R}^*(G)$ a weak solution *of the exterior boundary-value problem if, for all $\mathbf{\Phi} \in \mathring{C}(G)$, Eqn (4) and*

$$\|\mathbf{x}_0 \times \operatorname{rot} \mathbf{E} + i\kappa \mathbf{E}\|(G_a) < \infty \tag{5}$$

are valid, with $\mathbf{x}_0 = \mathbf{x}/|\mathbf{x}|$ and $\kappa = \omega\sqrt{(\varepsilon\mu)}$.

Equation (5) is Sommerfeld's radiation condition. We shall prove that the exterior boundary-value problem is uniquely solvable. In order to show that the solutions so defined are also solutions in the normal sense, one has to assume that the coefficients of the differential equations and $\mathbf{F}$ are sufficiently smooth and to apply regularity theorems. Let this be assumed in the following; we do not intend to enter into these problems.

If G is bounded and σ disappears, eigenvalue problems can be posed. We then introduce

Definition 3 A number λ is called an eigenvalue *of Maxwell's equations if an $\mathbf{E} \in \mathring{R} \cap D_{\varepsilon 0}$, $\mathbf{E} \neq \mathbf{0}$ exists with*

$$(\operatorname{rot} \mathbf{\Phi}, \mu^{-1} \operatorname{rot} \mathbf{E}) = \lambda (\mathbf{\Phi}, \varepsilon \mathbf{E})$$

for all $\mathbf{\Phi} \in \mathring{C}(G)$. $\mathbf{E}$ is then called the eigenfield *of λ.*

3 The damped case

In the damped case ($\sigma > 0$), one can easily prove the following estimate for an arbitrary domain:

$$|B(\mathbf{\Phi}, \mathbf{\Phi})| \geq p\, |\mathbf{\Phi}|_r \quad \text{for all} \quad \mathbf{\Phi} \in \mathring{R} \tag{6}$$

with $p > 0$. This estimate and the Riesz representation theorem, or rather its slight generalization (theorem of Lax and Milgram), imply immediately unique solubility of the boundary-value problem.

Let

$$M : L_2(G) \to \mathring{R}(G)$$

be the solution operator. Then we obtain

$$\|M\mathbf{F}\|_r \leq c\|\mathbf{F}\|$$

and

Theorem 1 *Let G be an arbitrary domain, $\sigma > 0$ and $\mathbf{F} \in L_2(G)$. Then, the boundary-value problem is uniquely soluble.*

When dealing with exterior boundary-value problems in section 7, we shall refer to this result and pass to the limit of vanishing dampening.

4 Boundary-value problems for bounded domains

Now let G be bounded and $\sigma = 0$. Then,

$$B(\mathbf{\Phi}, \mathbf{\Phi}) \geq p\ \|\mathbf{\Phi}\|_r^2 - c\ \|\mathbf{\Phi}\|^2 \tag{7}$$

holds for all $\mathbf{\Phi} \in \mathring{R}$ with positive constants p and c. It is also said that $B(\mathbf{\Phi}, \mathbf{E})$ is *coercive* over $\mathring{R}$. $B(\mathbf{\Phi}, \mathbf{E})$ is called *strongly coercive* if c can be taken to be zero. In the damped case, therefore, $B(\mathbf{\Phi}, \mathbf{E})$ is strongly coercive over $\mathring{R}$.

In the undamped case, one cannot deduce the unique solubility of our boundary-value problem. Here, Fredholm's alternative holds, as for elliptic boundary-value problems.

In order to show this, we first note that we can confine ourselves to the case $\operatorname{div} \mathbf{F} = 0$ without loss of generality. This follows from the orthogonal decomposition

$$L_2 = D_0 \oplus \overline{\operatorname{grad} \mathring{H}_1},$$

which goes back to H. Weyl and results from the definition of divergence in the weak sense.

We set

$$B_0(\mathbf{\Phi}, \mathbf{E}) = B(\mathbf{\Phi}, \mathbf{E}) + (1 + \omega^2)(\mathbf{\Phi}, \varepsilon\mathbf{E}).$$

Then, $B_0(\mathbf{\Phi}, \mathbf{E})$ is strongly coercive over $\mathring{R}$. Let M_0 be the solution operator which exists according to section 3:

$$M_0 : D_0 \to \mathring{R} \cap D_{\varepsilon 0}$$

and

$$M_0\varepsilon : D_{\varepsilon 0} \to D_{\varepsilon 0}.$$

Our boundary-value problem

$$B(\mathbf{\Phi}, \mathbf{E}) = (\mathbf{\Phi}, \mathbf{F})$$

or

$$B_0(\mathbf{\Phi}, \mathbf{E}) - (1+\omega^2)(\mathbf{\Phi}, \varepsilon\mathbf{E}) = (\mathbf{\Phi}, \mathbf{F})$$

is then equivalent to

$$\mathbf{E} - (1+\omega^2)M_0\varepsilon\mathbf{E} = M_0\mathbf{F}. \tag{8}$$

Therefore, we discuss Eqn (8) instead of Eqn (4).

First, we notice that it is sufficient to discuss this equation in $D_{\varepsilon 0}$, since $\mathbf{F} \in D_0$, $M_0F \in D_{\varepsilon 0}$. Now, let $\mathbf{E} \in D_{\varepsilon 0}$ be the solution. Then, from

$$M_0 : L_2 \to \mathring{R},$$

$\mathbf{E} \in \mathring{R}$ follows automatically.

If it is possible to show that the mapping $M_0\varepsilon : D_{\varepsilon 0} \to D_{\varepsilon 0}$ is compact, Fredholm's alternative holds for Eqn (8). One can easily see that the necessary and sufficient solution condition is $(\mathbf{L}, \mathbf{F}) = 0$ for all solutions $\mathbf{L} \in \mathring{R}$ of the homogeneous Eqn (4).

The real difficulty of our problem is the proof of the compactness of $M_0\varepsilon$. When the boundary ∂G is smooth, it can be shown that the solutions and their first derivatives are square integrable, and the selection theorem of Rellich may be used. Having a non-smooth boundary, a simple example in R^2 will show that generally the solutions do not belong to H_1 any more. Let $J_{2/3}(r)$ be the Bessel function, which is regular in the origin. Then,

$$u(r, \phi) := J_{2/3}(r) \cos 2\phi/3$$

solves the Helmholtz equation in a neighbourhood of the vertex $0 \leqslant \phi \leqslant 3\pi/2$. We define

$$\mathbf{E} = (\partial_2 u, -\partial_1 u).$$

$\mathbf{E}$ solves the homogeneous Maxwell equations in a neighbourhood of the vertex and satisfies the boundary condition. $\mathbf{E}$, $\operatorname{rot} \mathbf{E} \sim \Delta u \sim u$ and $\operatorname{div} \mathbf{E}$ are square integrable, but an arbitrary derivative is not (because of $J_{2/3}(r) \sim r^{2/3}$). The second with $N_{2/3}(r)$ analogously formed solution is itself no longer square integrable and, therefore does not appear in a L_2-theory.

In the general case, the compactness of M_0 follows from

$$\|M_0\mathbf{F}\|_r \leqslant c\,\|\mathbf{F}\|$$

and the following

Selection theorem: *Suppose* $\mathbf{E}_n \in \mathring{R}$, $\operatorname{div} \varepsilon\mathbf{E}_n \in L_2$ *and*

$$\|\operatorname{rot} \mathbf{E}_n\|^2 + \|\operatorname{div} \varepsilon\mathbf{E}_n\|^2 + \|\mathbf{E}_n\|^2 \leqslant 1.$$

Then, there exists a subsequence $\{\mathbf{E}_{n'}\}$ *and an* $\mathbf{E}\in L_2$ *with* $\|\mathbf{E}_{n'}-\mathbf{E}\|\to 0$.

This selection theorem can relatively easily be proved for bounded domains with smooth boundaries by partial integration. The proof can be carried over to domains which can be reduced to domains with smooth boundaries by reflection (for instance half-spheres). For more general domains the proof becomes difficult, and it is not yet clear in which generality the theorem is valid. So far, the most general proof was given by Weck for cone-like domains.

In the following, we always assume that for G or in the case of unbounded domains for $G\cap\{\mathbf{x}\mid |\mathbf{x}|<R_0\}$ the selection theorem is valid. Then, we have

Theorem 2 *Let* G *be bounded,* $\sigma=0$ *and* $\mathbf{E}\in L_2$. *Then, Fredholm's alternative is valid for the boundary-value problem, and the necessary and sufficient solution condition is* $(\mathbf{L},\mathbf{F})=0$ *for all solutions* $\mathbf{L}$ *of the homogeneous equation.*

If we have the selection theorem at our disposal, and, therefore, the compactness of the solution operator is clear, eigenvalue problems can be solved analogously to elliptic differential equations. We set

$$A=\varepsilon^{-1}\operatorname{rot}\mu^{-1}\operatorname{rot}$$

and $[\mathbf{F},\mathbf{G}]=(\mathbf{F},\varepsilon\mathbf{G})$. Then, we obtain

Theorem 3 *The eigenvalue problem* $A\mathbf{E}=\tau\mathbf{E}$ *has a countably infinite set of eigenvalues* $\tau_j\to\infty$ *and the corresponding eigenfields* $\mathbf{E}_j\in\mathring{R}\cap D_{\varepsilon 0}$, *where*

$$A\mathbf{E}_j=\tau_j\mathbf{E}_j,\qquad \tau_j\geqslant 0$$
$$[\mathbf{E}_j,\mathbf{E}_k]=\delta_{jk}.$$

Let $\mathbf{F}$ *be a field with* $\mathbf{F}\in\mathring{R}\cap D_{\varepsilon 0}$ *and* $f_j=[\mathbf{F},\mathbf{E}_j]$. *Then, the following results are valid:*

$$\left\|\mathbf{F}-\sum_{j=1}^{n} f_j\mathbf{E}_j\right\|_r\to 0,$$

$$\sum_{j=1}^{\infty} f_j^2=[\mathbf{F},\mathbf{F}],$$

$$\sum_{j=1}^{\infty}\tau_j f_j^2=(\operatorname{rot}\mathbf{F},\mu^{-1}\operatorname{rot}\mathbf{F}).$$

If, for $\tau_j\neq 0$, we define

$$\mathbf{H}_j=-i\mu^{-1}\operatorname{rot}\mathbf{E}_j/\sqrt{\tau_j},$$

then, for all $F \in \mu^{-1} \operatorname{rot} \mathring{R}$, we obtain

$$\mathbf{F} = \sum_{\tau_j > 0} (\mathbf{F}, \mu \mathbf{H}_j) \mathbf{H}_j$$

because, for $\tau = 0$, we have $\operatorname{rot} \mathbf{E}_0 = 0$. Thus, the zero eigenvalue does not appear.

5 The principle of unique continuation

In order to prove a uniqueness theorem for exterior boundary-value problems, we need the principle of unique continuation of solutions. Such a statement is obvious for analytic functions; it is, however, remarkable that it even holds for solutions of equations with non-analytic coefficients. The principle is valid, rather generally, for differential equations of second order. For equations of higher order or systems of equations, there are counter-examples.

In order to prove the principle of unique continuation for Maxwell's equations, we use an estimate which is due to Protter.

Let $a_{ik}(\mathbf{x}) \in C_2(G)$ be a positive definite matrix and

$$Lu = \sum_{i,k} a_{ik} \, \partial_i \, \partial_k u.$$

Let $u \in C_1(G)$ be twice piecewise continuously differentiable; u vanishes in the neighbourhood of the point $\mathbf{x}_0$. We then choose a sphere $K(\mathbf{x}_0, R) \subset G$ with $R < 1$, smooth u in the exterior of $|\mathbf{x} - \mathbf{x}_0| \leq R/2$ down to zero and call the smoothed function $\hat{u}$.

Let β be a positive number and

$$\phi(|\mathbf{x} - \mathbf{x}_0|) = \exp |\mathbf{x} - \mathbf{x}_0|^{-\beta}.$$

Then, there exist positive constants c and β_0 such that, for all $\beta \geq \beta_0$,

$$\beta^4 \int \frac{\phi^2 \, |\hat{u}|^2}{|\mathbf{x} - \mathbf{x}_0|^{2\beta+2}} + \beta^2 \int \phi^2 \, |\nabla \hat{u}|^2 \leq c \int \phi^2 \, |\mathbf{x} - \mathbf{x}_0|^{\beta+2} \, |L\hat{u}|^2$$

holds. Passing to the limit $\beta \to \infty$, this estimate yields the vanishing of u in $|\mathbf{x} - \mathbf{x}_0| < R/2$ if it satisfies the differential equation

$$Lu + \sum_i a_i \, \partial_i u + au = 0.$$

$u = 0$ in G then follows by a sphere-chain argument.

With regard to Maxwell's equations, one starts from $\operatorname{div} \varepsilon E = 0$ and $\operatorname{div} \mu H = 0$ and, by differentiation, gets a weakly-coupled elliptic system for E_1, E_2, E_3, H_1, H_2, H_3. By addition, an analogous estimate is obtained and we arrive at

Theorem 4 *Let* $\mathbf{E}, \mathbf{H} \in C_1(G)$ *be twice piecewise continuously differentiable solutions of the homogeneous Maxwell equations which vanish in the neighbourhood of a point* $\mathbf{x}_0 \in G$. *Let* $\varepsilon, \mu \in C_2(G)$. *Then,* $\mathbf{E}$ *and* $\mathbf{H}$ *vanish in* G.

6 The asymptotic behaviour at infinity

To prove the existence theorem for exterior boundary-value problems, we need an *a priori* estimate for the asymptotic behaviour of the solutions at infinity. Let $k=\omega\sqrt{(\mu\eta)}$, $\Phi(\mathbf{x},\mathbf{y})=e^{ik|\mathbf{x}-\mathbf{y}|}/4\pi\,|\mathbf{x}-\mathbf{y}|$ and $\mathbf{a}$ be an arbitrary vector. Then,

$$\mathbf{H}_1(\mathbf{a};\mathbf{x},\mathbf{y})=\nabla_x\times\mathbf{a}\Phi(\mathbf{x},\mathbf{y}),$$
$$\mathbf{E}_1(\mathbf{a};\mathbf{x},\mathbf{y})=i\nabla_x\times\mathbf{H}_1(\mathbf{a};\mathbf{x},\mathbf{y})/\omega\eta$$

are fundamental solutions (electric dipoles) of Maxwell's equations for homogeneous isotropic media. This follows from

Lemma 1 *Let* $\mathbf{y}$ *be fixed, then we have*

$$\operatorname{rot}\mathbf{E}_1(\mathbf{a};\mathbf{x},\mathbf{y})-i\omega\mu\mathbf{H}_1(\mathbf{a};\mathbf{x},\mathbf{y})=i\operatorname{rot}\mu^{-1}\delta(\mathbf{x}-\mathbf{y})/\omega\eta.$$

After simple calculation one obtains

Lemma 2 *Let* $\mathbf{y}$ *be fixed and* $\mathbf{x}_0=\mathbf{x}/|\mathbf{x}|$. *Then, for* $|\mathbf{x}|\to\infty$, *the following is valid uniformly in* $\mathbf{x}_0$:

$$\mathbf{H}_1(\mathbf{a};\mathbf{x},\mathbf{y})=-ik(\mathbf{a}\times\mathbf{x}_0)\Phi+O(|\mathbf{x}|^{-2})$$
$$\mathbf{E}_1(\mathbf{a};\mathbf{x},\mathbf{y})=i\omega\mu\mathbf{x}_0\times(\mathbf{a}\times\mathbf{x}_0)\Phi+O(|\mathbf{x}|^{-2})$$
$$\omega\eta(\mathbf{x}_0\times\mathbf{E}_1)-k\mathbf{H}_1=O(|\mathbf{x}|^{-2})$$
$$\omega\mu(\mathbf{x}_0\times\mathbf{H}_1)+k\mathbf{E}_1=O(|\mathbf{x}|^{-2}).$$

A representation of solutions in the whole R^3 for homogeneous isotropic media may be deduced from these lemmas.

Lemma 3 *Let* $\mathbf{J},\mathbf{K}\in\mathring{C}(R^3)$; $k=\omega\sqrt{(\mu\eta)}$ *with* $\operatorname{Re}k>0$, $\operatorname{Im}k\geq 0$ *and* $\mathbf{E}$ *a solution of Maxwell's equations with*

$$\|\mathbf{x}_0\times\operatorname{rot}\mathbf{E}+ik\,\mathbf{E}\|<\infty.$$

Then, we have

$$\mathbf{aE}(\mathbf{x})=(i/\omega\eta)\mathbf{aJ}(\mathbf{x})+\int_{R^3}\{\mathbf{K}(\mathbf{y})\mathbf{H}_1(\mathbf{a};\mathbf{y},\mathbf{x})-\mathbf{J}(\mathbf{y})\mathbf{E}_1(\mathbf{a};\mathbf{y},\mathbf{x})\}\,\mathrm{d}y,$$

$$\mathbf{a}\operatorname{rot}\mathbf{E}(\mathbf{x})=i\omega\int_{R^3}\{\mu\mathbf{J}(\mathbf{y})\mathbf{H}_1(\mathbf{a};\mathbf{y},\mathbf{x})+\eta\mathbf{K}(\mathbf{y})\mathbf{E}_1(\mathbf{a};\mathbf{y},\mathbf{x})\}\,\mathrm{d}y.$$

The required *a priori* estimates of the asymptotic behaviour of the solutions at infinity follow from Lemma 3. Since we intend later on to pass to the limit of vanishing damping, we also admit damped media and derive estimates uniform with respect to σ. Let G be an exterior domain in an inhomogeneous, anisotropic medium. In G_a, i.e., for $|\mathbf{x}|>R_0$, we assume the medium to be homogeneous and isotropic. Let $\mathbf{E}$ be the solution of the exterior boundary-value problem with $\|\mathbf{x}_0 \times \operatorname{rot} \mathbf{E} + ik\mathbf{E}\|(G_a)<\infty$. We set $k=\omega\sqrt{(\eta_0\mu_0)}$ with $\operatorname{Re} k=k_1>0$ and $\operatorname{Im} k\geqslant 0$. Then, we have

Theorem 5 *Let* $\kappa=\omega\sqrt{(\varepsilon_0\mu_0)}>0$, $c_0>0$, $R_1>R_0$. *Then, there exists a constant c and a compact set* $Z\Subset G_a$, *such that, for all solutions* $\mathbf{E}$ *for all* $\mathbf{x}$, $|\mathbf{x}|\geqslant R_1$ *and for all* σ_0 *with* $0\leqslant\sigma_0\leqslant c_0$, *the following is true*:

1. $|\mathbf{E}(\mathbf{x})|\leqslant c|\mathbf{x}|^{-1}\,\|\mathbf{E}\|(Z)$;
2. $|\mathbf{x}_0\times\operatorname{rot}\mathbf{E}(\mathbf{x})+ik\mathbf{E}(\mathbf{x})\,|\leqslant c\,|\mathbf{x}|^{-2}\,\|\mathbf{E}\|(Z)$.

7 Exterior boundary-value problems

After these preparations, we can prove the following theorem.

Theorem 6 *Let G be an exterior domain,* $\sigma=0$ *and* $\mathbf{F}\in L_2(G)$ *with* $\operatorname{supp}\mathbf{F}\in K(\mathbf{0},R_0)$ *and* $\operatorname{div}\mathbf{F}\in L_2(G)$. *Then, the exterior boundary-value problem is uniquely soluble.*

The uniqueness follows from a result of Kupradze and Rellich and the principle of unique continuation. In order to prove the existence of a solution, the following is needed.

Lemma 4 *Assume* $R_0<R'<R$. *Then, for all* $\mathbf{E}\in\mathring{R}^*$ *with* $\operatorname{rot}\mu^{-1}\operatorname{rot}\mathbf{E}\in L_2(G_R)$,

$$\|\mathbf{E}\|_r(G_{R'})\leqslant c\{\|\operatorname{rot}\mu^{-1}\operatorname{rot}\mathbf{E}\|(G_R)+\|\mathbf{E}\|(G_R)\}.$$

Now, we choose a positive null sequence $\{r_n\}$, set $\sigma_n=\omega\varepsilon r_n$ (thus, $\eta_n=(1+ir_n)\varepsilon$) and solve

$$\operatorname{rot}\mu^{-1}\operatorname{rot}\mathbf{E}_n-\omega^2\eta_n\mathbf{E}_n=\mathbf{F}$$

in the weak sense in $\mathring{R}$. We want to pass to the limit $r_n\to 0$ to obtain the required solution. Since, at infinity, $\mathbf{E}$ behaves like $1/|\mathbf{x}|$, i.e., it does not belong to $L_2(G)$, we choose the norm

$$|||\mathbf{E}|||=\|\mathbf{E}(\mathbf{x})/(1+|\mathbf{x}|^{3/4})\|(G)$$

and, thus, we can prove the existence of a limit **E** from the selection theorem and Theorem 5. **E** solves the exterior boundary-value problem. This procedure is called the method of limiting absorption.

8 The spectrum of Maxwell's operator in bounded domains

In investigating the time-dependent Maxwell equations, we need the spectral theorem for the Maxwell operator. In this section, therefore, we proceed to reformulate for our purposes the results of section 4; in particular, Theorem 3. Assume $\sigma = 0$. Then, the time-dependent Maxwell equations read

$$\left.\begin{aligned} -i\omega\varepsilon\mathbf{E} - \operatorname{rot}\mathbf{H} = \mathbf{J}, \\ -i\omega\mu\mathbf{H} + \operatorname{rot}\mathbf{E} = \mathbf{K}, \end{aligned}\right\} \tag{2}$$

or, in brief,

$$M\mathbf{U} - \omega\mathbf{U} = \mathbf{F}, \tag{9}$$

where

$$M = \begin{pmatrix} 0 & i\varepsilon^{-1}\operatorname{rot} \\ -i\mu^{-1}\operatorname{rot} & 0 \end{pmatrix}$$

and

$$\mathbf{U} = \begin{pmatrix} \mathbf{U}_1 \\ \mathbf{U}_2 \end{pmatrix} = \begin{pmatrix} \mathbf{E} \\ \mathbf{H} \end{pmatrix}, \qquad \mathbf{F} = -i\begin{pmatrix} \varepsilon^{-1}\mathbf{J} \\ \mu^{-1}\mathbf{K} \end{pmatrix}.$$

For $\omega \neq 0$, Eqn (2) and

$$\left.\begin{aligned} A\mathbf{E} - \omega^2\mathbf{E} = \varepsilon^{-1}(\operatorname{rot}\mu^{-1}\mathbf{K} - i\omega\mathbf{J}), \\ \mathbf{H} = i\mu^{-1}(\mathbf{K} - \operatorname{rot}\mathbf{E})/\omega \end{aligned}\right\} \tag{10}$$

coincide. In section 4, we have developed a solution theory for Eqn (10).

In order to determine the spectrum of M directly, we first have to fix the domain $D(M)$ and the range $W(M)$ of M. For the present, it suggests itself that we choose $\mathring{R} \times R$ as $D(M)$; but because of the decomposition of the Maxwell equations for $\omega = 0$, the null space would become infinite-dimensional, and this would not result in useful spectral formulae. Therefore, let us attempt to find out if it is not sufficient to let M operate in a smaller space.

We start from the orthogonal decomposition of L_2, which is valid for self-adjoint operators

$$L_2 = N(M) \oplus \overline{W(M)}.$$

Assume $\mathbf{U}=\mathbf{U}_N+\mathbf{U}_W$ and $\mathbf{F}=\mathbf{F}_N+\mathbf{F}_W$, where $\mathbf{U}_N, \mathbf{F}_N \in N(M)$ and $\mathbf{U}_W, \mathbf{F}_W \in W(M)$, respectively. Then, it follows from Eqn (9) that

$$-\omega \mathbf{U}_N = \mathbf{F}_N \qquad \text{in } N(M), \tag{11}$$

$$M\mathbf{U}_W - \omega \mathbf{U}_W = \mathbf{F}_W \qquad \text{in } \overline{W(M)}. \tag{12}$$

Equation (11) is trivally soluble; therefore, only Eqn (12) is still of interest, and this equation is soluble in $W(M)$ for $\omega=0$ as well. Obviously, $N(M)=\mathring{R}\cap R_0 \times R_0$. As the scalar product, we choose

$$[\mathbf{U}, \mathbf{V}] = (\mathbf{U}_1, \varepsilon \mathbf{V}_1) + (\mathbf{U}_2, \mu \mathbf{V}_2).$$

Then, according to Picard, the following orthogonal decompositions hold:

$$L_2 = \mathring{R}\cap R_0 \oplus \overline{\varepsilon^{-1} \operatorname{rot} R}, \qquad L_2 = R_0 \oplus \overline{\mu^{-1} \operatorname{rot} \mathring{R}}. \tag{13}$$

Thus, it suggests itself that we choose the range of M

$$W(M) = \varepsilon^{-1} \operatorname{rot} R \times \mu^{-1} \operatorname{rot} \mathring{R}$$

and the domain

$$D(M) = \mathring{R} \cap \varepsilon^{-1} \operatorname{rot} R \times R \cap \mu^{-1} \operatorname{rot} \mathring{R}.$$

From Eqn (13), it can easily be proved that $D(M)$ is dense in $\overline{W(M)}$. M is a self-adjoint operator.

Since the zero does not belong to the spectrum of M any more, Theorem 3 may be carried over directly. All eigenvalues of M are real. We set

$$M\mathbf{U}_j = -\lambda_j \mathbf{U}_j$$

and easily see that $-\lambda_j$ is an eigenvalue of $\overline{\mathbf{U}}_j$, i.e.,

$$M\overline{\mathbf{U}}_j = -\lambda_j \overline{\mathbf{U}}_j.$$

Thus, for $\tau_j \neq 0$, we set

$$\lambda_j = \sqrt{\tau_j}, \qquad \lambda_{-j} = -\lambda_j,$$

$$\mathbf{E}_{-j} = \overline{\mathbf{E}}_j = \mathbf{E}_j, \qquad \mathbf{H}_{-j} = \overline{\mathbf{H}}_j = -\mathbf{H}_j,$$

$$\mathbf{U}_j = \frac{1}{\sqrt{2}}\begin{pmatrix}\mathbf{E}_j \\ \mathbf{H}_j\end{pmatrix}, \qquad \mathbf{U}_{-j} = \overline{\mathbf{U}}_j.$$

Then, $\{\mathbf{U}_j\}$ is a complete orthogonal system in $W(M) = \overline{W(M)}$. An eigenfield $\mathbf{E}_0$ belonging to the eigenvalue $\tau=0$ (Theorem 3) is orthogonal to $\overline{\varepsilon^{-1} \operatorname{rot} R}$ (because $\mathbf{E}_0 \in \mathring{R} \cap R_0 \cap D_{\varepsilon 0}$ and Eqns (13)) and therefore does not appear any more.

Thus Theorem 3 implies

Theorem 7 *The eigenvalue problem*

$$M\mathbf{U} = \lambda \mathbf{U}, \qquad \mathbf{U} \in W(M),$$

has a countably infinite set of eigenvalues with no finite accumulation point and the corresponding eigenfields $\mathbf{U}_j \in D(M)$, *where*

$$M\mathbf{U}_j = \lambda_j \mathbf{U}_j, \qquad \lambda_j \neq 0,$$
$$[\mathbf{U}_j, \mathbf{U}_k] = \delta_{jk}.$$

For $\mathbf{F} \in W(M)$ *and* $f_j = [\mathbf{F}, \mathbf{U}_j]$, *we get*

$$\mathbf{F} = \sum_{-\infty}^{\infty} f_j \mathbf{U}_j.$$

This is the spectral theorem for M. In particular, for $\mathbf{F} \in D(M)$, we obtain

$$M\mathbf{F} = \sum_{-\infty}^{\infty} \lambda_j f_j \mathbf{U}_j.$$

9 Initial-value problems for bounded domains

In the following section, we deal with initial-value problems for the time-dependent Maxwell equations

$$\begin{pmatrix} \dot{\mathbf{E}} - \varepsilon^{-1} \operatorname{rot} \mathbf{H} \\ \dot{\mathbf{H}} + \mu^{-1} \operatorname{rot} \mathbf{E} \end{pmatrix} = \mathrm{e}^{-i\omega t} \mathbf{K}. \tag{14}$$

We write Eqn (14) in an abbreviated form as

$$\dot{\mathbf{U}} + iM\mathbf{U} = \mathrm{e}^{-i\omega t} \mathbf{K}. \tag{15}$$

The initial condition $\mathbf{U}(\mathbf{x}, 0) = \mathbf{U}^0$ and the boundary condition are added. For $\mathbf{K} \in N(M)$, Eqn (15) reduces to

$$\dot{\mathbf{U}} = \mathrm{e}^{-i\omega t} \mathbf{K}, \qquad \mathbf{U} \in N(M),$$

and is solved by

$$\mathbf{U} = i\, \mathrm{e}^{-i\omega t} \mathbf{K}/\omega - i\mathbf{K}/\omega + \mathbf{U}_N^0. \tag{16}$$

Thus, we are again led to the more interesting case $\mathbf{K} \in W(M)$. We deal with Eqn (15) as with an ordinary differential equation in the Banach space $W(M)$ and assume the tentative hypothesis

$$\mathbf{U}(\mathbf{x}, t) = \mathrm{e}^{-i\omega t} \mathbf{S}(\mathbf{x}) + \mathbf{T}(\mathbf{x}, t).$$

From this hypothesis, the following equations for $\mathbf{S}$ and $\mathbf{T}$ are deduced:

$$M\mathbf{S} - \omega \mathbf{S} = -i\mathbf{K}, \tag{17}$$

$$\dot{\mathbf{T}} + iM\mathbf{T} = \mathbf{0}, \qquad \mathbf{T}^0 = \mathbf{U}^0 - \mathbf{S}. \tag{18}$$

We set $k_j = [\mathbf{K}, \mathbf{U}_j]$. Then, Eqn (17) is solved by

$$\mathbf{S} = -i \sum_{-\infty}^{\infty} \frac{k_j}{\lambda_j - \omega} \mathbf{U}_j. \tag{19}$$

In the case of $\omega=\lambda_l$, k_l must, in addition, be zero; otherwise, we have to start with another hypothesis. Equation (18) can be solved formally by

$$\mathbf{T}=\mathrm{e}^{-iMt}\mathbf{T}^0.$$

The spectral theorem (Theorem 7) now permits us to calculate functions of M. Thus, we have

$$\mathbf{T}=\sum_{-\infty}^{\infty}\mathrm{e}^{-i\lambda_j t}[\mathbf{T}^0,\mathbf{U}_j]\mathbf{U}_j, \tag{20}$$

and the solution of the initial-value problem is given by Eqns (16), (19) and (20).

10 Initial-value problems for exterior domains

Finally, we are going to deal with the corresponding initial-value problems for exterior domains. We shall proceed in the same way, and, therefore, we have to determine the spectrum of M. Now, M has a purely continuous spectrum, and the spectral theorem

$$\mathbf{F}=\int_{-\infty}^{\infty}\mathrm{d}(P_\lambda\mathbf{F})$$

holds again.

In the case of bounded domains, we had

$$P_\lambda\mathbf{F}=\sum_{\lambda_j\le\lambda}[\mathbf{F},\mathbf{U}_j]\mathbf{U}_j.$$

To get P_λ for unbounded domains, we use the results of section 7. In section 7, we proved the existence of an outgoing solution of the exterior boundary-value problem by means of the method of limiting absorption. A second 'incoming' solution can be obtained in a similar way. We prescribe the inward radiation condition

$$\|\mathbf{x}_0\times\operatorname{rot}\mathbf{E}-ik\mathbf{E}\|(G_a)<\infty,$$

and only have to take a negative null sequence $\{r_n\}$.

Let $\mathbf{F}$ have compact support, let $\mathbf{U}_s^+[\mathbf{F}]$ be the outgoing solution and $\mathbf{U}_s^-[\mathbf{F}]$ be the incoming solution of

$$M\mathbf{U}-s\mathbf{U}=\mathbf{F}.$$

We then have (Stone's formula)

$$P_\lambda\mathbf{F}=\frac{1}{2\pi i}\int_{-\infty}^{\lambda}(\mathbf{U}_s^+[\mathbf{F}]-\mathbf{U}_s^-[\mathbf{F}])\,\mathrm{d}s. \tag{21}$$

and, for example, Eqn (20) reads

$$\mathbf{T}=\int_{-\infty}^{\infty} e^{-i\lambda t}\, d(P_\lambda \mathbf{T})$$

In the case of unbounded domains, the spectral theorem contains important special integral transformations of mathematical physics. For example, we get the Fourier sine transform,

$$f(x)=\sqrt{\left(\frac{2}{\pi}\right)}\int_0^\infty F(y)\sin yx\, dy,$$

$$F(y)=\sqrt{\left(\frac{2}{\pi}\right)}\int_0^\infty F(z)\sin zy\, dz,$$

provided we set $G=(0,\infty)$ and $Au=d^2u/dx^2$ with $u(0)=0$.

Many questions are still open in regard to the theory of wave propagation in unbounded domains. Solutions with finite energy are being sought. We want to learn more about their asymptotic behaviour over long periods of time, to discuss the energy flow, etc.

Bibliography

Agmon, S., *Lectures on Elliptic Boundary Value Problems*. Van Nostrand, Princeton, 1965.

Eidus, D. M., Principle of limiting absorption, *Mat. Sb.*, **57** (99), 13–44, 1962; also *AMS Transl.* (2), **47,** 157–191, 1965.

Kupradze, V. D., Dynamical problems in elasticity, in *Progress in Solid Mechanics* Vol III (Sneddon, I. N. and, Hill, R. eds), North-Holland, Amsterdam, 1–259, 1963.

Leis, R., Zur Theorie elektromagnetischer Schwingungen in anisotropen inhomogenen Medien, *Math. Z.*, **106,** 213–224, 1968.

Leis, R., Über die Eindeutige Fortsetzbarkeit der Lösungen der Maxwellschen Gleichungen in anisotropen inhomogenen Medien, *Bull. Polytech. Inst. Jassy*, XIV (XIII), Fasc. 3–4, 119–124, 1968.

Leis, R., Außenraumaufgaben zur Plattengleichung, *Arch. Rat. Mech. Anal.*, **35,** 226–233, 1969.

Leis, R., Zur Theorie elastischer Schwingungen in inhomogenen Medien, *Arch. Rat. Mech. Anal.*, **39,** 158–168, 1970.

Leis, R., Zur Theorie elastischer Schwingungen, *Berichte der Gesellschaft für Mathematik und Datenverarbeitung* BMFT-GMD-72, 1973.

Leis, R., Rand- und Eigenwertaufgaben in der Theorie elektromagnetischer Schwingungen, *Method. Verfahr. Math. Phys.*, **11,** 85–116, 1974; *see also ZAMM* **54,** T36–T40, 1974.

Leis, R., Zur Theorie der Plattengleichung. *Berichte der Gesellschaft für Mathematik und Datenverarbeitung* BMFT-GMD-101, 1975.

Müller, C., *Grundprobleme der mathematischen Theorie elektromagnetischer Schwingungen*, Springer, Heidelberg, 1957.

Müller, C. and Niemeyer, H., Greensche Tensoren und asymptotische Gesetze der elektromagnetischen Hohlraumschwingungen, *Arch. Rat. Mech. Anal.*, **7,** 305–348, 1961.

Picard, R., Ein Randwertproblem in der Theorie kraftfreier Magnetfelder, *ZAMP*, **27,** 169–180, 1976.

Picard, R., Zur Existenz des Wellenoperators bei Anfangsrandwertproblemen vom Maxwell-Typ, *Math. Z.*, **156,** 175–185, 1977.

Protter, M. H., Unique continuation for elliptic equations, *Trans. Amer. Math. Soc.*, **95,** 81–91, 1960.

Rellich, F., Über das asymptotische Verhalten der Lösungen von $\Delta u + \lambda u = 0$ in unendlichen Gebieten, *Jber. Dtsch. Math. Verein.*, **53,** 57–65, 1943.

Wacker, H. M., Existenz- und Eindeutigkeitssätze für die erste und zweite Randwertaufgabe des Außbenraumproblems der Gleichung elastischer Schwingungen, *Bonner Math. Schriften*, 31, 1968.

Weck, N., Maxwell's boundary value problem on Riemannian manifolds with nonsmooth boundaries, *J. Math. Anal. Appl.*, **46,** 410–437, 1974.

Wilcox, C. H., *Scattering Theory for the D'Alembert Equation in Exterior Domains*, Lecture Notes in Mathematics no. 442, Springer, Berlin, Heidelberg, New York, 1975.

Weyl, H., Die natürlichen Randwertaufgaben im Außenraum für Strahlungsfelder beliebiger Dimensionen und beliebigen Ranges, *Math. Z.*, **56,** 105–119, 1952.

Professor Dr Rolf Leis,
Institut für Angewandte Mathematik der Universität Bonn,
Wegelerstraße 10,
53 Bonn,
Bundesrepublik Deutschland

J.-L. Lions

Introductory remarks on asymptotic analysis of periodic structures

Introduction

In mechanics, physics, chemistry, engineering, in the study of composite materials, macroscopic properties of crystalline or polymeric structures, nuclear reactor design, etc., one is led to the study of *boundary-value problems in media with a periodic structure.*

If the period of the structure is small, say ε, as compared to the size of the region in which the system is to be studied, then one is led to the study of *an asymptotic expansion of the solution of a boundary-value problem in terms of* ε.

The simplest model one can consider along these lines is as follows. Let Y denote the unit cube in $\mathbb{R}^n_y$ ($n=2$ or 3 in most of the applications) and let us consider functions a_{ij} which satisfy the conditions

$$a_{ij} \in L^\infty(\mathbb{R}^n_y),\dagger \text{ with real values,} \tag{1}$$

$$a_{ij}(y_1+1, y_2, \ldots, y_n) = a_{ij}(y_1, y_2+1, y_3, \ldots, y_n) = \ldots = a_{ij}(y_1, \ldots, y_{n-1}, y_n+1) \quad \text{a.e. in } y,\ddagger \tag{2}$$

$$a_{ij}(y)\xi_i\xi_j \geqslant \alpha\xi_i\xi_i, \qquad \alpha>0, \quad \text{a.e. in } y.\S \tag{3}$$

We are given a material which is represented by the functions $a_{ij}(x/\varepsilon)$; we introduce the operator A^ε defined by

$$A^\varepsilon\varphi = -\frac{\partial}{\partial x_i}\left(a_{ij}(x/\varepsilon)\frac{\partial\varphi}{\partial x_j}\right) \tag{4}$$

and we consider, in a domain $\Omega \subset \mathbb{R}^n$, boundary-value problems associated with A^ε. To fix these ideas,¶ let us consider the Dirichlet

† Space of bounded and measurable functions.

‡ We will say that a_{ij} is Y-periodic.

§ Throughout this paper we will adopt a summation convention.

¶ The results are independent of the boundary conditions.

problem in Ω:

$$A^\varepsilon u_\varepsilon = f \quad \text{in} \quad \Omega, \qquad u_\varepsilon = 0 \quad \text{on} \quad \Gamma = \partial\Omega. \tag{5}$$

The problem is now to derive, if possible, an asymptotic expansion for u_ε in terms of ε.

In section 1, we give the so-called *homogenization formula* (or the formulae for the *effective coefficients*). There are several methods available to obtain and to prove these results. A very short presentation of these methods is given in section 2. Section 3 gives some details on the *method of multi-scales,* which has been developed during the past few years by A. Bensoussan, G. Papanicolaou and J. L. Lions and which is studied extensively in the book [1] by these authors.

In section 4, we show how to apply this technique to the problem of *perforated media with periodic holes.* The results of this section are announced here for the first time. Many developments will be given along these lines in several forthcoming publications by the same authors.

Bibliographical references are indicated in section 2. However, at this point, we also refer to the paper of E. Sanchez-Palencia [2].

1 Composite media

1.1 Setting the problem

Let u_ε be the solution of (5) in the Introduction. Because of (3), the operators A^ε are *uniformly* elliptic *in* ε, so that

$$u_\varepsilon \text{ remains in a bounded set of } H^1(\Omega) \text{ as } \varepsilon \to 0 \tag{1.1}$$

(where $H^1(\Omega)$ denotes the Sobolev space of functions which are square integrable in Ω together with their first-order derivatives).

However, passing to the limit in Eqn (5) is not trivial, since the limit of $a_{ij}(x/\varepsilon)(\partial u_\varepsilon/\partial x_j)$ will turn out not to be the product of the limits of $a_{ij}(x/\varepsilon)$ and of $\partial u_\varepsilon/\partial x_j$ (cf., Remark 1.5 below).

1.2 Homogenized coefficients

Let us introduce

$$A_1 = -\frac{\partial}{\partial y_i}\left(a_{ij}(y)\frac{\partial}{\partial y_j}\right) \tag{1.2}$$

and let us define χ^j by

$$A_1(\chi^j - y_j) = 0, \qquad \chi^j \text{ is } Y\text{-periodic}; \tag{1.3}$$

Eqn (1.3) defines χ^j up to an additive constant.

We then define

$$q_{ij} = \int_Y \left(a_{ij} - a_{ik}(y) \frac{\partial \chi^j}{\partial y_k} \right) \mathrm{d}y \tag{1.4}$$

and

$$\mathscr{A} = -q_{ij} \frac{\partial^2}{\partial x_i \, \partial x_j} \tag{1.5}$$

the operator $\mathscr{A}$ is the so-called *homogenized operator*.

Remark 1.1 One can prove that $\mathscr{A}$, as defined by (1.5), is an *elliptic operator*.

Remark 1.2 The coefficients q_{ij} (and the operator $\mathscr{A}$) *do not depend on Ω*.

1.3 Limit theorem

Let us consider the solution u_ε of problem (5) of the Introduction. Let u be the solution of

$$\mathscr{A}u = f \quad \text{in} \quad \Omega, \qquad u = 0 \quad \text{on} \quad \Gamma = \partial\Omega. \tag{1.6}$$

Then one can prove (and we shall give below some indications on the technique for proving this result) that

$$u_\varepsilon \to u \quad \text{weakly in a Sobolev space;} \tag{1.7}$$

more precisely,

$$u_\varepsilon \to u, \frac{\partial u_\varepsilon}{\partial x_i} \to \frac{\partial u}{\partial x_i} \quad \text{in} \quad L^2(\Omega) \textit{ weakly}. \tag{1.8}$$

Remark 1.3 One *cannot* replace the *weak* convergence in (1.8) by *strong* convergence. In order to obtain strong convergence, one has to introduce *corrector terms* as indicated in [1] (cf. also Remark 1.5 below).

Remark 1.4 If ε is small, one can replace the (difficult) computation of u_ε by the standard computation of u. For the numerical aspects of these questions, we refer to F. Bourgat and H. Lanchon [3], I. Babuska [4], J. L. Lions [5].

Remark 1.5 Let us set

$$a_{ij}^\varepsilon(x) = a_{ij}(x/\varepsilon); \tag{1.9}$$

when $\varepsilon \to 0$, a_{ij}^ε remains in a bounded set of $L^\infty(\Omega)$; one verifies that

$$a_{ij}^{\varepsilon} \to \mathcal{M}(a_{ij}) = \int_Y a_{ij}(y)\,dy \quad \text{in} \quad L^{\infty}(\Omega) \quad \text{weak star.} \tag{1.10}$$

If we had *strong* convergence in (1.8), we would have

$$a_{ij}^{\varepsilon} \frac{\partial u_{\varepsilon}}{\partial x_j} \to \mathcal{M}(a_{ij}) \frac{\partial u}{\partial x_j} \quad \text{in} \quad L^2(\Omega) \text{ weakly}$$

and, for the limit problem, we would obtain

$$-\mathcal{M}(a_{ij}) \frac{\partial^2 u}{\partial x_i\, \partial x_j} = f \quad \text{in} \quad \Omega, \qquad u = 0 \quad \text{on} \quad \Gamma. \tag{1.11}$$

Since, in general, the operators $\mathcal{A}$ and $\mathcal{M}(a_{ij})(\partial^2/\partial x_i\, \partial x_j)$ are *distinct*, this observation proves that (1.8) cannot be improved† and it shows how *one should modify the simple average* $\mathcal{M}(a_{ij})$ *by formula* (1.5) *in order to obtain the effective coefficients of the material.*

2 General indications of the methods available

One method relies on the *calculus of variations* and was developed by de Giorgi and his group (cf., E. de Giorgi and S. Spagnolo [6], de Giorgi [7]); this is the generalized convergence (G-convergence); it is not restricted to *periodic* coefficients, but it is restricted to *symmetric* partial differential operators and does not give *asymptotic expansions.*

The most suitable method for obtaining the complete formulae, in the case of *highly varying periodic coefficients,* is the method of asymptotic expansions developed in [1], *using multi-scale methods.* Some indications regarding this method are given in section 3 below. This method is entirely general, and it is useful for all kind of partial differential operators, stationary or of evolution, linear or non-linear.

In the justification of the method of multi-scales, one can meet difficulties with the regularity of the coefficients and with boundary conditions which are not of Dirichlet's type. These difficulties have been solved by the *energy method* of Tartar [8], which is also used in [1] in conjunction with asymptotic expansions. (cf., also N. S. Bakbalov [9].)

For second-order problems, elliptic or parabolic, one can also use *probabilistic arguments* (cf., [1]).

For hyperbolic problems, the homogenization should be refined using a more complete spectral decomposition based on *Bloch Waves,* (cf., [1], Chap. 4‡).

† One can actually prove weak convergence of u_{ε} and of $\partial u_{\varepsilon}/\partial x_i$ in a space $L^{2+\eta}(\Omega)$, $\eta > 0$.

‡ There are probably connections between this method and the operational methods of V. P. Maslov [11].

3 The multi-scale method

3.1 Notations

We set

$$y = x/\varepsilon. \tag{3.1}$$

The operator $\partial/\partial x_j$ applied to a function of x and of $y = x/\varepsilon$ becomes $(\partial/\partial x_j) + \varepsilon^{-1}(\partial/\partial y_j)$; therefore,

$$A^\varepsilon = \varepsilon^{-2}A_1 + \varepsilon^{-1}A_2 + \varepsilon^0 A_3, \tag{3.2}$$

$$A_1 = -\frac{\partial}{\partial y_i}\left(a_{ij}(y)\frac{\partial}{\partial y_j}\right) \tag{3.3}$$

(cf., (1.1)),

$$A_2 = -\frac{\partial}{\partial y_i}\left(a_{ij}(y)\frac{\partial}{\partial x_j}\right) - \frac{\partial}{\partial x_i}\left(a_{ij}(y)\frac{\partial}{\partial y_j}\right), \tag{3.4}$$

$$A_3 = -\frac{\partial}{\partial x_i}\left(a_{ij}(y)\frac{\partial}{\partial x_j}\right). \tag{3.5}$$

3.2 Asymptotic expansion

We are looking for u_ε in the form

$$u_\varepsilon = u_0 + \varepsilon u_1 + \varepsilon^2 u_2 + \dots, \tag{3.6}$$

where

$$\left.\begin{aligned} &u_j = u_j(x, y),\\ &u_j \text{ is defined for } x \in \Omega,\ y \in \mathbb{R}^n,\\ &u_j \text{ is } Y\text{-periodic in } y. \end{aligned}\right\} \tag{3.7}$$

(We do not worry, for the time being, about the boundary conditions.)

We replace u_ε by the expansion (3.6) in Eqn (5) of the Introduction; we use (3.2) and we identify the various powers of ε. Thus, we obtain

$$A_1 u_0 = 0, \tag{3.8}$$

$$A_1 u_1 + A_2 u_0 = 0, \tag{3.9}$$

$$A_1 u_2 + A_2 u_1 + A_3 u_0 = f. \tag{3.10}$$

etc. These are the *equations of the expansion.*

3.3 Resolution of the equations of the expansion

We shall use the following simple lemma.

Lemma *We consider in Y the equation*

$$A_1\Phi = F, \tag{3.11}$$

with the boundary conditions

$$\Phi \text{ is } Y\text{-periodic.} \tag{3.12}$$

In order for this problem to admit a solution (defined up to an additive constant) it is necessary and sufficient that

$$\int_Y F(y)\, dy = 0. \tag{3.13}$$

Application From (3.8), where we think of x as a parameter, it follows that

$$u_0 = u(x) \quad \text{does not depend on } y. \tag{3.14}$$

Using (3.14), (3.9) reduces to

$$A_1 u_1 = \frac{\partial a_{ij}}{\partial y_i}(y)\frac{\partial u}{\partial x_j}(x). \tag{3.15}$$

Using functions χ^j defined in (1.2), (1.3), (and, say, (1.4)), it follows from (3.15) that†

$$u_1 = -\chi^j(y)\frac{\partial u}{\partial x_j}(x) + \tilde{u}_1(x). \tag{3.16}$$

We now consider (3.10), which we write as

$$A_1 u_2 = f - A_2 u_1 - A_3 u_0.$$

Using (3.13), we see that u_2 exists iff

$$\int_Y (f - A_2 u_1 - A_3 u_0)\, dy = 0. \tag{3.17}$$

Using (3.14) (3.16), one easily verifies that (3.17) *becomes*

$$\mathcal{A}u = f \tag{3.18}$$

when $\mathcal{A}$ is given by (1.4) (1.5). *The homogenized operator is obtained* (formally) *in this manner*: *it is a direct consequence of the expansion* (*the* '*Ansatz*') (3.6) *and of the Lemma* (*Fredholm's alternative*).

† Since $\int_Y (\partial a_{ij}/\partial y_i)(y)\, dy = 0$, condition (3.13) is satisfied.

Boundary condition It is obvious that we should take

$$u_0 = u = 0 \quad \text{on} \quad \Gamma. \tag{3.19}$$

Formula (3.16) shows that, *in general, it will be impossible to achieve*

$$u_1(x, y) = 0 \quad \text{on} \quad \Gamma(y = x/\varepsilon).$$

Hence, *in order to obtain higher-order expansions*, (i) one can use (3.6) for *interior expansions*; (ii) one has to introduce *boundary layers* near Γ. The study of boundary layers in this context does not seem to be completed yet.

Remark 3.1 One can justify the above expansion using the *maximum principle.*

For other boundary conditions, such as Neumann's boundary condition, energy methods are better for the justification of the first term of this expansion.

4 Perforated media with periodic holes

4.1 A model problem

Let us consider the cube Y with a 'hole' $\mathcal{O}$; $\mathcal{O}$ is actually an arbitrary open set such that

$$\bar{\mathcal{O}} \subset Y.$$

We consider, in $\mathbb{R}^n$, the set of all translated cubes Y and all holes $\mathcal{O}$ by integers, so as to obtain a covering of $\mathbb{R}^n$ by τY (the set of all translations of Y) and a set of holes $\mathcal{O}$.

We make next an homothety ε. We obtain a covering $\varepsilon\tau Y$ of $\mathbb{R}^n$ and a set of 'small' holes $\varepsilon\tau\mathcal{O}$, arranged in a periodic manner.

In this material, we consider $\Omega \subset \mathbb{R}^n$; more precisely, let Ω be an open set of $\mathbb{R}^n$ with boundary Γ. We define

$$\Omega_\varepsilon = \Omega \cap \varepsilon(\tau Y - \tau\mathcal{O}) \tag{4.1}$$

i.e.,

$$\Omega_\varepsilon = \Omega - \text{'periodic holes'};$$

$$\left.\begin{aligned} &\Gamma_\varepsilon = \Gamma \cap \varepsilon(\tau Y - \tau\mathcal{O}), \qquad \Gamma = \partial\Omega, \\ &S_\varepsilon = \text{boundary of all the holes contained (in whole or in} \\ &\qquad \text{part) in } \Omega. \end{aligned}\right\} \tag{4.2}$$

The boundary of Ω_ε is $\Gamma_\varepsilon \cup S_\varepsilon$.

We consider now the *model problem.*† Let u_ε be the solution of

$$-\Delta u_\varepsilon = f \quad \text{in} \quad \Omega_\varepsilon, \tag{4.3}$$

where f is given in Ω; more precisely,

$$f \in L^2(\Omega), \tag{4.4}$$

and with u_ε subject to the boundary condition

$$u_\varepsilon = 0 \quad \text{on} \quad \partial\Omega_\varepsilon. \tag{4.5}$$

We want to study the behaviour of u_ε as $\varepsilon \to 0$.

Remark 4.1 One could consider as well, instead of (4.3), the equation

$$A^\varepsilon u_\varepsilon = f \quad \text{in} \quad \Omega, \tag{4.6}$$

where A^ε is given as before.

Remark 4.2 One can also study other boundary conditions.

4.2 Asymptotic expansion

We are looking for u_ε in the form

$$u_\varepsilon = u_0(x, y) + \varepsilon u_1(x, y) + \ldots + \varepsilon^j u_j(x, y) + \ldots, \tag{4.7}$$

where

$$\left.\begin{array}{l} u_j(x, y) \quad \text{is defined for} \quad x \in \Omega_\varepsilon,\ y \in Y - \mathcal{O}; \\ u_j(x, y) \quad \text{is } Y\text{-periodic in} \quad y; \\ u_j(x, y) = 0 \quad \text{for} \quad y \in \partial\mathcal{O}. \end{array}\right\} \tag{4.8}$$

Remark 4.3 It follows from the last condition (4.8) that

$$\sum \varepsilon^j u_j(x, x/\varepsilon) = 0 \quad \text{for} \quad x \in S_\varepsilon,$$

(assuming convergence, of course!). *There are no boundary layers in the neighbourhood of S_ε.*

We now proceed with the identification of various powers of ε. We observe that

$$\left.\begin{array}{l} -\Delta = -\varepsilon^{-2}\Delta_y - 2\varepsilon^{-1}\Delta_{xy} - \Delta_x, \\ \Delta_{xy} = \dfrac{\partial^2}{\partial x_j\, \partial y_j} \end{array}\right\} \tag{4.9}$$

† Much more realistic models can be considered using the techniques given below.

(this is a particular case of (3.2)). Thus,

$$-\Delta_y u_0 = 0, \tag{4.10}$$

$$-\Delta_y u_1 - 2\Delta_{xy} u_0 = 0, \tag{4.11}$$

$$-\Delta_y u_2 - 2\Delta_{xy} u_1 - \Delta_x u_0 = f, \tag{4.12}$$

etc.

4.3 Resolution of the equations of the expansion

The difference between the situation described in section 3 and the present one lies in the fact that (compare to Lemma of section 3.3)

$$\left.\begin{aligned} &-\Delta_y \Phi = F \quad \text{in} \quad Y - \mathcal{O}, \\ &\Phi \ \text{ is } Y\text{-periodic}, \\ &\Phi = 0 \quad \text{on} \quad \partial\mathcal{O}, \end{aligned}\right\} \tag{4.13}$$

always admit a unique solution (if $\mathcal{O}$ is not empty!). Therefore, (4.10) implies

$$u_0 = 0. \tag{4.14}$$

Thus, (4.11) reduces to $-\Delta_y u_1 = 0$, which implies

$$u_1 = 0, \tag{4.15}$$

and (4.12) reduces to

$$-\Delta_y u_2 = f(x). \tag{4.16}$$

We define $w(y)$ by

$$\left.\begin{aligned} &-\Delta_y w(y) = 1 \quad \text{in} \quad Y - \mathcal{O}, \\ &w \text{ is } Y\text{-periodic and } w = 0 \text{ on } \partial\mathcal{O}. \end{aligned}\right\} \tag{4.17}$$

Then,

$$u_2(x, y) = w(y) f(x). \tag{4.18}$$

Conclusion

$$u_\varepsilon(x) = \varepsilon^2 f(x) w(y) + \ldots \tag{4.19}$$

One can prove that

$$\| u_\varepsilon - \varepsilon^2 f(x) w(y) \|_{L^\infty(\Omega_\varepsilon)} \leq C\varepsilon^3 \tag{4.20}$$

under suitable regularity assumptions on f (cf., A. Bensoussan, J. L. Lions and G. Papanicolaou [11]).

Remark 4.4 One can give a probabilistic proof of (4.20) (cf., [11]).

Remark 4.5 The above computations extend to operators of any order.

Remark 4.6 One can prove, by energy estimates (cf., D. Cioranescu [12]), the following. Let us define

$$\tilde{u}_\varepsilon = \begin{Bmatrix} u_\varepsilon & \text{in} & \Omega_\varepsilon, \\ 0 & \text{in} & \Omega - \Omega_\varepsilon. \end{Bmatrix} \tag{4.21}$$

Then $\tilde{u}_\varepsilon \to 0$ and $\partial \tilde{u}_\varepsilon / \partial x_i \to 0$ in $L^2(\Omega)$ weakly.

5 Some extensions

As already indicated, the above methods extend to evolution problems and to non-linear problems (cf., [1]), where one also solves the case of coefficients $a_{\alpha\beta}(x, x/\varepsilon, x/\varepsilon^2, \ldots, x/\varepsilon^N)$ and where the $a_{\alpha\beta}(x, y, z, \ldots)$ are bounded in x, periodic in $y, z, \ldots$ (with different periods). Applications to problems of elasticity are given in G. Duvaut [13].

Problems of homogenization appear in transport theory (cf., E. Larsen [14] and A. Bensoussan, J. L. Lions and G. Papanicolaou [15]).

References

1 Bensoussan, A., Lions, J.-L. and Papanicolaou, G., *Asymptotic Methods in Periodic Structures*, North-Holland, Amsterdam, 1978.

2 Sanchez-Palencia, E., Comportement local et macroscopique d'un type de milieux physiques hétérogènes, *Int. J. Eng. Sci.*, **12,** 331–351, 1976.

3 Bourgat, F. and Lanchon, H., Application of the homogenization method to composite materials with periodic structures, LABORIA Report, IRIA F-78150, Le Chesnay, 1976.

4 Babuska, I., Homogenization approach in engineering, Technical Report, University of Maryland, 1974.

5 Lions, J.-L., Remarques sur les aspects numériques de la méthode d'homogénéisation dans les milieux composites, IRIA-Novosibirsk Colloquium 1976, Novosibirsk, 1978.

6 de Giorgi, E. and Spagnolo, S., Sulla convergenza degli integrali dell' energia per operatori del 2° ordine, *Boll. U.M.I.*, **8,** 339–411, 1973.

7 de Giorgi, E., Γ-convergenze e G-convergenze, *Boll. U.M.I.*, (5), **14**-A, 213–220, 1977.

8 Tartar, T., Some remarks on homogenization, in *Functional Analysis and Numerical Analysis* (H. Fujita, ed.), J.S.P.S., 1978, pp. 469–481.

9 Bakbalov, N. S., Average characteristic of bodies with periodic structure, *Dokl. Akad. Nauk USSR*, **218,** 1046–1048, 1974.

10 Bensoussan, A., Lions, J.-L. and Papanicolaou, G., Asymptotic expansions in perforated media, to appear.

11 Maslov, V. P., *Operational Methods*, Mir, Moscow, 1976 (translated and revised from the 1973 Russian edition).

12 Cioranesnu, D., Homogenisation dans milieux perfous, Thesis, Paris, 1977.

13 Duvaut G., Loi de comfortement élastique homogénéisé, in *Functional Analysis and Numerical Analysis* (H. Fugita, ed.), J.S.P.S., 1978, pp. 71–81.

14 Larsen, E., Neutron transport and diffusion in inhomogeneous media, *Int. J. Math. Phys.*, **16,** 1421–1427, 1975.

15 Bensoussan, A., Lions, J.-L. and Papanicolaou, G., Boundary layers and homogenization of transport processes, *J. Publ. R.I.M.S. Kyoto*, 1978.

Professor J.-L. Lions
Collège de France,
Paris,
France

E. Meister and F.-O. Speck

Some multidimensional Wiener–Hopf equations with applications

1 Introduction

In the book by Noble [1], the famous Sommerfeld problem, viz., the diffraction of a plane wave by a half-plane, is solved by means of the Fourier transform and the Wiener–Hopf technique. This method was invented in 1931 by Wiener and Hopf [2] in order to solve certain types of integral equations of the convolution type on the half-line, viz.,

$$(W\varphi)(t) := \varphi(t) - \int_0^\infty k(t-\tau)\varphi(\tau)\,\mathrm{d}\tau = f(t), \qquad t \in \mathbb{R}_+, \tag{1.1}$$

by means of function theoretic methods.

Systematic investigations for $k \in L^1(\mathbb{R})$ in spaces $\mathfrak{X} = L^p(\mathbb{R}_+)$, $1 \leqslant p \leqslant \infty$, and certain subspaces of $L^\infty(\mathbb{R}_+)$ were undertaken by Kreĭn [3]. They were generalized by Gohberg and Kreĭn [4] to the case of systems of Wiener–Hopf integral equations (1.1) on $\mathbb{R}_+$ and their discrete analogues, the Toeplitz equations

$$\sum_{k=1}^{\infty} a_{j-k}\xi_k = \eta_j \qquad (j = 1, 2, \ldots), \tag{1.2}$$

where the Toeplitz matrix $(a_{j-k})_{j,k\in\mathbb{N}}$ with $\{a_j\}_{j\in\mathbb{Z}} \in l_1$ and the right-hand side $\{\eta_j\}_{j\in\mathbb{N}} \in l_p$, $1 \leqslant p < \infty$, are given and the sequence $\{\xi_k\}_{k\in\mathbb{N}} \in l_p$ is sought. For a general account of the theory and of some methods for their approximate solution, consult the book by Gohberg and Feldman [5].

Introducing the concepts of the *defect numbers* $\alpha(W) := \dim N(W)$, $N(W)$ being the kernel or *null space* of the Wiener–Hopf operator W in Eqn (1.1), $\beta(W) := \operatorname{codim} R(W)$, $R(W)$ being the *range* of W, and the *index* $\operatorname{ind} W = \alpha - \beta$, the, now classical, result of the Kreĭn–Gohberg theory runs as follows [5].

Theorem 1 *Let k be $\in L^1(\mathbb{R})$ and $f \in \mathfrak{X}$ be given functions. Then W is a Fredholm–Noether operator, i.e., normally soluble with finite defect* *numbers, if and only if (iff) the* symbol *of W*

$$\Phi_W(\xi) := 1 - \hat{k}(\xi) \neq 0 \quad \text{on} \quad \dot{\mathbb{R}}_\xi := \mathbb{R}_\xi \cup \{\infty\}, \tag{1.3}$$

where

$$\hat{k}(\xi) = (Fk)(\xi) := \frac{1}{\sqrt{2\pi}} \int_{-\infty}^{\infty} e^{ix\xi} k(x) \, dx$$

denotes the Fourier transform. The defect numbers are then given by

$$\alpha(W) = \max(-\kappa, 0) \tag{1.4}$$

$$\beta(W) = \max(0, \kappa) \tag{1.5}$$

such that ind $W = -\kappa(1-\hat{k}(\cdot)) := -(1/2\pi)[\arg(1-\hat{k}(\xi))]_{-\infty}^{\infty}$, *where* $\kappa = \kappa(1-\hat{k}(\cdot))$ *is the* winding number *of the symbol. Moreover, all elements* $\varphi_0 \in N(W)$ *belong to* $\cap\,\mathfrak{X} = L^1(\mathbb{R}_+) \cap C_0(\dot{\mathbb{R}}_+)$ *and there is a basis forming a* d-chain, *i.e.,* $\varphi_{0,1}, \ldots, \varphi_{0,|\kappa|} \in \cap\,\mathfrak{X}$ *in case of* $-\kappa > 0$, *where* $\varphi_{0,1}, \ldots, \varphi_{0,|\kappa|-1}$ *are also absolutely continuous and*

$$\varphi_{0,j+1} = \frac{d}{dt}\varphi_{0,j} \quad \text{for} \quad j = 1, \ldots, |\kappa| - 1,$$

$$\varphi_{0,j}(+0) = 0 \quad \text{for} \quad j = 1, \ldots, |\kappa| - 2,$$

but

$$\varphi_{0,|\kappa|-1}(+0) \neq 0.$$

The crucial step in proving this theorem is the *factorization* of the symbol into

$$\Phi_W(\xi) = 1 - \hat{k}(\xi) = a_-(\xi)\left(\frac{\xi - i}{\xi + i}\right)^{\kappa} a_+(\xi), \tag{1.6}$$

where $a_\pm(\xi)$ are $\neq 0$ on $\dot{\mathbb{R}}_\xi$ and are holomorphically extendable into the upper and lower half-planes of the complex $\zeta = \xi + i\eta$-plane, respectively, where $|a_\pm(\zeta)|$ are bounded from below and above by positive constants.

The Kreĭn–Gohberg theory has been extended to (scalar) integro-differential equations of the type

$$(A\varphi)(t) := \sum_{j=0}^{m} \left\{ a_j(t)(D^j\varphi)(t) + b_j(t) \cdot \int_0^\infty c_j(\tau) k_j(t-\tau)(D^j\varphi)(\tau) \, d\tau \right\}$$
$$= f(t), \qquad t \in \mathbb{R}_+, \tag{1.7}$$

by Gerlach in his thesis [6]. He assumed $k_j \in L^1(\mathbb{R})$, a_j, b_j, $c_j \in C([0, \infty])$, $f \in \mathfrak{X}$ to be given functions and $\varphi \in \mathfrak{X}$ or $\mathfrak{X}_0^m$ to be found. $\mathfrak{X}^m$ and $\mathfrak{X}_0^m$ denote the Banach subspaces of $\mathfrak{X}$ such that $D^j\varphi = d^j/dt^j \in \mathfrak{X}$ for $j = 0, \ldots, m$ and, additionally, $(D^\mu\varphi)(+0) = 0$ for $\mu = 0, \ldots, m-1$ in the case of $\mathfrak{X}_0^m$, respectively. Kremer [7] derived a Fredholm–Noether theory for the more general Wiener–Hopf integral equation on the half-line:

$$\begin{aligned}(A\varphi)(t) := a(t)\varphi(t) + \frac{b(t)}{\pi i}\int_0^\infty \frac{\varphi(\tau)\,\mathrm{d}\tau}{\tau - t} + c(t)\int_0^\infty k(t-\tau)\varphi(\tau)\,\mathrm{d}\tau \\ = f(t), \qquad t \in \mathbb{R}_+, \end{aligned} \tag{1.8}$$

where $k \in L^1(\mathbb{R})$; a, b, $c \in C([0, \infty])$; $f \in L^2(\mathbb{R}_+)$ are given and $\varphi \in L^2(\mathbb{R}_+)$ is unknown. Generalizing the class of admitted symbols, $\Phi(t, \xi)$ being the Fourier transforms, in the distributional (γ') sense, of certain L^∞-functions, Dudučava recently established [8, 9] Fredholm theorems for equations of the form

$$\begin{aligned}(A\varphi)(t) := \sum_{j=1}^{N} \left\{ \beta_j a_j(t) c_j(t) \varphi(t) + \int_{-\infty}^{\infty} a_j(t) c_j(\tau) k_j(t-\tau)\varphi(\tau)\,\mathrm{d}\tau \right\} \\ = f(t), \qquad t \in \mathbb{R}_+, \end{aligned} \tag{1.9}$$

in the spaces $L^p(\mathbb{R}_+)$, $1 < p < \infty$, where a_j, c_j are piecewise continuous on $\ddot{\mathbb{R}} := [-\infty, \infty]$ and the Fourier transforms $\hat{k}_j(\xi)$ of the k_j are composed of piecewise combinations $\sum_{\nu=1}^{N_j} \chi_{E_{j,\nu}}(\xi)[c_{j,\nu} + \hat{k}_{j,\nu}(\xi)]$, $c_{j,\nu} \in \mathbb{C}$, and $k_{j,\nu} \in L^1(\mathbb{R})$ of elements of the Wiener algebra, the $E_{j,\nu}$ denoting measurable disjoint subsets covering $\mathbb{R}$ for each j. The classical Wiener–Hopf technique has also been applied to numerous problems in mathematical physics, particularly to mixed boundary-value problems in diffraction theory. Besides the above-mentioned book by Noble [1], Talenti [10] more recently gave a detailed review of the subject in 1973.

Much less is known about higher-dimensional integral equations of the Wiener–Hopf type:

$$(Wu)(x) := u(x) - \int_G k(x-y)u(y)\,\mathrm{d}y = v(x), \qquad x \in G, \tag{1.10}$$

where $k \in L^1(\mathbb{R}^n)$, $n \geq 2$, $v \in L^p(G)$, $1 \leq p \leq \infty$, or is an element of a subspace of $L^\infty(G)$ with a measurable set $G \subset \mathbb{R}^n$, mes $G = \infty$, are given and $u \in L^p(G)$ is unknown. The problems arising become even more complicated if one admits strongly singular kernel-functions like those of the Cauchy principal-value type or Calderón–Mikhlin type. The factorization procedure plays a dominant part in the local theory of elliptic boundary-value problems, as has been shown by Višik and Eskin [11–13], Dikanskiĭ [14, 15] and others.

Before we enter into the discussion of certain classes of multidimensional integral equations on semi-infinite domains $G \subset \mathbb{R}^n$, we shall formulate some mixed boundary-value problems arising in the theory of diffraction of electromagnetic and elastic waves.

2 Some multidimensional Wiener–Hopf equations occurring in diffraction theory

2.1 Diffraction of scalar waves by a plane screen

Generalizing the above mentioned Sommerfeld half-plane problem, Radlow [16, 17] studied the problem of diffraction of an incoming plane scalar wave by a quarter-plane in $\mathbb{R}^3$-space:

Let $\Psi_{\text{inc}}(x_1, x_2, x_3, t) = \operatorname{Re}[\Phi_{\text{inc}}(x_1, x_2, x_3)e^{-i\omega t}]$ be an incoming wave-field falling upon the screen $\Sigma := \{(x_1, x_2, x_3) \in \mathbb{R}^3 : x_3 = 0, x_1, x_2 \geq 0\}$. The following boundary-value problem for the scattered field amplitude $\Phi(x_1, x_2, x_3)$ then arises†:

Problem *Find $\Phi \in C^2(\mathbb{R}^3 \backslash \Sigma)$ being a solution to Helmholtz's equation*

$$(\Delta_3 + k^2)\Phi = 0 \quad \text{in} \quad \mathbb{R}^3 \backslash \Sigma \tag{2.1}$$

with the boundary conditions

$$\Phi(x_1, x_2, x_3)\big|_\Sigma = -\Phi_{\text{inc}}(x_1, x_2, x_3)\big|_\Sigma \tag{2.2a}$$

or

$$\frac{\partial}{\partial x_3}\Phi(x_1, x_2, x_3)\big|_{\mathring{\Sigma}} = -\frac{\partial}{\partial x_3}\Phi_{\text{inc}}(x_1, x_2, x_3)\big|_{\mathring{\Sigma}} \tag{2.2b}$$

where $\mathring{\Sigma} = \Sigma \backslash \partial\Sigma$. Additionally, Φ has to satisfy the asymptotic edge conditions

$$\Phi = O(1),\ \operatorname{grad}\Phi = O(r^{-1-\alpha}) \quad \text{as} \quad r = \sqrt{(x_1^2 + x_2^2 + x_3^2)} \to 0, \quad 0 \leq \alpha < 1, \tag{2.3a}$$

$$\Phi = O(1),\ \operatorname{grad}\Phi = O(\rho_j^{-\beta}) \quad \text{as} \quad \rho_1 = \sqrt{(x_2^2 + x_3^2)} \quad \text{or} \quad \rho_2 = \sqrt{(x_1^2 + x_3^2)} \to 0, \qquad 0 \leq \beta < 1, \tag{2.3b}$$

and the radiation conditions

$$\Phi,\ \operatorname{grad}\Phi = O\left(\frac{e^{-\operatorname{Im} kr}}{r}\right) \tag{2.4a}$$

$$\text{as} \quad r \to \infty$$

$$\left(\frac{\partial}{\partial r} - ik\right)\Phi = O\left(\frac{e^{-\operatorname{Im} kr}}{r}\right) \tag{2.4b}$$

where $k \neq 0$, $\operatorname{Re} k \geq 0$, $\operatorname{Im} k \geq 0$.

† Because of the analogy with problems coming from classical physics in this context, Φ and k have another sense than being a symbol function and a convolution kernel; these, here and in the following section exceptionally, will be denoted by σ and γ.

Applying the two-dimensional Fourier transform to Φ with respect to (x_1, x_2), viz.,

$$(F_{12}\Phi(x_1, x_2, x_3))(\xi_1, \xi_2) = \hat{\Phi}(\xi_1, \xi_2, x_3)$$
$$:= \iint_{\mathbb{R}^2_{x_1,x_2}} e^{i(x_1\xi_1 + x_2\xi_2)} \Phi(x_1, x_2, x_3)\, d(x_1, x_2), \qquad (2.5)$$

the asymptotic conditions lead to the following representation of the transformed function:

$$\hat{\Phi}_D(\xi_1, \xi_2, x_3) = a_D(\xi_1, \xi_2) \exp[-|x_3|\, \gamma(\xi_1, \xi_2)] \qquad (2.6a)$$

in the case of the Dirichlet problem and

$$\hat{\Phi}_N(\xi_1, \xi_2, x_3) = \pm a_N(\xi_1, \xi_2) \exp[-|x_3| \gamma(\xi_1, \xi_2)] \qquad (2.6b)$$

in the case of the Neumann problem, respectively, where $\gamma(\xi_1, \xi_2) := \sqrt{(\xi_1^2 + \xi_2^2 - k^2)}$ is defined in such a way that $\mathrm{Re}\, \gamma \geqslant 0$ for all $(\xi_1, \xi_2) \in \mathbb{R}^2$ behaving asymptotically like $|\xi|$ as $|\xi| := \sqrt{(\xi_1^2 + \xi_2^2)} \to \infty$. Taking the boundary conditions into consideration, we arrive at the following relations due to the continuity of Φ and $\partial\Phi/\partial x_3$ off the screen Σ:

$$a_D(\xi_1, \xi_2) = -\hat{\Phi}^{++}_{\mathrm{inc},0}(\xi_1, \xi_2) + \hat{\Phi}_0^{-+}(\xi_1, \xi_2) + \hat{\Phi}_0^{--}(\xi_1, \xi_2) + \hat{\Phi}_0^{+-}(\xi_1, \xi_2) \qquad (2.7a)$$

and

$$\gamma \cdot a_N(\xi_1, \xi_2) = -\frac{\partial}{\partial x_3} \hat{\Phi}^{++}_{\mathrm{inc},0}(\xi_1, \xi_2) + \frac{\partial}{\partial x_3} \hat{\Phi}_0^{-+}(\xi_1, \xi_2)$$
$$+ \frac{\partial}{\partial x_3} \hat{\Phi}_0^{--}(\xi_1, \xi_2) + \frac{\partial}{\partial x_3} \hat{\Phi}_0^{+-}(\xi_1, \xi_2), \qquad (2.7b)$$

respectively.

Here, we have introduced the unknown F-transforms of the restrictions of the functions $\Phi(x_1, x_2, 0)$ and $(\partial/\partial x_3)\Phi(x_1, x_2, 0)$, respectively, to the second $(-+)$, third $(--)$ and fourth $(+-)$ quadrant, i.e., for instance,

$$\hat{\Phi}_0^{-+}(\xi_1, \xi_2) = (F\chi_{\mathbb{R}^2_{-+}} \cdot \Phi(x_1, x_2, 0))(\xi_1, \xi_2) \qquad (2.8)$$

with the characteristic function $\chi_{\mathbb{R}^2_{-+}}$ of the second quadrant $\mathbb{R}^2_{-+}$. Introducing the unknown jumps of $\partial\Phi_D/\partial x_3$ and Φ_N across the screen, $2I^{++}(x_1, x_2)$ and $2J^{++}(x_1, x_2)$, respectively, we arrive at the following Wiener–Hopf-type functional relations in the transform plane

$$\frac{1}{\gamma} \cdot \hat{I}^{++}(\xi_1, \xi_2) - \hat{\Phi}_0^{-+}(\xi_1, \xi_2) - \hat{\Phi}_0^{--}(\xi_1, \xi_2) - \hat{\Phi}_0^{+-}(\xi_1, \xi_2) = -\hat{\Phi}^{++}_{\mathrm{inc},0}(\xi_1, \xi_2) \qquad (2.9a)$$

and

$$\gamma \cdot \hat{J}^{++}(\xi_1, \xi_2) - \frac{\partial}{\partial x_3} \hat{\Phi}_0^{-+}(\xi_1, \xi_2) - \frac{\partial}{\partial x_3} \hat{\Phi}_0^{--}(\xi_1, \xi_2) - \frac{\partial}{\partial x_3} \hat{\Phi}_0^{+-}(\xi_1, \xi_2)$$

$$= -\frac{\partial}{\partial x_3} \hat{\Phi}_{\text{inc},0}^{++}(\xi_1, \xi_2). \quad (2.9b)$$

Now, writing

$$\hat{P} = \hat{P}_{++} := FP_{++}F^{-1} = F\chi_{\mathbb{R}^2_{++}} \cdot F^{-1}$$

and

$$\hat{Q} := I - \hat{P} = \hat{P}_{-+} + \hat{P}_{--} + \hat{P}_{+-} = F\chi_{\mathbb{R}^2 \setminus \mathbb{R}^2_{++}} \cdot F^{-1}$$

for the continuous linear projectors on the spaces $FL^1(\mathbb{R}^2)$, we arrive at the *two-part* or *paired Wiener–Hopf equations*

$$\frac{1}{\gamma} \hat{P}\hat{I}^{++} - \hat{Q}\hat{\Phi}_0 = -\hat{P}\hat{\Phi}_{\text{inc},0} \quad (2.10a)$$

and

$$\gamma \hat{P}\hat{J}^{++} - \hat{Q}\frac{\partial}{\partial x_3} \hat{\Phi}_0 = -\hat{P}\frac{\partial}{\partial x_3} \hat{\Phi}_{\text{inc},0} \quad (2.10b)$$

respectively, or at the equivalent Wiener–Hopf equations

$$\hat{P}\frac{1}{\gamma} \hat{P}\hat{I}^{++} = -\hat{P}\hat{\Phi}_{\text{inc},0} \quad (2.11a)$$

and

$$\hat{P}\gamma\hat{P}\hat{J}^{++} = \hat{P}\frac{\partial}{\partial x_3} \hat{\Phi}_{\text{inc},0}. \quad (2.11b)$$

Taking the inverse F-transform, we may write

$$PF^{-1}\frac{1}{\gamma} FPI^{++} = -P\Phi_{\text{inc}}\big|_{x_3=0}, \quad (2.12a)$$

$$PF^{-1}\gamma FPJ^{++} = -P\frac{\partial}{\partial x_3} \Phi_{\text{inc}}\big|_{x_3=0}. \quad (2.12b)$$

The left-hand sides involve the translation-invariant operators

$$A := F^{-1}\frac{1}{\gamma} \cdot F$$

with the symbol $\sigma_A = 1/\gamma$ and

$$A^{-1} = F^{-1}\gamma \cdot F$$

with the symbol $\sigma_{A^{-1}} = \gamma$ in the notation of Hörmander [18].

It is easily seen that the diffraction problems treated here may be generalized to the case of an arbitrary plane screen $\Sigma \subset \mathbb{R}^2_{x_1,x_2}$, which may also consist of disjoint parts. We only have to replace the projector $P = P_{++}$ for the first quadrant by $P_\Sigma := \chi_\Sigma \cdot$ given by the characteristic function of the screen Σ. The problem of the unique existence of a solution to the boundary-value problems is then equivalent to the unique continuous invertibility of the Wiener–Hopf operator $W_\Sigma := P_\Sigma A P_\Sigma$ on the range $P_\Sigma L^1$ or $P_\Sigma \mathfrak{X}$, where $\mathfrak{X}$ is a suitable Banach space of functions on $\mathbb{R}^2_{x_1,x_2}$. It can be solved on the basis of the theory developed by Devinatz and Shinbrot [31], (cf. Chap. 3).

2.2 Diffraction of waves by dielectric wedges

Now, let us assume that the $\mathbb{R}^3$-space is divided into four wedges given by $G_j := \{(x_1, x_2, x_3) \in \mathbb{R}^3, (x_1, x_2) \in \Gamma_j, x_3 \in \mathbb{R}\}$, where Γ_j denotes the jth quadrant, and is filled by dielectric media with equal permeabilities $\mu_j = \mu$ but different dielectric constants ε_j and conductivities $\sigma_j > 0$. Without loss of generality, we assume a source (or dipole) located in the interior of the first wedge. The electric vector of this primary electromagnetic field (E_{pr}, H_{pr}) will be polarized in the x_3-direction; thus, we assume the electric field to be of the form $E(x_1, x_2, x_3) = (0, 0, \Phi(x_1, x_2))$ and the magnetic field of the form $H(x_1, x_2, x_3) = (H_1(x_1, x_2), H_2(x_1, x_2), 0)$. The splitting of the total electromagnetic field into the primary one and a scattered field in the first wedge G_1 results in the decomposition $\Phi_{\text{tot}\,1}(x_1, x_2) = \Phi_{pr}(x_1, x_2) + \Phi_1(x_1, x_2)$. We take $\Phi_j(x_1, x_2)$ for the total x_3-component of the electric vector in the other wedges G_j, $j \neq 1$. Without going into details, as has been done by Latz [19, 22], Radlow [20], Kuo and Plonus [21], Kraut and Lehmann [23], and Rawlins [24], the introduction of the specialized field vectors into Maxwell's equations leads to the following *transmission* or *interface problem* for Helmholtz's equation.

Find $\Phi_j(x_1, x_2) \in C^2(\mathring{\Gamma}_j)$, $j = 1, \ldots, 4$, satisfying

$$(\Delta_2 + k_j^2)\Phi_j(x_1, x_2) = 0 \quad \text{in} \quad \mathring{\Gamma}_j, \tag{2.13}$$

where $k_j^2 := \omega^2 \varepsilon_j \mu + i\omega\sigma_j\mu$, and the transmission conditions written in polar coordinates (j counted modulo 4)

$$\Phi_j - \Phi_{j-1} = \left\{ \begin{array}{lll} -\Phi_{pr}(x_1, 0), & x_1 > 0, & j = 1, \\ \Phi_{pr}(0, x_2), & x_2 > 0, & j = 2, \\ & 0, & j = 3, 4, \end{array} \right\} \tag{2.14a}$$

and

$$\frac{\partial \Phi_j}{\partial n_j} - \frac{\partial \Phi_{j-1}}{\partial n_j} = \left\{ \begin{array}{lll} -i\omega\mu H_{pr1}(x_1, 0), & x_1 > 0, & j = 1, \\ i\omega\mu H_{pr2}(0, x_2), & x_2 > 0, & j = 2, \\ & 0, & j = 3, 4. \end{array} \right\} \tag{2.14b}$$

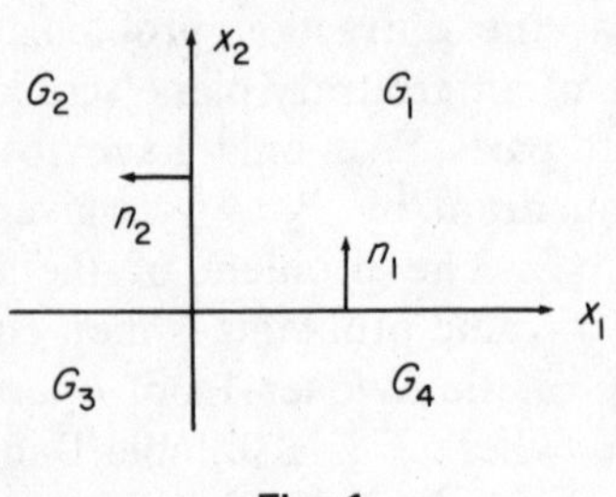

Fig. 1

Additionally, Φ_j have to satisfy the asymptotic conditions

$$\left.\begin{aligned} \Phi_j &= O(1) \quad \text{as} \quad r=\sqrt{(x_1^2+x_2^2)}\to 0,\\ \operatorname{grad}\Phi_j &= O(r^{-\alpha_j}),\ \Delta\Phi_j = O(r^{-1-\alpha_j}) \quad \text{as} \quad r\to 0, \end{aligned}\right\} \tag{2.15}$$

$$\Phi_j, \operatorname{grad}\Phi_j = O(\mathrm{e}^{-qr}) \quad \text{as} \quad r\to\infty, \tag{2.16}$$

where $q := \min(\omega\sigma_j\mu) > 0$ for $j = 1, \ldots, 4$.

These conditions allow us to apply the two-dimensional Fourier transform F, resulting in the following functional relation for

$$\hat{\Phi}_j(\xi_1, \xi_2) = (F\Phi_j(x_1, x_2))(\xi_1, \xi_2) = (FP_j\Phi(x_1, x_2))(\xi_1, \xi_2)$$

with the projectors $P_j\Phi := \chi_{\Gamma_j} \cdot$ restricting functions in $\mathbb{R}^2$ to the jth quadrant:

$$\hat{\Phi}_j(\xi_1, \xi_2) = \frac{z_j(\xi_1, \xi_2)}{\xi_1^2+\xi_2^2-k_j^2}, \qquad (\xi_1, \xi_2)\in\mathbb{R}^2. \tag{2.17}$$

The numerator contains the one-dimensional transforms of the boundary values of Φ_j on $\partial\Gamma_j$, e.g.,

$$\begin{aligned} z_1(\xi_1, \xi_2) = &-(F_2\Phi_{1,x_1}(0+, x_2))(\xi_2) + i\xi_1(F_2\Phi_1(0+, x_2))(\xi_2)\\ &-(F_1\Phi_{1,x_2}(x_1, 0+))(\xi_1) + i\xi_2(F_1\Phi_1(x_1, 0+))(\xi_1), \end{aligned} \tag{2.18}$$

where F_j denotes partial Fourier transformation with respect to the jth variable.

After adding all z_j, we obtain, due to transmission conditions (2.14a, b),

$$\begin{aligned} \sum_{j=1}^{4} z_j(\xi_1, \xi_2) &= -i\xi_1 F_2[\Phi_{\text{pr}}(0, x_2)](\xi_2) - i\omega\mu F_2[H_{\text{pr}2}(0, x_2)](\xi_2)\\ &\quad - i\xi_2 F_1[\Phi_{\text{pr}}(x_1, 0)](\xi_1) + i\omega\mu F_1[H_{\text{pr}1}(x_1, 0)](\xi_1)\\ &= z(\xi_1, \xi_2), \end{aligned} \tag{2.19}$$

which is known by the data of the primary field. Replacing the numerators due to Eqn (2.17), we arrive at $-|\xi|^2 := \xi_1^2 + \xi_2^2 -$:

$$\sum_{j=1}^{4} (|\xi|^2 - k_j^2)\hat{\Phi}_j(\xi_1, \xi_2) = z(\xi_1, \xi_2), \tag{2.20}$$

which is an example for a *four-part composite Wiener–Hopf equation*

$$\sum_{j=1}^{4} A_j(\hat{P}_j\hat{\Phi}) = z. \tag{2.21}$$

After dividing (2.20) by $n(\xi_1, \xi_2; k) := |\xi|^2 - k^2$, we get

$$\sum_{j=1}^{4} \left(1 + \frac{k^2 - k_j^2}{n(\xi_1, \xi_2; k)}\right)\hat{\Phi}_j(\xi_1, \xi_2) = \frac{z(\xi_1, \xi_2)}{n(\xi_1, \xi_2; k)} \tag{2.22}$$

or, after applying the projectors $\hat{P}_l$, $l = 1, \ldots, 4$,

$$\hat{\Phi}_l + \sum_{j=1}^{4} (k^2 - k_j^2)\hat{P}_l n^{-1}(\xi_1, \xi_2; k)\hat{\Phi}_j = \hat{P}_l z \cdot n^{-1}. \tag{2.23}$$

For small values of $k^2 - k_j^2$, this system becomes soluble by means of the iteration scheme corresponding to Banach's fixed-point principle in $L^2(\mathbb{R}_\xi^2)$. This has been done by Meister and Latz [25] and by Kraut and Lehman [23]. The four orthogonal projectors $\hat{P}_j$ acting on the Hilbert space $L^2(\mathbb{R}^2_{\xi_1,\xi_2})$ may then be written down explicitly; e.g.,

$$\hat{P}_1 = \tfrac{1}{2}(I + S_1)\tfrac{1}{2}(I + S_2), \tag{2.24}$$

and with similar representations for $\hat{P}_2$, $\hat{P}_3$ and $\hat{P}_4$ by changing signs in front of the *partial Hilbert transforms,*

$$(S_1\hat{\Phi})(\xi_1, \xi_2) := \frac{1}{\pi i}\int_{-\infty}^{\infty} \frac{\hat{\Phi}(\tau_1, \xi_2)\,d\tau_1}{\tau_1 - \xi_1} \tag{2.25a}$$

and

$$(S_2\hat{\Phi})(\xi_1, \xi_2) := \frac{1}{\pi i}\int_{-\infty}^{\infty} \frac{\hat{\Phi}(\xi_1, \tau_2)\,d\tau_2}{\tau_2 - \xi_2}. \tag{2.25b}$$

Latz [22] extended the investigations in order to include any finite number N of dielectric wedges with different material constants $\varepsilon_j, \sigma_j > 0$. He managed to show that there always exist complex k such that (2.23) contains a contracting operator. He showed that it is sufficient to choose

$$k \in \left\{\kappa = k^2 \in \mathbb{C} : 0 \leqslant \left|\frac{k_j^2 - k^2}{k^2}\right| < 1 \quad \text{for all} \quad j = 1, \ldots, N \quad \text{and} \quad 0 \leqslant |\text{Re}\, k| \leqslant \text{Im}\, k\right\}.$$

After a retransformation to the original (x_1, x_2)-space, Eqn (2.23) turns into a system of convolution equations:

$$\Phi_l(x_1, x_2) + \sum_{j=1}^{4} (k^2 - k_j^2)\chi_{\Gamma_l}(x_1, x_2)\frac{1}{4i}\iint_{\Gamma_j} H_0^{(1)}(k\sqrt{((x_1-y_1)^2+(x_2-y_2)^2)})$$
$$\cdot\, \Phi_j(y_1, y_2)\, \mathrm{d}(y_1, y_2) = \chi_{\Gamma_l}(x_1, x_2)\cdot(F^{-1}(z\cdot n^{-1}))(x_1, x_2)$$
$$\text{for } l = 1, 2, 3, 4. \quad (2.26)$$

In the case of $k = k_1 = k_d$ and $k_2 = k_3 = k_4 = k_v$, where $1 < |k_d/k_v| < \sqrt{2}$, the iteration procedure has been applied and approximate solutions have been calculated by Kuo and Plonus in [21], Kraut and Lehman [23] and, recently, by Rawlins [24].

2.3 Diffraction of electromagnetic waves by three-dimensional semi-infinite bodies

Let G_j, $j = 1, \ldots, N$, denote a finite number of closed domains decomposing $\mathbb{R}^3$ with boundaries ∂G_j being piecewise smooth such that Gauss' theorem may be applied to any finite portion $G_j(R) := G_j \cap K_R(0)$, where $K_R(0)$ is a ball of radius R with centre at the origin O. We assume that an electromagnetic primary field (E_{pr}, H_{pr}) with harmonic time-dependence is generated by sources confined to one, let's say the first, region G_1. The N regions G_j are assumed to be filled by different dielectric absorbing materials characterized by the coefficients $\mu_j = \mu$, ε_j and $\sigma_j > 0$, as in section 2.2 above.

The following problem is considered by Latz [26].

Problem *Find the total electromagnetic field* (E_{tot}, H_{tot}) *in the whole* $\mathbb{R}^3$*-space. Separating the primary field, the following conditions are to be satisfied by the scattered field* (E, H) *in* G_j. *Let* E_j, H_j *be* $\in (C^1(\mathring{G}_j) \cap C(\bar{G}_j))^3$ *such that Maxwell's equations hold:*

$$\left.\begin{aligned} \operatorname{curl} E_j - i\omega\mu H_j &= 0 \\ \operatorname{curl} H_j + i\omega\eta_j E_j &= 0 \end{aligned}\right\} \quad \text{in} \quad \mathring{G}_j, \qquad (2.27)$$

where $\eta_j := \varepsilon_j + i\sigma_j/\omega$, $\omega > 0$. *On the common non-void boundaries* $\partial G_j \cap \partial G_k$, *the transmission conditions hold:*

$$\begin{aligned} (E_j - E_k) \wedge n_{jk}(x) &= A^0(x), \\ (H_j - H_k) \wedge n_{jk}(x) &= B^0(x), \end{aligned} \qquad (2.28)$$

where $n_{jk}(x)$ *denotes the unit normal vector pointing outward from* G_k *into* G_j *for neighbouring regions and* $A^0(x) := -E_{pr}(x) \wedge n(x)$ *on* $\partial G_2 \cap \partial G_1$ *and* $= E_{pr}(x) \wedge n(x)$ *on* $\partial G_N \cap \partial G_1$ *and, similarly,* $B^0(x)$. *Besides these interface conditions, the asymptotic conditions will hold:*

$$\text{curl } E, \text{ curl } H \in L^2_{\text{loc}}(\mathbb{R}^3) \tag{2.29}$$

and

$$E, \frac{\partial}{\partial x_l} E, H, \frac{\partial}{\partial x_l} H = O(\mathrm{e}^{-qr}) \quad \textit{for} \quad l = 1, 2, 3 \tag{2.30}$$

as $r := \sqrt{(x_1^2 + x_2^2 + x_3^2)} \to \infty$ *with some* $q := \min_j (\mu\sigma_j\omega) > 0$.

In order to transform this problem into an N-part Wiener–Hopf problem for vector-valued transform functions, we shall apply the three-dimensional Fourier transform F and shall use the following version of Gauss' theorem:

$$\int_B (\langle R, \text{curl } S\rangle) - (\langle S, \text{curl } R\rangle) \, \mathrm{d}(x_1, x_2, x_3) = -\oint_B \langle R \wedge S, n\rangle \, \mathrm{d}o \tag{2.31}$$

for $R, S : B \to \mathbb{C}^3$ continuous and possessing integrable curl on the bounded domain B which has a piecewise smooth boundary ∂B. Now insert $C\mathrm{e}^{ix\xi} = C\mathrm{e}^{i(x_1\xi_1 + x_2\xi_2 + x_3\xi_3)}$ with a constant $\mathbb{C}^3$-valued vector C for R and $E-$ or $H-$ for S. Taking into account Maxwell's equations, this leads to

$$\int_{G_j} (\mathrm{e}^{ix\xi}[\langle C, i\omega\mu H_j\rangle + \langle E_j, C \wedge (i\xi_1, i\xi_2, i\xi_3)\rangle] \, \mathrm{d}(x_1, x_2, x_3)$$

$$= -\int_{\partial G_j} \mathrm{e}^{ix\xi} \langle C \wedge E_j, n_j\rangle \, \mathrm{d}o_x \tag{2.32}$$

and a similar equation with E_j and H_j interchanged. After choosing C to be one of the cartesian unit vectors and after introducing the matrix

$$\mathbf{Z}(\xi_1, \xi_2, \xi_3) := \begin{pmatrix} 0 & -i\xi_3 & i\xi_2 \\ i\xi_3 & 0 & -i\xi_1 \\ -i\xi_2 & i\xi_1 & 0 \end{pmatrix}, \tag{2.33}$$

we may add all N equations (2.32) and recognize that the boundary terms on the right-hand sides adjust exactly for the transmission conditions in such a way as to allow for retransformation of the boundary integral along ∂G_1 into a domain integral over G_1 with the primary field components inserted ($\mu = 1$ for convenience):

$$\int_{\mathbb{R}^3} (\mathrm{e}^{ix\xi}[i\omega H(x) - \mathbf{Z}(\xi)E(x)] \, \mathrm{d}(x_1, x_2, x_3)$$

$$= -\int_{G_1} \mathrm{e}^{ix\xi}[i\omega H_{\text{pr}}(x) - \mathbf{Z}(\xi)E_{\text{pr}}(x)] \, \mathrm{d}(x_1, x_2, x_3) \tag{2.34a}$$

and, similarly,

$$\int_{\mathbb{R}^3} e^{ix\xi}[-i\omega\eta E(x) - \mathbf{Z}(\xi)H(x)]\, d(x_1, x_2, x_3)$$
$$= -\int_{G_1} e^{ix\xi}[-i\omega\eta_1 E_{pr}(x) - \mathbf{Z}(\xi)H_{pr}(x)]\, d(x_1, x_2, x_3), \quad (2.34b)$$

where we have $\eta\,|_{G_j} = \eta_j$, $E\,|_{G_j} = E_j$ and $H\,|_{G_j} = H_j$, $j = 1, \ldots, N$. After multiplying the first equation by $\mathbf{Z}(\xi)$ and the second by $i\omega$ and adding both, we obtain, $\mathbf{I}$, denoting the 3×3-unit matrix,

$$\int_{\mathbb{R}^3} e^{ix\xi}[-\mathbf{Z}^2(\xi) + \omega^2\eta\mathbf{I}]E(x)\, d(x_1, x_2, x_3)$$
$$= -\int_{G_1} e^{ix\xi}[-\mathbf{Z}^2(\xi) + \omega^2\eta_1\mathbf{I}]E_{pr}(x)\, d(x_1, x_2, x_3), \quad (2.35)$$

which, in terms of the Fourier transforms, means that

$$\sum_{j=1}^{N} (-\mathbf{Z}^2(\xi) + \omega^2\eta_j \cdot \mathbf{I})\hat{E}_j(\xi) = -(-\mathbf{Z}^2(\xi) + \omega^2\eta_1 \cdot \mathbf{I})\hat{E}_{pr}(\xi) \quad \text{on} \quad \mathbb{R}^3_\xi. \quad (2.36)$$

Setting $\kappa_j := \omega^2\eta_j$, $j = 1, \ldots, N$, and κ arbitrary within $\mathbb{C}^+ = \{\zeta \in \mathbb{C} : \operatorname{Im} \zeta > 0\}$, we may multiply Eqn (2.36) by

$$(-\mathbf{Z}^2(\xi) + \kappa\mathbf{I})^{-1} = x(|\xi|^2 - \kappa)^{-1} \cdot \begin{pmatrix} \xi_1^2 - \kappa & \xi_1\xi_2 & \xi_1\xi_3 \\ \xi_1\xi_2 & \xi_2^2 - \kappa & \xi_2\xi_3 \\ \xi_1\xi_2 & \xi_2\xi_3 & \xi_3^2 - \kappa \end{pmatrix} \quad (2.37)$$

and arrive at

$$\sum_{j=1}^{N} (\mathbf{I} + (\kappa_j - \kappa)(-\mathbf{Z}^2(\xi) + \kappa)^{-1})\hat{E}_j(\xi)$$
$$= -(\mathbf{I} + (\kappa_1 - \kappa)(-\mathbf{Z}^2(\xi) + \kappa))^{-1})\hat{E}_{pr}(\xi) \quad \text{on} \quad \mathbb{R}^3. \quad (2.38)$$

This is a special N-part composite Wiener–Hopf equation on $(L^2(\mathbb{R}^3_\xi))^3$, specializing to $\kappa = \kappa_1$, the right-hand side of Eqn (2.37) just gives $\hat{E}_{pr}(\xi)$. Applying consecutively the N projectors $\hat{P}_l := FP_lF^{-1} = F\chi_{G_l} \cdot F^{-1}$, we obtain, alternatively, the following N-system:

$$\hat{E}_l(\xi) + \sum_{j=1}^{N} (\kappa_j - \kappa)\hat{P}_l(-\mathbf{Z}^2(\xi) + \kappa\mathbf{I})^{-1}\hat{E}_j(\xi)$$
$$= -(\delta_{l1} \cdot \mathbf{I} + (\kappa_1 - \kappa)\hat{P}_l(-\mathbf{Z}^2(\xi) + \kappa\mathbf{I})^{-1})\hat{E}_{pr}(\xi) \quad \text{for} \quad l = 1, \ldots, N. \quad (2.39)$$

Now, if we write

$$\mathbf{K}(\xi; \kappa) = \mathbf{K}_1(\xi; \kappa) \cdot K_2(\kappa), \quad (2.40a)$$

where

$$\mathbf{K}_1(\xi;\kappa):=\kappa(-\mathbf{Z}^2(\xi)+\kappa)^{-1} \tag{2.40b}$$

$$K_2(\kappa):=\sum_{j=1}^{N}\frac{(\kappa-\kappa_j)}{\kappa}\hat{P}_j, \tag{2.40c}$$

we obtain an estimate for $|\mathbf{K}_1(\xi_i\kappa)v|$ for vectors $v\in\mathbb{C}^3$ and matrices $\mathbf{K}_1\in\mathbb{C}^{3\times 3}$: $\langle\mathbf{K}_1v,\mathbf{K}_1v\rangle=\langle\mathbf{K}_1^*\mathbf{K}_1v,v\rangle\leqslant\lambda_{\max}\langle v,v\rangle$, where $\lambda_{\max}$ is the greatest eigenvalue of the Hermitean matrix $\mathbf{K}_1^*\mathbf{K}_1$. Here, we have the following eigenvalues:

$$\lambda_1(\xi;\kappa)=\lambda_2(\xi;\kappa)=|\kappa|^2\,||\xi|^2-\kappa|^{-2},\qquad \lambda_3=1. \tag{2.41}$$

$\lambda_3=1$ is the greatest eigenvalue for all $\xi=(\xi_1,\xi_2,\xi_3)\in\mathbb{R}^3$ if $|\kappa|^2\leqslant||\xi|^2-\kappa|^2$, which is equivalent to $0\leqslant|\xi|^2(|\xi|^2-2\operatorname{Re}\kappa)$, which will be the case for $\operatorname{Re}\kappa\leqslant 0$. Choosing

$$\kappa\in\mathbb{C}_0:=\{z\in\mathbb{C}:\operatorname{Re}z<0\quad\text{and}\quad|z-\kappa_j|<|z|,\,j=1,\ldots,N\}$$

to be a non-empty set, we have

$$\left|\mathbf{K}_1(\xi;\kappa)\frac{\kappa-\kappa_j}{\kappa}v\right|^2_{\mathbb{R}^3}\leqslant\left|\frac{\kappa-\kappa_j}{\kappa}\right|^2|v|^2_{\mathbb{R}^3}<|v|^2_{\mathbb{R}^3} \tag{2.42}$$

for all $j=1,\ldots,N$, which results in the following estimate of $(L^2(\mathbb{R}^3))^3$-norms:

$$\begin{aligned}&\left\|\sum_{j=1}^{N}(\kappa_j-\kappa)(-\mathbf{Z}^2(\xi)+\kappa\mathbf{I})^{-1}\hat{P}_j\hat{E}(\xi)\right\|^2\\&=\left\|\mathbf{K}_1(\xi;\kappa)\sum_{j=1}^{N}\frac{\kappa-\kappa_j}{\kappa}\hat{P}_j\hat{E}(\xi)\right\|^2\\&\leqslant\left\|\sum_{j=1}^{N}\frac{\kappa-\kappa_j}{\kappa}\hat{P}_j\hat{E}(\xi)\right\|=\sum_{j=1}^{N}\left|\frac{\kappa-\kappa_j}{\kappa}\right|^2\|\hat{P}_j\hat{E}(\xi)\|^2<\sum_{j=1}^{N}\|\hat{P}_j\hat{E}(\xi)\|^2=\|\hat{E}(\xi)\|^2\end{aligned} \tag{2.43}$$

due to the orthogonality of $\hat{P}_j\hat{E}$ and $\hat{P}_1\hat{E}$ for $j\neq 1$. Thus, Eqn (2.37) may be solved by the fixed-point principle uniquely in $(L^2(\mathbb{R}^3))^3$. A Fourier inverse transform applied to Eqn (2.38) would lead to an N-system of singular integral equations involving a matrix $(F_\xi^{-1}\mathbf{K}(\xi;\kappa))(x)$ of translation invariant operators whose kernels behave like singular Riesz kernels. In the special case of $\kappa=0$, they are products of such Riesz transforms.

2.4 Diffraction of elastic waves by three-dimensional semi-infinite bodies

The above investigations may also be applied to the study of the behaviour of elastic waves propagating in $\mathbb{R}^3$-space, which is divided into N

semi-infinite regions $\bar{G}_j$ filled by different homogeneous isotropic absorbing materials. Following an orally communicated idea by Latz, we again assume that Gauss' theorem may be applied to every bounded subdomain $\bar{G}_j(R) := \bar{G}_j \cap K_R(0)$. Let a primary field characterized by the displacement vector field $U_{pr}(x)$, the stress tensor field $\mathbf{S}_{pr}(x)$, and the body force field $K_{pr}(x)$ be given in $\bar{G}_1$ with the density ρ_1, the Lamé coefficients λ_1, μ_1, and the damping coefficient η_1. Our problem then is the following one.

Problem *Find* $U_j(x) \in C^2(\mathring{G}_j) \cap C^1(\bar{G}_j') \cap C(\bar{G}_j)$, $j = 1, \ldots, N$, $\mathbf{S}(x) = (\sigma_{kl}(x))_{k,l=1,2,3} \in C^1(\bar{G}_j') \cap C^1(\bar{G}_j)$, *where* $\bar{G}_j'$ *denotes the closed regions with the exception of edges and vertices of their boundaries. These fields shall satisfy the equations of a harmonically time-dependent motion*

$$div\ \mathbf{S}_j + (\omega^2 \rho_j + i\omega\eta_j) U_j + K_j = 0, \tag{2.44}$$

where $K_j = 0$ *for* $j \neq 1$. *Hooke's law is assumed to hold, i.e.,*

$$\mathbf{S}_j = \lambda_j\ div\ U_j \cdot (\delta_{kl})_{k,l=1,2,3} + \mu_j \left(\frac{\partial U_k^{(j)}}{\partial x_l} + \frac{\partial U_l^{(j)}}{\partial x_k} \right)_{k,l=1,2,3}. \tag{2.45}$$

Across the common non-empty boundaries $\partial G_j \cap \partial G_{j-1}$, $j = 1, \ldots, N$, *the total displacement vectors and surface tensions shall behave continuously, i.e.,*

$$U_{j,\text{tot}} \big|_{\partial G_{j-1}} = U_{j-1,\text{tot}} \big|_{\partial G_{j-1}} \tag{2.46a}$$

and

$$\mathbf{S}_{j,\text{tot}} n_j \big|_{\partial G_j} + \mathbf{S}_{j-1,\text{tot}} n_{j-1} \big|_{\partial G_{j-1}} = 0, \tag{2.46b}$$

where n_j *denotes the outward-directed unit normal vector to* ∂G_j *such that* $n_{j-1} = -n_j$ *at common boundary points.*

In the neighbourhood of edges and vertices, we claim that $\mathbf{S}_j$, $\nabla \mathbf{S}_j$, $\nabla \otimes U_j$ *behave in such a way that the integrals occurring in the following context converge absolutely. Additionally, as* $r \to \infty$ *we claim that*

$$U_j, \nabla \otimes U_j, \mathbf{S}_j, \text{div}\ \mathbf{S}_j = O(e^{-qr}) \tag{2.47}$$

with some $q > 0$.

We are going to derive an N-part composite Wiener–Hopf system for 10-vectors having

$$U_h^{(j)}, h = 1, 2, 3, \sigma_{kl}^{(j)}, \qquad 1 \leq k \leq l \leq 3, \quad \text{and} \quad \theta_j := \text{div}\ U_j$$

as components.

Let $\mathbf{T}$ denote a symmetric continuous bounded tensor field with div $\mathbf{T} \in L^1(\bar{G})$ and assume the existence of a continuously-differentiable bounded vector field with $\nabla \otimes V := (\partial V_l / \partial x_k)_{k,l=1,2,3} \in L^1(\bar{G})$, $\bar{G}$ a bounded regular domain; then, the following integral theorems hold:

$$\int_B \operatorname{div} \mathbf{T} \, \mathrm{d}x = \oint_{\partial B} \mathbf{T}n \, \mathrm{d}o, \tag{2.48}$$

$$\int_B \nabla \otimes V \, \mathrm{d}x = \oint_{\partial B} n \otimes V \, \mathrm{d}o, \tag{2.49}$$

where $n \otimes V := (n_k V_l)_{k,l=1,2,3}$ denotes the tensor product of V and the surface normal vector n.

Now we choose $\mathbf{T} := \mathbf{S}\mathrm{e}^{ix\xi}$ and obtain

$$\int_B \mathrm{e}^{ix\xi}[\operatorname{div} \mathbf{S} + i\mathbf{S} \cdot \xi] \, \mathrm{d}x = \oint_{\partial B} \mathrm{e}^{ix\xi}\mathbf{S}n \, \mathrm{d}o_x. \tag{2.50}$$

Taking $\mathbf{S} = \mathbf{S}_{\text{tot}}$ and $B = \bar{G}_j \cap K_R(0)$ and letting $R \to \infty$, we obtain the following result due to the relations (2.46b) and (2.47):

$$\sum_{j=1}^{N} \int_{\bar{G}_j} \mathrm{e}^{ix\xi}[\operatorname{div} \mathbf{S}_j + i\mathbf{S}_j\xi] \, \mathrm{d}x + \int_{\bar{G}_1} \mathrm{e}^{ix\xi}[\operatorname{div} \mathbf{S}_{\text{pr}} + i\mathbf{S}_{\text{pr}}\xi] \, \mathrm{d}x = 0. \tag{2.51}$$

Again, denoting the three-dimensional Fourier transform by F and $P_j\mathbf{S}_j = \chi_{\bar{G}_j} \cdot \mathbf{S}_j$, we obtain

$$\sum_{j=1}^{N} (FP_j \operatorname{div} \mathbf{S} + i(FP_j\mathbf{S})\xi) = -FP_1 \operatorname{div} \mathbf{S}_{\text{pr}} - i(FP_1\mathbf{S}_{\text{pr}})\xi. \tag{2.52}$$

Inserting the left-hand side of Eqn (2.44), we obtain

$$-\sum_{j=1}^{N} (\omega^2\rho_j + i\omega\eta_j)FP_jU + i\sum_{j=1}^{N} (FP_j\mathbf{S}) \\ = -FP_1 \operatorname{div} \mathbf{S}_{\text{pr}} - i(FP_1\mathbf{S}_{\text{pr}})\xi + FP_1K_1 =: FP_1R_{\text{pr}}, \tag{2.53}$$

where the vector field R_{pr} is known in $\bar{G}_1$.

Now we insert $U_{j,\text{tot}}\mathrm{e}^{ix\xi}$ for V in Eqn (2.49) and obtain

$$\int_{\bar{G}_j} \mathrm{e}^{ix\xi}[\nabla \otimes U_{j,\text{tot}} + i\xi \otimes U_{j,\text{tot}}] \, \mathrm{d}x = \oint_{\partial G_j} \mathrm{e}^{ix\xi}(n_j \otimes U_{j,\text{tot}}) \, \mathrm{d}o_x. \tag{2.54}$$

Summing from 1 to N, splitting $U_{1,\text{tot}}$ into $U_1 + U_{\text{pr}}$ and taking into account the relations (2.46a), we arrive at

$$\sum_{j=1}^{N} \int_{\bar{G}_j} \mathrm{e}^{ix\xi}[\nabla \otimes U_j + i\xi \otimes U_j] \, \mathrm{d}x = -\int_{\bar{G}_1} \mathrm{e}^{ix\xi}[\nabla \otimes U_{\text{pr}} + i\xi \otimes U_{\text{pr}}] \, \mathrm{d}x. \tag{2.55}$$

Now, after adding the equations of the transposed tensor fields $(\nabla \otimes U_j)^{\mathrm{T}}$ and $i(\xi \otimes U_j)^{\mathrm{T}}$, respectively, we make use of Hooke's law and obtain

$$\sum_{j=1}^{N} \left(\frac{1}{\mu_j} FP_j\mathbf{S} - \frac{\lambda_j}{\mu_j} FP_j\theta\mathbf{I} + iFP_jU \otimes \xi + i\xi \otimes FP_jU \right) = -FP_1(\nabla \otimes U_{\mathrm{pr}} + (\nabla \otimes U_{\mathrm{pr}})^{\mathrm{T}}) - iF(P_1U_{\mathrm{pr}} \otimes \xi + \xi \otimes P_1U_{\mathrm{pr}}) =: FP_1\mathbf{W}_{\mathrm{pr}}, \quad (2.56)$$

which is a known tensor field. Here, we have introduced the notation $\theta := \operatorname{div} U$.

The final equation we are looking for is derived from the most common form of Gauss theorem, viz.,

$$\int_{\bar{G}_j} \operatorname{div}(U_{j,\mathrm{tot}} \mathrm{e}^{ix\xi})\, \mathrm{d}x = \int_{\bar{G}_j} \mathrm{e}^{ix\xi}[\operatorname{div} U_{j,\mathrm{tot}} + i\langle U_{j,\mathrm{tot}}, \xi\rangle]\, \mathrm{d}x = \int_{\partial G_j} \mathrm{e}^{ix\xi}\langle U_{j,\mathrm{tot}}, n_j\rangle\, \mathrm{d}o_x, \quad (2.57)$$

Splitting $U_{1,\mathrm{tot}}$ into U_1 and U_{pr} and summing, we obtain

$$\sum_{j=1}^{N} (FP_j\theta + i\langle FP_jU, \xi\rangle) = -FP_1\theta_{\mathrm{pr}} - i\langle FP_1U_{\mathrm{pr}}, \xi\rangle =: FP_1r, \quad (2.58)$$

which is a known function.

The equations (2.52), (2.56) and (2.58) constitute the announced system of Wiener–Hopf equations for 10-vectors $(FP_j\theta, FP_jU, FP_j\mathbf{S})$. The corresponding N 10×10-matrices of functions on $\mathbb{R}^3_\xi$ are only partially occupied by elements different from zero. A careful investigation gives the following N matrices $\mathbf{M}^{(j)}(\xi; \lambda_j, \mu_j)$, from which we write down one representative:

$$\mathbf{M}(\xi; \lambda, \mu) = \begin{pmatrix} 1 & i\xi_1 & i\xi_2 & i\xi_3 & 0 & 0 & 0 & 0 & 0 & 0 \\ 0 & M_{22} & 0 & 0 & i\xi_1 & i\xi_2 & i\xi_3 & 0 & 0 & 0 \\ 0 & 0 & M_{33} & 0 & 0 & i\xi_1 & 0 & i\xi_2 & i\xi_3 & 0 \\ 0 & 0 & 0 & M_{44} & 0 & 0 & i\xi_1 & 0 & i\xi_2 & i\xi_3 \\ -\lambda/\mu & 2i\xi_1 & 0 & 0 & \mu^{-1} & & & & & \\ 0 & i\xi_2 & i\xi_1 & 0 & & \mu^{-1} & & & 0 & \\ 0 & i\xi_3 & 0 & i\xi_1 & & & \mu^{-1} & & & \\ -\lambda/\mu & 0 & 2i\xi_2 & 0 & & & & \mu^{-1} & & \\ 0 & 0 & i\xi_3 & i\xi_2 & & & 0 & & \mu^{-1} & \\ -\lambda/\mu & 0 & 0 & 2i\xi_3 & & & & & & \mu^{-1} \end{pmatrix} \quad (2.59)$$

where $M_{ll} = -(\rho_l\omega^2 + i\omega\eta_l)$ for $l = 2, 3, 4$.

After taking suitable linear combinations of the equations of this Wiener–Hopf system, it is possible to decouple the system in such a way that there results a 3×3 system containing only the unknown three components of the displacement vectors $U^{(j)}$, while the six components of the stress tensors may be calculated from the remaining six equations. The matrices corresponding to the equations of motion

$$\mu_j \Delta U_j + (\lambda_j + \mu_j)\,\text{grad div}\, U_j + (\rho_j\omega^2 + i\eta_j\omega)U_j + K = 0$$

are given by

$$\mathbf{A}_j := \begin{pmatrix} \mu_j\,|\xi|^2 + (\lambda_j+\mu_j)\xi_1^2 - (\omega^2\rho_j + i\omega\eta_j) & (\lambda_j+\mu_j)\xi_1\xi_2 & (\lambda_j+\mu_j)\xi_1\xi_3 \\ (\lambda_j+\mu_j)\xi_2\xi_1 & \mu_j\,|\xi|^2 + (\lambda_j+\mu_j)\xi_2^2 - (\omega^2\rho_j + i\omega\eta_j) & (\lambda_j+\mu_j)\xi_2\xi_3 \\ (\lambda_j+\mu_j)\xi_3\xi_1 & (\lambda_j+\mu_j)\xi_3\xi_2 & \mu_j\,|\xi|^2 + (\lambda_j+\mu_j)\xi_3^2 - (\omega^2\rho_j + i\omega\eta_j) \end{pmatrix}, \tag{2.60}$$

the determinant of which is

$$\det \mathbf{A}_j = (\mu_j\,|\xi|^2 - (\omega^2\rho_j + i\eta_j\omega))^2((\lambda_j + 2\mu_j)\,|\xi|^2 - (\omega^2\rho_j + i\omega\eta_j)). \tag{2.61}$$

One can show that the matrices $\mathbf{A}_j$ are unitarily equivalent to the diagonal matrices

$$\mathbf{D}_j = \begin{pmatrix} \mu_j\,|\xi|^2 - (\omega^2\rho_j + i\eta_j\omega) & 0 & 0 \\ 0 & \mu_j\,|\xi|^2 - (\omega^2\rho_j + i\eta_j\omega) & 0 \\ 0 & 0 & (\lambda_j + 2\mu_j)\,|\xi|^2 - (\omega^2\rho_j + i\eta_j\omega) \end{pmatrix} \tag{2.62}$$

where we have $\mathbf{D}_j = \mathbf{T}^*\mathbf{A}_j\mathbf{T}$ with the orthogonal matrix

$$\mathbf{T} = (e_1, e_2, e_3) = \begin{pmatrix} \dfrac{-\xi_2}{\sqrt{(\xi_1^2+\xi_2^2)}} & \dfrac{-\xi_1\xi_3}{|\xi|\sqrt{(\xi_1^2+\xi_2^2)}} & \dfrac{\xi_1}{|\xi|} \\ \dfrac{\xi_1}{\sqrt{(\xi_1^1+\xi_2^2)}} & \dfrac{-\xi_2\xi_3}{|\xi|\sqrt{(\xi_1^2+\xi_2^2)}} & \dfrac{\xi_2}{|\xi|} \\ 0 & \dfrac{\sqrt{(\xi_1^2+\xi_2^2)}}{|\xi|} & \dfrac{\xi_3}{|\xi|} \end{pmatrix} \tag{2.63}$$

where $\mathbf{T}$ is independent of j!

Similar equations may be derived in the aperiodic time-dependent case, where we have to replace $i\omega$ by s with $\operatorname{Re} s > s_0$ after applying a one-dimensional Laplace transform with respect to t.

3 On general Wiener–Hopf operators

3.1 Formulation of the problem and some examples

In 1964, Shinbrot [27] started his investigations on *general Wiener–Hopf operators* in Hilbert spaces while Devinatz [28] was concerned with similar studies in Toeplitz operators. Their work was published in several papers [29–34] and continued by Reeder [35, 36] and Pellegrini [37, 38], attaining a certain final stage. These authors consider operators of the type

$$T_p(A) = PA\,|_{R(P)} \tag{3.1}$$

where P is always an orthogonal projector acting on a separable Hilbert space $\mathcal{H}$ and A is a linear and injective operator with domain $D(A)$ and range $R(A)$.

All operators are assumed to be linear in the following, but they do not necessarily belong to $\mathcal{L}(\mathcal{H})$, the Banach space of all bounded operators on $\mathcal{H}$. A is said to be *invertible* (on $\mathcal{H}$) iff A is a homeomorphism of $\mathcal{H}$.

First, we give a few examples of special Wiener–Hopf operators (WHO*s*).

1 The WH integral operator occurring in Eqn (1.1) belongs to this type when it is defined as follows:

$$(T_p(A)\varphi)(t) := \lambda\varphi(t) - \int_0^\infty k(t-\tau)\varphi(\tau)\,d\tau, \qquad t \geq 0, \tag{3.2}$$

for $\varphi \in L^2(\mathbb{R}_+)$, $\lambda \in \mathbb{C}$, $k \in L^1(\mathbb{R})$, $D(A) = \mathcal{H} = L^2(\mathbb{R})$, $A := \lambda I - k*$ with the *convolution operator*

$$(k*u)(t) := \int_{-\infty}^{+\infty} k(t-\tau)u(\tau)\,d\tau \tag{3.3}$$

and the *space-projector* P defined by

$$(Pu)(t) := \begin{Bmatrix} u(t), & t \geq 0, \\ 0, & t < 0. \end{Bmatrix} \tag{3.4}$$

It is a well-known classical result that A is invertible iff $\inf_{\xi \in \mathbb{R}} |\lambda - \hat{k}(\xi)| > 0$, $\hat{k}$ denoting the Fourier transform of k.

Further examples are easily established by replacing $\mathbb{R}_+$ by an arbitrary measurable set E, the invertible operator $\lambda I - k*$ by a translation invariant operator $F^{-1}\Phi(\cdot) \cdot F$, $\Phi \in L^\infty$, essinf $|\Phi(\cdot)| > 0$ or, more generally, $|\Phi(\xi)| > 0$ for almost every ξ, as in the case of classical WH equations of the first kind ($\lambda = 0$).

2 We should mention the discrete analogue of (3.2), the Toeplitz operators, as introduced in Eqn (1.2). Here we have $\mathcal{H} = L^2(T)$, the Hilbert space of all complex-valued square integrable functions on the unit-circle T or, equivalently, of all 2π-periodic functions. P denotes the linear continuous projector from $\mathcal{H}$ on to the Hardy space H^2, defined by

$$(P\varphi)(\theta) := \sum_{k=0}^{\infty} \varphi_k e^{ik\theta} \tag{3.5}$$

for all

$$\varphi(\theta) := \sum_{k=-\infty}^{\infty} \varphi_k e^{ik\theta}, \quad \text{where} \quad \sum_{k=-\infty}^{\infty} |\varphi_k|^2 < \infty.$$

The operator A is defined by an arbitrary $a(\theta) \in L^\infty(T)$ such that

$$(A\varphi)(\theta) := a(\theta)\varphi(\theta). \tag{3.6}$$

The corresponding Toeplitz operators were investigated in this context by, for example, Devinatz [28, 34].

3 In the one-dimensional case, one is led to a close connection with singular integral operators, since the following representation for the *Cauchy* (or *Hilbert*) *operator* on $L^2(\mathbb{R})$ holds:

$$(S\varphi)(t) := \frac{1}{\pi i} \int_{-\infty}^{\infty} \frac{\varphi(\tau)\, d\tau}{\tau - t} = (F^{-1} \operatorname{sgn}(\cdot) \cdot F\varphi)(t). \tag{3.7}$$

From this, it follows that

$$\tfrac{1}{2}(I \pm S) = F^{-1}\chi_{\mathbb{R}\pm} \cdot F =: P_\pm \tag{3.8}$$

are orthogonal projectors on $L^2(\mathbb{R})$. Taking this into consideration, we immediately see that the following equations are equivalent:

$$a(t)\varphi(t) + b(t)(S\varphi)(t) = f(t), \qquad t \in \mathbb{R}, \tag{3.9}$$

and

$$(P_- + \Phi(t)P_+)\varphi(t) = g(t), \qquad t \in \mathbb{R}, \tag{3.10}$$

respectively, where we have set

$$\Phi(t) := (a-b)^{-1}(a+b)(t), \qquad g(t) := (a-b)^{-1}(t)f(t),$$

if these functions exist. Equation (3.10) is also equivalent [27] to

$$P_+(\Phi \cdot)\varphi = P_+\Phi \cdot P_+\varphi = P_+g. \tag{3.11}$$

After introducing the *space-projector* $P := F^{-1}P_+F = \chi_{\mathbb{R}_+} \cdot$ and the translation invariant operator $A := F^{-1}\Phi \cdot F$, the equivalence of the singular equation (3.9) to the above mentioned types is obvious, viz.,

$$T_P(A)\hat{\varphi} = P\hat{g}. \tag{3.12}$$

4 In the higher-dimensional case, there exist not only situations analogous to the equivalence between the general Wiener–Hopf equation (3.1) and the special types of Eqns (3.2) to (3.4), but there are many more kinds of singular operators: operators of the Calderón–Zygmund–Mikhlin type (CMOs), whose combinations do not result in projectors, or operators with strongly singular behaviour on submanifolds and which behave as identity operators in the orthogonal directions. Let us look at the following example.

$$Lu(x) = (a_0 I + a_1 S_1 + a_2 S_2 + a_{12} S_1 S_2) u(x) = v(x), \qquad x \in \mathbb{R}^2 \tag{3.13}$$

on $L^2(\mathbb{R}^2)$, where $a_0, \ldots, a_{12} \in L^\infty(\mathbb{R}^2)$ and $v \in L^2(\mathbb{R}^2)$ are given functions and S_j denote the *partial Cauchy transforms* along the jth coordinate line. Introducing the projectors as in Eqn (3.8), we may rewrite Eqn (3.13) as a four-part Wiener–Hopf equation:

$$\begin{aligned} Wu = [&(a_0 + a_1 + a_2 + a_{12}) P_{++} + (a_0 + a_1 - a_2 - a_{12}) P_{+-} \\ &+ (a_0 - a_1 + a_2 - a_{12}) P_{-+} + (a_0 - a_1 - a_{12} + a_{12}) P_{--}] u = v. \end{aligned}$$

with $P_{-+} := F^{-1} \chi_{\mathbb{R}_-}(\xi_1) \cdot \chi_{\mathbb{R}_+}(\xi_2) \cdot F$ as an example.

As far as the authors know, such *multiple-part composite WH equations* have not yet been treated by means of the general theory, apart from special cases in which some of the multipliers in front of the projectors are equal or absent.

One is led to a large class of multiple-part composite WH equations by taking combinations (sums and products) of operators consisting of (a) $F^{-1} \Phi \cdot F$ with Φ continuous for $\xi \neq 0$ and homogeneous of degree zero, (b) Cauchy-operators along arbitrary lines in $\mathbb{R}^n$-space, and (c) multipliers $a \in L^\infty(\mathbb{R}^n)$. These then generalize operators treated by Simonenko [39].

Problems The main problems which arise for general Wiener-Hopf equations are the following:

(i) *Under which conditions on A and P acting on a Hilbert space $\mathcal{H}$ – or, more generally, on a given Banach space $\mathfrak{X}$ – do there exist solutions of*

$$T_P(A)\varphi = f \tag{3.14}$$

for $f \in R(P)$ given and when are they unique?

(ii) *Is there an equivalence between the unique solubility of Eqn* (3.14) *and the possibility of factorizing A, analogous to the classical WH equation?*

(iii) *What do the representations of the solution look like by orthogonal series expansions in the case of a Hilbert space $\mathcal{H}$ given by eigenvalue problems?*

3.2 The unique solubility of general WHOs with invertible A

Here, we wish to present the general idea of factorization. We assume A to be a positive operator invertible on $\mathscr{H}$, i.e., $(Au, u) \geq \delta \|u\|^2$ for all $u \in \mathscr{H}$ and $\delta > 0$. The equation

$$T_P(A)\varphi := PAP\varphi = f \tag{3.15}$$

is equivalent to

$$(Q + AP)u = v \tag{3.16}$$

where $Q = I - P$ is the complementary continuous projector on $\mathscr{H}$ and $Pv = f$. Since $A > 0$ and $A^{-1} \in \mathscr{L}(\mathscr{H})$, both operators $A^{1/2}$ and $A^{-1/2}$ belong to $\mathscr{L}(\mathscr{H})$.

We try to find a factorization

$$A = A_- A_+, \tag{3.17}$$

where $A_+, A_- \in \mathscr{L}(\mathscr{H})$ with $R(A_+P) = R(P)$ and $R(A_-Q) = R(Q)$, both invertible on $\mathscr{H}$. If such a factorization exists, we may write Eqn (3.16) in the form

$$(Q + AP)u = A_-(A_-^{-1}Q + A_+P)u = v, \tag{3.18}$$

which is obviously uniquely soluble by

$$u = A_-QA_-^{-1}v + A_+^{-1}PA_-^{-1}v \tag{3.19}$$

from which the solution of (3.15) follows:

$$\varphi = Pu = A_+^{-1}PA_-^{-1}f. \tag{3.20}$$

The factors in Eqn (3.17) may be calculated by means of the formulae

$$A_+ := UA^{1/2} \quad \text{and} \quad A_- := (UA^{-1/2})^{-1}, \tag{3.21}$$

where U is the unitary operator given by

$$Uw = \sum_j (w, \varphi_j^+)\psi_j^+ + \sum_k (w, \varphi_k^-)\psi_k^-; \tag{3.22}$$

$\{\psi_j^+\}$, $\{\psi_k^-\}$, $\{\varphi_j^+\}$ and $\{\varphi_k^-\}$ denote complete orthonormal systems of the ranges of P, Q, $A^{1/2}P$, and $A^{-1/2}Q$, respectively [27]. Summarizing, we obtain

Theorem 2 *Let A be a positive and invertible operator on a separable Hilbert space $\mathscr{H}$ and let P be an orthogonal projector. Then the WH Eqn (3.15) is uniquely soluble for all $f \in R(P)$ by Eqn (3.20).*

But positivity of A is not necessary for $T_P(A)$ to be invertible on $R(P)$, as the following lemmas prove.

Lemma 1 *Let A be a* strongly elliptic *operator, i.e., $\operatorname{Re} A \geqslant \delta > 0$, which means that $\operatorname{Re}(Au, u) \geqslant \delta \|u\|^2$ for all $u \in \mathcal{H}$; then $T_P(A)$ is invertible on $R(P)$.*

This can be understood immediately by taking $u = Pu$, $\|u\| = 1$ and estimating

$$\|T_P(A)u\| \geqslant |(T_P(A)u, u)| = |(Au, u)| \geqslant \delta \|u\|^2 = \delta. \tag{3.23}$$

In the case of translation invariant operators $A = F^{-1}\Phi \cdot F$ on $\mathcal{H} = L^2(\mathbb{R}^n)$, the strong ellipticity is equivalent to $\operatorname{Re} \Phi(\xi) \geqslant \delta > 0$ for a.e. $\xi \in \mathbb{R}^n$.

It is obvious that multiplication by a fixed complex number $a \neq 0$ does not destroy the invertibility of $T_P(A)$; in other words, even more generally the following lemma holds.

Lemma 2 *Let A, H be linear and invertible on $\mathcal{H}$, P an orthogonal projector such that $R(HP) = R(P)$ and $T_P(AH)$ invertible; then $T_P(A)$ is also invertible on $R(P)$.*

These results give the proof of the sufficiency part of

Theorem 3 [31, Theorem 3] *Let A be invertible on $\mathcal{H}$; then $T_P(A)$ is invertible on $R(P)$ iff there exists a linear invertible operator H on $\mathcal{H}$ with $R(HP) = R(P)$ and such that AH is strongly elliptic on $\mathcal{H}$.*

To prove the necessity of the condition, one has to make use of the polar decomposition of A into $\tilde{A}B := A^{*-1}B^*B := A^{*-1}CU^*UC$, where $C = (A^*A)^{1/2}$ and U as in Eqn (3.22), setting C there instead of $A^{1/2}$. H is given by

$$H = (P\tilde{A}P + Q\tilde{A}Q)B. \tag{3.24}$$

The second main result of the theory of general WHOs is the necessity of the existence of a factorization, given by

Theorem 4 [31, Theorem 5] *Let A be invertible on $\mathcal{H}$; then $T_P(A)$ is invertible on $R(P)$ iff there are invertible operators $A_\pm$ such that $R(A_+P) = R(P)$, $R(A_-Q) = R(Q)$, with $Q := I - P$, and $A = A_-A_+$.*

Corollary

$$[T_P(A)]^{-1} = A_+^{-1}PA_-^{-1}. \tag{3.25}$$

It is easy to prove the sufficiency of the conditions as in the case of positive A treated above. A short proof of the necessity was given by

Pellegrini [38, Theorem 1.3], who makes use of the representation

$$A_+ u = \sum_j (u, \psi_j^+)\varphi_j^+ + \sum_k (u, \psi_k^-)\psi_k^- \tag{3.26}$$

with complete orthonormal systems $\{\psi_j^+\}$, $\{\psi_k^-\}$ and $\{\varphi_j^+\}$ on $R(A^*P)$, $R(Q)$, and $R(P)$, respectively.

3.3 The case of unbounded operators A or A^{-1} and orthogonal series expansions

In the last section, the numerical range of the operator A occurs:

$$W(A) = \left\{\frac{(Au, u)}{\|u\|^2} : 0 \neq u \in D(A)\right\} \tag{3.27}$$

assuming $D(A) = \mathcal{H}$. Now, we shall consider the more general case of a dense subset $D(A) \subset H$ and an unbounded, but in general injective, A, i.e., $A^{-1}: R(A) \to D(A)$ exists as a not necessarily bounded operator.

The first papers on this topic were due to Shinbrot [29, 30] who assumes

$$W(A) \subset (0, \infty) \tag{3.28}$$

instead of the strong positivity of the numerical range. He uses the method of modifying $\mathcal{H}$ in such a way that he gets an equivalent equation

$$T_{\mathscr{P}}(\mathscr{A})\varphi = f \tag{3.29}$$

involving other spaces $\mathcal{H}_\pm$ instead of the original WH equation $T_P(A)u = v$ on $\mathcal{H}$. The new problem is to characterize these generally different function spaces for φ and f.

Without using the modified operator, Shinbrot proved previously [27] the possibility of factorizing

$$A = A_- A_+, \tag{3.30}$$

where

$$R(A_+P) = R(P),\ R(A_-^{-1}Q) = R(Q), \tag{3.31}$$

involving operators not necessarily bounded. Some interesting examples can be found in Shinbrot's paper [30].

Generalizations to the case of A with a positive real part (*accretive operator*) were studied by the same author in [32, 33] by means of analytic operator families. But according to Reeder [36], the proofs are only true for a bounded A. Finally, Pellegrini [37] obtained the following result in a simpler way.

Let $A:D(A)\to R(A)\subset\mathcal{H}$ be not necessarily bounded, then define

$$\begin{aligned}&\mathcal{H}_+:=\mathrm{cl}\,\{D(A),\|.\|_+\}, && \|u\|_+:=\|Au\|_{\mathcal{H}},\\ &P_+:\mathcal{H}_+\to\mathrm{cl}\,\{R(A)\cap D(A),\|.\|_+\}, && Q_+=I_+-P_+\end{aligned}\tag{3.32}$$

as orthogonal projectors. Furthermore, define the operator

$$\mathcal{A}:\mathcal{H}_+\to\mathcal{H}\tag{3.33}$$

by extension of A from $P(A)$ to $\mathcal{H}_+$ and the modified WHO

$$P\mathcal{A}\,|_{R(P_+)}:R(P_+)\to\mathcal{H}\tag{3.34}$$

being linear and bounded.

Theorem 5 *Let $R(P)\cap D(A)$ and $R(Q)\cap D(A^{*-1})$ be dense subsets of $R(P)$ and $R(Q)$, respectively, and let dim $R(Q)=$ dim $\mathcal{H}$ or A be bounded. Then the following statements are equivalent.*

(i) *$P\mathcal{A}\,|_{R(P_+)}$ is invertible.*
(ii) *There exists a $\delta>0$ and an invertible $L\in\mathcal{L}(\mathcal{H},\mathcal{H}_+)$ such that $R(LQ)=R(Q_+)$, $R(L^*P_+)=R(P)$ and $\mathrm{Re}\,L\mathcal{A}\geqslant\delta>0$.*
(iii) *There exist invertible operators $\mathcal{A}_+\in\mathcal{L}(\mathcal{H}_+,\mathcal{H})$, $\mathcal{A}_-\in\mathcal{L}(\mathcal{H})$ such that $R(\mathcal{A}_+P_+)=R(P)$, $R(\mathcal{A}_-^{-1}Q)=R(Q)$ and $\mathcal{A}=\mathcal{A}_-\mathcal{A}_+$.*

Now, let us briefly examine the problem of representing the solutions of WH equations by means of orthogonal series expansions. The principle may be displayed by

Theorem 6 [32, Theorem 4.1] *Let A be a bounded self-adjoint and strongly positive operator and let $\{\psi_j\}$ be a sequence such that $\mathrm{sp}\,\{\psi_j\}=R(P)$ and $\{A^{1/2}\psi_j\}$ is orthonormal. Then the solution u of the WH equation $T_P(A)u=v$ for $v=Pv$ is given by*

$$u=\sum_j(v,\psi_j)\psi_j.\tag{3.35}$$

Proof Inserting the last expression for u, we obtain

$$T_P(A)u=PA^{1/2}\sum_j(A^{-1/2}v,A^{1/2}\psi_j)A^{1/2}\psi_j=PA^{1/2}\mathbb{P}A^{-1/2}v,\tag{3.36}$$

where $\mathbb{P}$ is the orthogonal projector onto $R(A^{1/2}P)=N(PA^{1/2})^{\perp}$, i.e.,

$$T_P(A)u=PA^{1/2}A^{-1/2}v=v.\tag{3.37}$$

Remark Extensions to the case of positive operators $A:D(A)\to R(A)$ have been investigated mainly by Reeder [36] by means of biorthogonal systems.

3.4 A problem of representation

We have discussed the *Wiener–Hopf problem* concerning a fixed, though arbitrary, orthogonal projector P on $\mathcal{H}$; there is, however, a very remarkable result by Pellegrini [38] concerning the *global Wiener–Hopf problem* for all $T_P(A)$ when A is one fixed injective linear operator with a dense domain $D(A) \subset \mathcal{H}$ and P is an arbitrary orthogonal projector. Before giving the result, we introduce

Definition 1 *The operator A possesses a* Wiener–Hopf factorization (strong WH factorization) *with respect to P, if $A = A_- A_+$, where A_+ is bounded and injective, such that $R(A_+P) \subset R(P)$ and where A_- is injective, such that $D(A_-) = R(A_+)$, $R(A_-^{-1}Q) \subset R(Q)$ holds (both factor operators are invertible) with $R(A_+P) = R(P)$, $R(A_-^{-1}Q) = R(Q)$.*

Theorem 7 [38, Theorems 2.1 and 3.1] *Let A be a bounded linear operator on the separable Hilbert space $\mathcal{H}$; then the following statements are equivalent.*

(i) *$T_P(A)$ is injective (invertible) for all orthogonal projectors P on $\mathcal{H}$.*
(ii) *A possesses a (strong) WH factorization with respect to all P.*
(iii) *There is a complex number $\lambda_0 \neq 0$ such that $\operatorname{Re} \lambda_0 A > 0 (\geqslant \delta > 0)$.*

This theorem solves the following problem of representation:

Problem *Describe all invertible operators A on $\mathcal{H}$, whose WHOs $T_P(A)$ are invertible for all orthogonal P!*

Since there do not exist finite-dimensional space-projectors $P = \chi_E \cdot$ corresponding to measurable sets E with mes $E > 0$, the following problem, in the case of $\mathcal{H} = L^2(\mathbb{R}^n)$, seems to sound reasonable:

Problem *What does the characterization of all translation-invariant operators $A = F^{-1}\Phi_A \cdot F \in \mathcal{L}(L^2(\mathbb{R}^n))$ by properties of their symbol $\Phi_A \in \mathcal{L}^\infty(\mathbb{R}^n)$ look like, such that $T_P(A)$ is an injective, or even invertible, or at least Fredholm operator for all or a fixed space-projector P?*

The classical situation for $A = \lambda I - k*$ with $k \in L^1(\mathbb{R})$, $P = \chi_{\mathbb{R}_+}$ (cf., Theorem 1) could lead to the following:

Conjecture *Let $P = \chi_E \cdot$ with a measurable $E \subset \mathbb{R}^n$, $A = F^{-1}\Phi_A \cdot F$ with $0 \neq \Phi_A \in C(\dot{\mathbb{R}}^n)$, $\dot{\mathbb{R}}^n := \mathbb{R}^n \cup \{\infty\}$ being homotopic to the constant 1 in this class; then $T_P(A)$ is Fredholmian on $R(P) \subset L^2(\mathbb{R}^n)$.*

That this is not generally true can be shown by considering three-dimensional WHOs on wedge-shaped regions making use of results by Doughs and Howe [47].

4 Wiener–Hopf and Toeplitz operators on a quadrant. Bisingular integral operators

4.1 Wiener–Hopf operators on a half-space and related regions

In 1960, Goldenstein and Gohberg [40] studied the simplest multidimensional WH integral equation, viz.,

$$(W\varphi)(t) := \varphi(t) - \int_{\mathbb{R}^n_+} k(t-\tau)\varphi(\tau)\,d\tau = f(t), \qquad t \in \mathbb{R}^n_+, \tag{4.1}$$

where $\mathbb{R}^n_+ := \{t = (t_1, \ldots, t_n) \in \mathbb{R}^n : t_1 \geqslant 0\}$ denotes the *first half-space* ($n \geqslant 2$). As in the one-dimensional case, described in Theorem 1, the authors make use of the factorization of the symbol

$$\Phi_W(\xi) := 1 - \hat{k}(\xi) = 1 - (Fk)(\xi) = \Phi_-(\xi)\Phi_+(\xi), \tag{4.2}$$

where $k \in L^1(\mathbb{R}^n)$, the factors being given by

$$\Phi_\pm(\xi) := \exp\{\hat{P}_{\mathbb{R}^n_\pm} \log(1 - \hat{k}(\xi))\}, \tag{4.3}$$

the projectors $\hat{P}_{\mathbb{R}^n_\pm} := F\chi_{\mathbb{R}^n_\pm} \cdot \bar{F}^1$ operating on the Wiener algebra $\mathbb{C} \oplus FL^1(\mathbb{R}^n)$.

An important result consists in the fact that the necessary condition $\Phi_W(\xi) \neq 0$ on $\dot{\mathbb{R}}^n$ already implies that the *partial winding numbers* of $\Phi_W(\xi)$ are zero. This is equivalent to the property of Φ_W being homotopic to the constant 1 in the class of non-vanishing elements of the Wiener algebra.

For Banach spaces $\mathfrak{X}$ of functions defined on $\mathbb{R}^n_+$, similar to those mentioned in Theorem 1, we then arrive at

Theorem 8 *The WH equation* (4.1) *is uniquely soluble for an arbitrary* $f \in \mathfrak{X}$ *iff* $\Phi_W(\xi) \neq 0$ *on* $\dot{\mathbb{R}}^n$. *The solution may be written in the form*

$$\varphi(t) = f(t) + \int_{\mathbb{R}^n_+} \gamma(t, \tau) f(\tau)\,d\tau, \qquad t \in \mathbb{R}^n_+, \tag{4.4}$$

where $\gamma(t, \tau)$ *is given by*

$$\gamma(t, \tau) := \gamma_+(t-\tau) + \gamma_-(\tau - t) + \int_{\mathbb{R}^n_+} \gamma_+(t-s)\gamma_-(\tau - s)\,ds \tag{4.5}$$

for $t, \tau \in \mathbb{R}^n_+$, *and*

$$1 + \int_{\mathbb{R}^n_\pm} \gamma_\pm(t) e^{\pm i\xi t}\,dt = \Phi_\pm^{-1}(\xi). \tag{4.6}$$

Remark Take, for simplicity, $n = 2$ and choose $\Phi_W \in W_1 \otimes W_2$ as an element of the tensor product of the one-dimensional Wiener algebras; then Φ_W is continuous on $(\dot{\mathbb{R}})^2$, which is homeomorphic to the surface of the torus, instead of $(\mathbb{R}^2)^{\cdot}$, which is homeomorphic to that of the Riemannian sphere. In the former case, condition (4.3) is not sufficient to guarantee the invertibility of $T_{\mathbb{R}^2_{++}}(F^{-1}\Phi_W \cdot F)$; one has to require, additionally, the vanishing of the partial winding numbers of $\Phi_W(\xi) = \Phi_W(\xi_1, \xi_2)$. This will be discussed later on.

From the viewpoint of the general Fredholm–Noether theory, two ways emerged independently in the sixties but lead, in many cases, to the same results.

(a) The *Banach-algebraic approach* – in this context, mainly founded by the research work of Breuer and Cordes [41] and Breuer [42], which proved later on to be of significant importance to the general theory of pseudo-differential operators (cf., for example, [43]).
(b) The *local theory* founded by Simonenko [44] and continued by his work on convolutions in cones [39] and other papers.

Without touching the delicate question of priority and without going into detailed discussions of the closely related work by R. Seeley, V. S. Rabinovič and others, we want now to outline the main ideas of Simonenko's work.

Let $\overline{\mathbb{R}^n}$ denote a definite compactification of $\mathbb{R}^n$ with an extension of the Lebesgue measure by the value zero to subsets of the infinitely distant points $\overline{\mathbb{R}^n} - \mathbb{R}^n$. We may think, for example, of the *one-point-compactification* $\dot{\mathbb{R}}^n$ or the *ray-compactification* $\tilde{\mathbb{R}}^n$. When we denote the classes of all linear bounded, Fredholm, and compact operators, by $\mathscr{L}(\mathfrak{X})$, $\mathscr{F}(\mathfrak{X})$, and $\mathscr{C}(\mathfrak{X})$ respectively, operating on the Banach space $\mathfrak{X} = L^p(\mathbb{R}^n)$ $(1 < p < \infty, n \in \mathbb{N})$, we shall write for $A, B \in \mathscr{L}(\mathfrak{X})$

$$A \sim B \quad \text{iff} \quad A - B \in \mathscr{C}(\mathfrak{X}), \tag{4.7}$$

and $A \in \mathscr{F}(\mathfrak{X})$ iff there exist $R_l, R_r \in \mathscr{L}(\mathfrak{X})$ such that $R_l A \sim AR_r \sim I$. Then we have

Definition 2 (i) $A \in \mathscr{L}(L^p(\mathbb{R}^n))$ *is said to be* of local type (with respect to $\overline{\mathbb{R}^n}$), *i.e.,* $A \in \Lambda(L^p(\mathbb{R}^n))$ *iff, for all continuous functions* $\omega \in C(\overline{\mathbb{R}^n})$, *the commutators with A are compact, i.e.,* $[\omega\cdot, A] \sim 0$.

(ii) $A, B \in \Lambda$ *are said to be* locally equivalent at $x_0 \in \overline{\mathbb{R}^n}$, *i.e.,* $A \overset{x_0}{\approx} B$, *iff, for all* $\varepsilon > 0$, *there exists a neighbourhood* $U(x_0)$ *such that*

$$\inf_{V \sim 0} \|\chi_U \cdot (A - B) - V\|_{\mathscr{L}} < \varepsilon. \tag{4.8}$$

(iii) $A \in \mathscr{L}$ *is said to be* locally Fredholmian at x_0, i.e., $A \in \mathscr{F}_{x_0}$, *iff there exist a neighbourhood* $U(x_0)$ *and two operators* $R_{l,x_0}, R_{r,x_0} \in \mathscr{L}$, *such that*

$$R_{l,x_0} A\, \chi_U \cdot \sim \chi_U \cdot AR_{r,x_0} \sim 0 \tag{4.9}$$

where $\chi_U \cdot$ again denotes the operator of multiplication by the characteristic function of U.

With these notations, we may formulate the following theorems established by Simonenko (*loc. cit.*).

Theorem 9 *Let $A \in \Lambda$ with respect to $\overline{\mathbb{R}^n}$. Then $A \in \mathcal{F}$ iff $A \in \mathcal{F}_{x_0}$ for all $x_0 \in \overline{\mathbb{R}^n}$.*

Theorem 10 *Let $A, B \in \Lambda$ with respect to $\overline{\mathbb{R}^n}$ and $A \overset{x_0}{\simeq} B$ for a fixed $x_0 \in \overline{\mathbb{R}^n}$. Then $A \in \mathcal{F}_{x_0}$ iff $B \in \mathcal{F}_{x_0}$.*

These preparations lead to the following concept

Definition 3 *$A \in \Lambda(L^p(\mathbb{R}^n))$ is called a* generalized convolution operator (with respect to $\overline{\mathbb{R}^n}$) *iff, for any $x_0 \in \overline{\mathbb{R}^n}$, there is a translation invariant operator of local type, A_{x_0}, such that $A \overset{x_0}{\simeq} A_{x_0}$.*

Examples (1) For $\overline{\mathbb{R}^n} = \dot{\mathbb{R}}^n$, one may admit the singular integral operators of the Calderón–Zygmund–Mikhlin type (CMOs):

$$Au(x) := a(x)u(x) + \int_{\mathbb{R}^n} \frac{f\left(x, \dfrac{x-y}{|x-y|}\right)}{|x-y|^n} u(y)\,\mathrm{d}y, \tag{4.10}$$

where $a \in C(\dot{\mathbb{R}}^n)$ and $f \in C(\dot{\mathbb{R}}^n, L^q(S^{n-1}))$ with $1 < q < \infty$ and $\int_{S^{n-1}} f(x, \theta)\,\mathrm{d}\theta = 0$ holding for all $x \in \dot{\mathbb{R}}^n$. (This is not true for $\widetilde{\mathbb{R}^n}$ replacing $\dot{\mathbb{R}}^n$, as shown by Speck [45].)

(2) Taking $\overline{\mathbb{R}^n} = \widetilde{\mathbb{R}^n}$, one may admit integral operators of the L^1-type:

$$Bu(x) := a(x)u(x) + \int_{\mathbb{R}^n} k(x, x-y)u(y)\,\mathrm{d}y, \tag{4.11}$$

where $a \in C(\widetilde{\mathbb{R}^n})$ and $k \in C(\widetilde{\mathbb{R}^n}, L^1(\mathbb{R}^n))$.

Remarks The *local theory* described here has been introduced by Simonenko [44] in more general versions, namely:

(a) $\mathfrak{X} = L^p(X)$, where X is a compact Hausdorff space carrying a measure;

(b) $\mathfrak{X}$ is a Banach space of vector-valued functions having components of the above mentioned kind and the operators A, etc., then have to be replaced by operator-matrices,

(c) the concept of local equivalence is replaced by that of *local quasi-equivalence.* This is defined as follows. Given two homomorphic Hausdorff spaces X, Y, related by $\varphi: X \to Y$, and a non-distorting transformation $(T_\varphi f)(x) := f(\varphi(x))$ for all $f \in L^p(Y)$, then the relation

$$T_{\varphi^{-1}} P_U A P_U T_\varphi P_V \overset{y_0}{\simeq} P_V B P_V \tag{4.12}$$

holds. This relation may be written shortly as $A \overset{x_0}{\approx} \overset{y_0}{\approx} B$ for $A \in \Lambda(L^p(X))$ and $B \in \Lambda(L^p(Y))$, $y_0 = \varphi(x_0)$, and homeomorphic neighbourhoods $U = U(x_0)$ and $V = V(y_0)$, respectively.

The last point is of major interest to us. We refer to the results obtained by Simonenko [39].

Theorem 11 *Let $\Gamma_j \subset \mathbb{R}^n$, $j = 1, \ldots, N$, $n \geq 2$, be a finite number of smooth cones, i.e., the intersections $\partial\Gamma_j \cap S^{n-1}$ of their boundaries with the unit-sphere are smooth $(n-2)$-manifolds, such that*

$$\sum_{j=1}^{N} P_j := \sum_{j=1}^{N} \chi_{\Gamma_j} \cdot = I \quad \text{on} \quad L^p(\mathbb{R}^n).$$

Let the N linear bounded operators A_j be generalized convolution operators in the sense of (4.11). Then the following assertion holds. The N-part composite Wiener–Hopf operator $W_N := \sum_{j=1}^{N} A_j P_j$ is Fredholmian iff

$$\operatorname{ess\,inf} |\Phi_{W_N}(x_0, \xi)| \geq \delta > 0, \tag{4.13}$$

where the symbol Φ_{W_N} is given by

$$\Phi_{W_N}(x_0, \xi) := \left\{ \begin{array}{lll} a_j(x_0) & \text{for} & x_0 \in \mathbb{R}^n \cap \Gamma_j, \\ a_j(x_0) + \hat{k}_j(x_0, \xi) & \text{for} & x_0 \in (\widetilde{\mathbb{R}^n} - \mathbb{R}^n) \cap \tilde{\Gamma}_j,\ \xi \in \dot{\mathbb{R}}^n. \end{array} \right\} \tag{4.14}$$

($\tilde{\Gamma}_j$ denotes the closure of Γ_j in the $\widetilde{\mathbb{R}^n}$ – topology.)

Proof First of all, one shows that all operators involved are of local type with respect to $\widetilde{\mathbb{R}^n}$. Then the following statements are equivalent:

$$W_N \in \mathscr{F} \quad \text{and} \quad \sum_{j=1}^{N} a_j(x_0)I + k_j(x_0, \cdot) * \in \mathscr{F}_{x_0} \quad \text{for all} \quad x_0 \in \widetilde{\mathbb{R}^n}.$$

Due to the smoothness of the cones Γ_j at every point $x_0 \neq 0$, at most two terms in the sum are involved. For finite points x_0, we have local equivalence to the operator $\sum_{j=1}^{N} a_j(x_0)I$, which yields the necessary and sufficient condition for the local Fredholm property. Taking $x_0 \in \widetilde{\mathbb{R}^n} - \mathbb{R}^n$, the local quasi-equivalence to a half-space WH problem assures the second part of the condition by Goldenstein and Gohberg's results.

4.2 The problem of the quadrant

Let us consider the following WH integral equation on the first quadrant $\mathbb{R}^2_{++} := \{x = (x_1, x_2) \in \mathbb{R}^2 : x_1 \geq 0, x_2 \geq 0\}$ with the corresponding projector $P_{++} := \chi_{\mathbb{R}^2_{++}}$.

$$(W_{++}u)(x) := \lambda u(x) - \int_{\mathbb{R}^2_{++}} k(x-y)u(y)\,dy = v(x), \tag{4.15}$$

for $v \in L^p(\mathbb{R}^2_{++})$. In the special case of $p=2$, the invertibility of $W_{++} = T_{P_{++}}(\lambda I - k*)$ for $\operatorname{Re} e^{i\alpha}[\lambda - \hat{k}(\xi)] \geq \delta > 0$ on $\mathbb{R}^n_\xi$ is known from section 3. Here, we seek more general conditions and those in the case of $p \neq 2$. Without loss of generality, we assume $\lambda = 1$ and factorize the symbol $\Phi(\xi) \neq 0$ within the two-dimensional Wiener algebra into

$$\Phi(\xi) := 1 - \hat{k}(\xi) = \Phi_{++}(\xi)\Phi_{+-}(\xi)\Phi_{-+}(\xi)\Phi_{--}(\xi), \tag{4.16}$$

where we have, for example,

$$\Phi_{++}(\xi) = 1 + \hat{k}_{++}(\xi) = \exp\{\hat{P}_{++} \log(1 - \hat{k}(\xi))\}, \tag{4.17}$$

where $k_{++} \in L^1(\mathbb{R}^2_{++})$.

While Eqn (4.16) constitutes no WH factorization of Φ for P_{++}, we easily obtain two different ones for two half-plane operators, viz., $(\Phi_{-+}\Phi_{--})(\xi)(\Phi_{++}\Phi_{+-})(\xi)$ corresponding to the *right WH operator* $T_{P_R}(A)$ and $(\Phi_{+-}\Phi_{--})(\xi)(\Phi_{++}\Phi_{-+})(\xi)$ corresponding to the *upper WH operator* $T_{P_U}(A)$, respectively. Their inverses exist according to Goldenstein and Gohberg [40] and are given by

$$[T_{P_R}(A)]^{-1} = F^{-1}\Phi_{++}^{-1} \cdot \Phi_{+-}^{-1} \cdot FP_R F^{-1}\Phi_{-+}^{-1} \cdot \Phi_{--}^{-1} \cdot F \tag{4.18}$$

and a similar formula for $[T_{P_U}(A)]^{-1}$. Now, making use of a result by Simonenko [39] ensuring that the operator $\chi_E \cdot h * \chi_G \cdot$ is compact on $L^p(\mathbb{R}^2)(1<p<\infty)$ for quadrants E and $G \subset \mathbb{R}^2$ lying opposite on the same diagonal, we immediately conclude that

$$TW_{++} = I + V_1 \quad \text{and} \quad W_{++}T = I + V_2, \tag{4.19}$$

where V_1, V_2 are compact and the regularizer T is given by

$$T = P_{++}([T_{P_R}(A)]^{-1} + [T_{P_U}(A)]^{-1} - A^{-1})P_{++}. \tag{4.20}$$

This yields the result obtained by Strang [46]:

Theorem 12 *Let $k \in L^1(\mathbb{R}^2)$ be such that $|\lambda - \hat{k}(\xi)| \geq \delta > 0$ for all $\xi \in \dot{\mathbb{R}}^n$. Then the WH operator W_{++} in Eqn (4.15) is Fredholmian with index zero, having a regularizer T given by Eqn (4.20).*

Remarks The construction of the regularizer T for W_{++} carries over to the case of an arbitrary sector. Apart from formula (4.20), the result is contained in Simonenko's paper [39] – based on quasi-equivalence – and was re-obtained by Douglas and Howe [47] by their tensor-product method. More recently, Dudučava [48, 49] succeeded in including the case of piecewise continuous non-vanishing symbols $\Phi_W(\xi)$ on $\mathbb{R}^2$, the lines of discontinuities being parallel to the ξ_1- and ξ_2-axis.

Kraut [50] treated the case of a WH integral equation on an n-dimensional hyperoctant by projection of the Fourier transformed space $L^2(\mathbb{R}^n_\xi)$ on to the 2^n Hardy subspaces. Invertibility is proved in this case by a condition which follows from the half-plane condition for general WHOs.

Latz [51] confines himself to special $L^1(\mathbb{R}^n)$-kernels given by

$$k(x) := C^n_\alpha(\mu)\,|x|^{-n/2+\alpha} H^{(1)}_{n/2-\alpha}(\mu\,|x|), \tag{4.21}$$

where

$$C^n_\alpha(\mu) := \frac{1}{\Gamma(\alpha)}\,\mathrm{e}^{-i\pi(\alpha-1)}(4\pi)^{1-\alpha}(4i)^{-1}(\mu/2\pi i)^{n/2-\alpha}\,\mathrm{e}^{i(\pi/2)(n/2-\alpha)} \tag{4.22}$$

and Hankel's function of the first kind is involved. The Fourier transform then has the relatively simple form

$$\hat{k}(\xi) = (\mu^2 - |\xi|^2)^{-\alpha} \in L^\infty(\mathbb{R}^n_\xi). \tag{4.23}$$

The quadrant $\mathbb{R}^2_{++}$ may now be replaced by an arbitrary domain $G \subset \mathbb{R}^n$, $n \geqslant 2$.

Kremer [52] studied the case of *non-normal* WH integral equations on the quadrant, viz., he allows the symbol $\Phi(\xi)$ to have a finite number of zero-lines of finite order in each of the variables ξ_1 and ξ_2. His method is algebraic and adapted to that of Douglas and Howe by showing that the Fredholm properties of the quarter-plane operator W_{++} are equivalent to the invertibility of two families of one-dimensional WHOs. Since he has no C^*-algebras involved, he makes essential use of the nuclearity of the algebras there.

4.3 Two-dimensional Toeplitz operators

The discrete analogue to the half-space WH problem treated by Goldenstein and Gohberg [40] is the infinite system of linear algebraic equations

$$\sum_{j\in\mathbb{Z}^n_+} a_{k-j}\xi_j = \eta_k \quad \text{for} \quad k \in \mathbb{Z}^n_+ := \{(k_1, \ldots, k_n) \in \mathbb{Z}^n, k_\nu \geqslant 0\}, \tag{4.24}$$

where the n-fold matrix $a = (a_j)_{j\in\mathbb{Z}^n} \in l^1(\mathbb{Z}^n)$ and the sequence $(\eta_k)_{k\in\mathbb{Z}^n_+} \in l^p(\mathbb{Z}^n_+)(1 \leqslant p \leqslant \infty)$ – or subspaces of $l^\infty(\mathbb{R}^n_+)$ – are given. Generalizing the results for the one-dimensional case, one has, according to Goldenstein and Gohberg [40], the following facts.

Let the symbol

$$\Phi(\zeta) = \Phi(\zeta_1, \ldots, \zeta_n) := \sum_{k\in\mathbb{Z}^n} a_k \zeta^k$$

$$= \sum_{(k_1,\ldots,k_n)\in\mathbb{Z}^n} a_{k_1\ldots k_n} \zeta_1^{k_1} \ldots \zeta_n^{k_n} \tag{4.25}$$

be non-vanishing on the n-torus $\{\zeta \in \mathbb{C}^n : |\zeta_\nu| = 1 \text{ for } \nu = 1, \ldots, n\}$ and let the first partial index $\kappa_1 \neq 0$,where

$$\kappa_\nu := -\frac{1}{2\pi}[\arg \Phi(\zeta)\,|_{\zeta_\nu = e^{i\varphi}}]_{\varphi=0}^{2\pi}, \qquad \nu = 1, \ldots, n. \tag{4.26}$$

Then Eqn (4.24) is uniquely soluble for every right-hand side $(\eta_k)_{k \in \mathbb{Z}_+^n}$. Due to homotopy considerations, the indices κ_ν are independent of ζ_μ, $\mu \neq \nu$. In the other cases – for $\kappa_1 \neq 0$ – the inhomogeneous equation still remains normally soluble, but the defect numbers then are $(0, \infty)$ and $(\infty, 0)$, respectively.

Strang [46] discusses the case of Toeplitz operators on the quarter-plane and makes use of the factorization within the discrete Wiener algebra of functions on the torus. His formulae run completely parallel to those in Eqns (4.18) to (4.20). Prior to Strang, Simonenko [53] had obtained similar results. We should emphasize that here only the Fredholm–Noether property is implied by the symbol condition.

4.4 Bisingular integral equations of Cauchy principal-value type

In higher dimensions ($n \geq 2$), there are completely different types of strongly singular integral operators. Besides those of the Calderón–Zygmund–Mikhlin type, with kernels having poles at fixed points, one may consider tensor-products of lower-dimensional operators such that lines, curves, submanifolds, and cartesian products of these may occur. Singular integral operators of this kind have mainly been studied by Russian mathematicians during the last decade. We shall not give a review of the whole present status, but shall point out the main ideas in some relatively simple cases.

Let us assume that we have a pair Γ_1, $\Gamma_2 \subset \mathbb{C}$ of simple Lyapunov contours and let us treat the case studied by Pilidi [54] and Pilidi and Sazonov [55] with Hoelder-continuous coefficients prescribed:

$$\begin{aligned}(Au)(t_1, t_2) := {} & a_0(t_1, t_2)u(t_1, t_2) + a_1(t_1, t_2)(S_1 u)(t_1, t_2) \\ & + a_2(t_1, t_2)(S_2 u)(t_1, t_2) + a_{12}(t_1, t_2)(S_{12} u)(t_1, t_2)\end{aligned} \tag{4.27}$$

in $L^p(\Gamma_1 \times \Gamma_2)(1 < p < \infty)$. Here the *partial Cauchy operators* and the *total Cauchy operator* are involved:

$$(S_1 u)(t_1, t_2) := \frac{1}{\pi i}\int_{\Gamma_1} \frac{u(\tau_1, t_2)\,d\tau_1}{\tau_1 - t_1}, \qquad (t_1, t_2) \in \Gamma_1 \times \Gamma_2 \tag{4.28a}$$

$$(S_2 u)(t_1, t_2) := \frac{1}{\pi i}\int_{\Gamma_2} \frac{u(t_1, \tau_2)\,d\tau_2}{\tau_2 - t_2}, \qquad (t_1, t_2) \in \Gamma_1 \times \Gamma_2 \tag{4.28b}$$

and

$$(S_{12}u)(t_1, t_2) := (S_1(S_2u))(t_1, t_2) = (S_2(S_1u))(t_1, t_2). \tag{4.28c}$$

To treat such operators – and generalizations of them – Pilidi [54] generalized Simonenko's theory [44] of operators of local type to tensor products along the following line.

Let $\mathscr{C}_\nu$, Λ_ν denote the classes of compact operators and those of local type (cf., Definition 2) on $L^2(\Gamma_\nu)$ and set $\Lambda := \Lambda_1 \otimes \Lambda_2$ for the closure of all linear-combinations $\sum_{j=1}^{N} c_j A_j^{(1)} A_j^{(2)}$, where $A_j^{(\nu)} \in \Lambda_\nu$ in the norm of $\mathscr{L}(L^2(\Gamma_1 \times \Gamma_2))$. Then we have

Definition 4 (a) *Two operators* $A, B \in \Lambda$ *are called* 1-*equivalent iff* $A - B \in \mathscr{C}_1 \otimes \Lambda_2$.

(b) $A \in \Lambda$ *is called* 1-Fredholmian, *i.e.*, $A \in \mathscr{F}_1$, *iff there are* R_l, $R_r \in \Lambda$ *such that* $R_l A \overset{1}{\sim} AR_r \overset{1}{\sim} I$.

(c) *A and B are called* locally 1-equivalent at $t_1^0 \in \Gamma_1$ *iff, for all* $\varepsilon > 0$, *there exists a neighbourhood* $U(t_1^0) \subset \Gamma_1$ *such that*

$$\inf_{T \in \mathscr{C}_1 \otimes \Lambda_2} \|(A - B)(P_U \otimes I) - T\|_{\mathscr{L}} < \varepsilon, \tag{4.29}$$

(d) $A \in \Lambda$ *is called* locally 1-Fredholmian at $t_1^0 \in \Gamma_1$ *iff there is a neighbourhood* $U(t_1^0) \subset \Gamma_1$ *and a pair of operators* R_{l,t_1^0}, $R_{r,t_1^0} \in \Lambda$ *such that the following relations hold*:

$$R_{l,t_1^0} A(P_U \otimes I) \overset{1}{\sim} (P_U \otimes I) A R_{r,t_1^0} \overset{1}{\sim} P_U \otimes I. \tag{4.30}$$

Remark Analogous notations are defined for the second variable $t_2 \in \Gamma_2$.

The main result by Pilidi [54] is the following:

Theorem 13 *An operator* $A \in \Lambda$ *is Fredholmian on* $L^2(\Gamma_1 \times \Gamma_2)$ *iff it is locally* 1-*Fredholmian at every point* $t_1^0 \in \Gamma_1$ *and it is locally* 2-*Fredholmian at every point* $t_2^0 \in \Gamma_2$.

To prove the necessity, it is important to know that the Fredholm property of A implies that its regularizer is an element of Λ. Theorem 10 of the Simonenko theory cited in section 4.1 is carried over as well in the following way.

Theorem 14 *Let* $A, B \in \Lambda$ *and* $A \overset{1,t_1^0}{\sim} B$ *at a point* $t_1^0 \in \Gamma_1$. *Then A and B are locally* 1-*Fredholmian at* t_1^0 *or are not, at the same time.*

On this basis, Pilidi [56] and Pilidi and Sazonov [55] proved

Theorem 15 *Let $A \in \Lambda$ be given by Eqn (4.27), where $\Gamma_1, \Gamma_2 \subset \mathbb{C}$ are simple Lyapunov contours and a_0, a_1, a_2, and a_{12} are continuous functions on $\Gamma_1 \times \Gamma_2$. Then A is Fredholmian iff the following operators are invertible:*

$$(a_0 \pm a_2)(., t_2)I + (a_1 \pm a_{12})(., t_2)S_1 \quad on \quad L^2(\Gamma_1)$$

and

$$(a_0 \pm a_1)(t_1, .)I + (a_2 \pm a_{12})(t_1, .)S_2 \quad on \quad L^2(\Gamma_2)$$

for all $t_2 \in \Gamma_2$ and for all $t_1 \in \Gamma_1$, respectively. This is the case iff

(i) $\inf|\Phi(t_1, t_2, \xi_1, \xi_2)| > 0$ *for* $(t_1, t_2) \in \Gamma_1 \times \Gamma_2$ *and* $(\xi_1, \xi_2) \in \ddot{\mathbb{R}}_1 \times \ddot{\mathbb{R}}_2$,

(ii) $[\arg \Phi(t_1, t_2^0, \xi_1, \xi_2^0)]_{t_1 \in \Gamma_1} = 0$ *for every fixed* $t_2^0 \in \Gamma_2$, $\xi_2^0 \in \ddot{\mathbb{R}}$ *and all* $\xi_1 \in \ddot{\mathbb{R}}$,

(iii) $[\arg \Phi(t_1^0, t_2, \xi_1^0, \xi_2)]_{t_2 \in \Gamma_2} = 0$ *for every fixed* $t_1^0 \in \Gamma_1$, $\xi_1^0 \in \ddot{\mathbb{R}}$ *and all* $\xi_2 \in \ddot{\mathbb{R}}$,

where the symbol is defined by

$$\Phi(t_1, t_2, \xi_1, \xi_2) := a_0(t_1, t_2) + a_1(t_1, t_2) \operatorname{sgn} \xi_1 + a_2(t_1, t_2) \operatorname{sgn} \xi_2 + a_{12}(t_1, t_2) \operatorname{sgn} \xi_1 \operatorname{sgn} \xi_2 \quad on \quad \Gamma_1 \times \Gamma_2 \times \ddot{\mathbb{R}} \times \ddot{\mathbb{R}}. \tag{4.31}$$

Corollary [55] *Let $A \in \Lambda$ be a Fredholmian bisingular operator of type (4.27) and R_1, R_2 denote the partial 1- and 2-regularizers, respectively; then a regularizer for A is given by*

$$R := R_1 + R_2 - R_1 A R_2 \tag{4.32}$$

and the index of the Fredholm operator A may be calculated according to the formula

$$\operatorname{ind} A = [\kappa_1(a_{++}) - \kappa_1(a_{--})][\kappa_2(a_{+-}) - \kappa_2(a_{-+})] \tag{4.33}$$

with the partial winding-numbers κ_ν with respect to $t_\nu \in \Gamma_\nu$ of the functions

$$a_{\pm\pm}(t_1, t_2) = (a_0 \pm a_1 \pm a_2 \pm a_{12})(t_1, t_2), \qquad (t_1, t_2) \in \Gamma_1 \times \Gamma_2. \tag{4.34}$$

Remarks The conditions of Theorem 15 to ensure the Fredholm property of operator (4.27) may be extended to incorporate the following situations (cf., Dudučava [48]):

(i) replace Γ_1, Γ_2 by two finite systems of Lyapunov contours and/or Lyapunov arcs in $\mathbb{C}$ each;

(ii) replace $L^2(\Gamma_1 \times \Gamma_2)$ by $L^{\vec{p}}(\Gamma_1 \times \Gamma_2; \vec{\rho}) := L^{p_1}(\Gamma_1; \rho_1) \times L^{p_2}(\Gamma_2; \rho_2)$ with non-negative weight-functions ρ_ν and $1 < p_1, p_2 < \infty$;

(iii) take systems of equations (4.27) with continuous matrix-valued coefficients $a_0, \ldots, a_{12}$ on $\Gamma_1 \times \Gamma_2$;

(iv) admit piecewise continuous coefficients;

(v) replace the B-spaces $L^{\vec{p}}(\Gamma_1 \times \Gamma_2; \vec{\rho})$ by certain Sobolev spaces $H^{s,p}(\Gamma_1 \times \Gamma_2)$ or even Frechét spaces $H^{\infty}(\Gamma_1 \times \Gamma_2)$ for sufficiently smooth coefficients (cf., Dudučava [49]) making use of certain regularity properties of the solutions to $Au = v$ according Pilidi and Sazonov [55], originating in *a priori* estimates for A;

(vi) Pilidi [54] and Pilidi and Sazonov [55] also treated operators $A \in \Lambda := \Lambda_1(\dot{\mathbb{R}}^m) \otimes \Lambda_2(\dot{\mathbb{R}}^n)$, $m, n \geq 2$, with the properties

$$A \overset{1,x}{\sim} A_x \in Q_m \otimes \Lambda(\dot{\mathbb{R}}^n) \quad \text{for all} \quad x \in \dot{\mathbb{R}}^m$$

and

$$A \overset{2,y}{\approx} B_y \in \Lambda(\dot{\mathbb{R}}^m) \otimes Q_n \quad \text{for all} \quad y \in \dot{\mathbb{R}}^n,$$

where Q_m denotes multiplication by functions from $C(\mathbb{R}^m - \{0\})$ and homogeneous of degree zero. They proved

Theorem 16 *The bisingular operator $A \in \Lambda_1(\dot{\mathbb{R}}^m) \otimes \Lambda_2(\dot{\mathbb{R}}^n)$ is Fredholmian iff its* partial symbols $\hat{A}_x(\xi) \in \Lambda(\dot{\mathbb{R}}^n)$ *for all $x \in \dot{\mathbb{R}}^m$ and $\xi \in S^{m-1}$ and $\hat{B}_y(\eta) \in \Lambda(\dot{\mathbb{R}}^m)$ for all $y \in \dot{\mathbb{R}}^n$ and $\eta \in S^{n-1}$. Here $\hat{A}_x(\xi)$ and $\hat{B}_y(\eta)$ denote the images under the canonical isomorphisms $Q_m \otimes \Lambda_2(\dot{\mathbb{R}}^n) \to C(S^{m-1}) \otimes \Lambda_2(\dot{\mathbb{R}}^n)$, and $\Lambda_1(\dot{\mathbb{R}}^m) \otimes Q_n \to \Lambda_1(\dot{\mathbb{R}}^m) \otimes C(S^{n-1})$ with invertible n-dimensional and m-dimensional CMOs in the last n and the first m variables, respectively.*

Remark Taking a pair of Lyapunov contours $\Gamma_1, \Gamma_2 \subset \mathbb{C}$ and the subspaces $L^2_{\pm}(\Gamma_\nu)$ of L^2-functions on Γ_ν extendable to holomorphic functions in the interior Δ^+_ν and the exterior domains Δ^-_ν, respectively, one may introduce the four projectors

$$\tilde{P}_{\pm\pm} := \tfrac{1}{4}(I \pm S_1)(I \pm S_2) \tag{4.35}$$

with the partial Cauchy operators S_ν along Γ_ν and may treat the *problems of linear conjugacy*

$$G_{++} \cdot \tilde{P}_{++}u + G_{+-} \cdot \tilde{P}_{+-}u + G_{-+} \cdot \tilde{P}_{-+}u + G_{--} \cdot \tilde{P}_{--}u = g \tag{4.36}$$

with continuous functions $G_{\pm\pm}$ on $\Gamma_1 \times \Gamma_2$ and a given $g \in L^2(\Gamma_1 \times \Gamma_2)$. Research work in this field has been done by Kakičev [57] and others (cf., also the book by Przeworska-Rolewicz [58]!).

5 Some results concerning composite WHOs involving generalized translation-invariant operators

5.1 Relations between WHOs and Banach's fixed-point theorem

Let us consider first the *special* single WHO

$$T_P(A) = PA\,|_{R(P)}, \tag{5.1}$$

where $A = F^{-1}\Phi \cdot F$ denotes a translation-invariant operator on $\mathcal{H} = L^2(\mathbb{R}^n)$ with a symbol $\Phi \in L^\infty(\mathbb{R}^n)$ and a space-projector $P = \chi_E \cdot$ involving a measurable set $E \subset \mathbb{R}^n$. In this case, one of the main results for general WHOs (cf., Lemma 1, Theorems 3 and 5 in Section 3) is a consequence of Banach's fixed-point theorem, viz.,

Theorem 17 *Let there exist an $\alpha \in [0, 2\pi)$ and a $\delta > 0$ such that $\mathrm{Re}\, e^{i\alpha}\Phi(\xi) \geq \delta > 0$ for almost all $\xi \in \mathbb{R}^n$; then, $T_P(A)$ is invertible.*

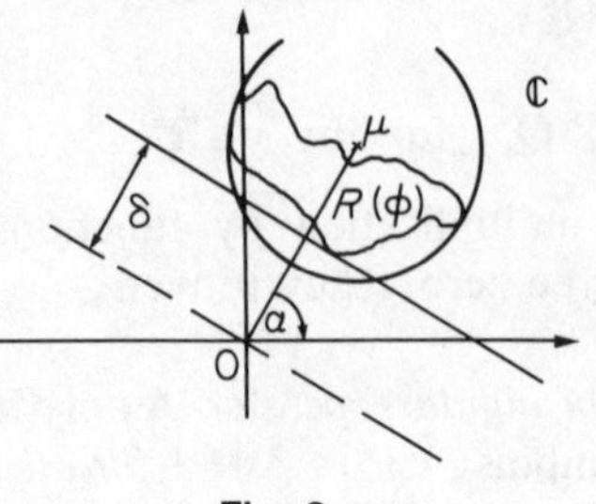

Fig. 2

Remark The idea for the proof of this theorem may be found in papers by Latz [22] and Busenberg [59]. One has just to write

$$T_P(A) = \mu P(I - \mu^{-1}(\mu I - A))|_{R(P)} \tag{5.2}$$

with a suitably chosen $\mu \in \mathbb{C}$; this equation implies the estimate

$$\|\mu^{-1}(\mu I - A)\|_{\mathcal{L}} = |\mu|^{-1}\,\|\mu - \Phi(\xi)\|_{L^\infty} < 1. \tag{5.3}$$

Corollary (cf., Gerlach and Latz [60]) *Let $A = k *$ with a kernel function $k \in L^1(\mathbb{R}^n)$. Then the following relation for the spectrum of $T_P(A)$ holds:*

$$\sigma(T_P(A)) \subset \bigcap \{\mathrm{ch}(\hat{h}(\dot{\mathbb{R}}^n)) : h \in L^1(\mathbb{R}^n),\ h - k\,|_{E-E} = 0\} \tag{5.4}$$

where ch *denotes the convex hull and*

$$E - E := \{z \in \mathbb{R}^n : z = x - y,\ x, y \in E\}.$$

It is easy to see that in the case of the translation invariant $A = F^{-1}\Phi \cdot F$ on $L^2(\mathbb{R}^n)$, the following conditions are equivalent:

$$\mathrm{Re}\, e^{i\alpha}\Phi(\xi) \geq \delta > 0 \tag{5.5}$$

for suitably chosen α, δ;

$$\|\mu^{-1}(\mu - \Phi(\cdot))\|_{L^\infty} < 1 \tag{5.6}$$

for suitably chosen $\mu \in \mathbb{C}$. Written in a different form,

$$\mathrm{Re}\, e^{i\alpha} A \geq \delta > 0 \tag{5.7}$$

and finally

$$\|\mu^{-1}(\mu I - A)\|_{\mathscr{L}} < 1. \tag{5.8}$$

We want to abbreviate the above by $m(A) < 1$, making use of the definition

$$m(A) := \inf_{\mu \in \mathbb{C}} m(A; \mu) := \inf_{\mu \in \mathbb{C}} \|\mu^{-1}(\mu I - A)\|_{\mathscr{L}}. \tag{5.9}$$

The conditions (5.7) and (5.8) are even equivalent in the following case.

Lemma 3 *Let $A \in \mathscr{L}(L^2(\mathbb{R}^n))$ be a normal or normaloid operator, i.e., $\sup\{|\lambda| : \lambda \in W(A)\} = \|A\|$ holds according to Halmos [61; §173]; then (5.7) and (5.8) are equivalent.*

Proof We have the following chain of equivalent relations:

$$\operatorname{Re} e^{i\alpha} A \geq \delta > 0 \quad \text{or} \quad 0 \notin \overline{W(A)},$$

due to the convexity of the numerical range, or

$$W(A) \subset K_r(\mu) \quad \text{and} \quad r < |\mu| \quad \text{for a pair} \quad r \geq 0, \qquad \mu \in \mathbb{C},$$

and, finally,

$$\sup_{\|u\|=1} |((\mu I - A)u, u)| = \|\mu I - A\|_{\mathscr{L}} \leq r < |\mu| \tag{5.10}$$

for suitably chosen $r \geq 0$, $\mu \in \mathbb{C}$.

Now we are in a position to modify the main theorem of Devinatz and Shinbrot (cf., Theorem 3) in the following sense:

Theorem 18 *Let A be an invertible operator on the separable Hilbert space $\mathscr{H}$ and P an orthogonal projector. Then $T_P(A)$ is invertible on $R(P)$ iff there exists an invertible operator $\tilde{H}$ on $\mathscr{H}$ such that $R(\tilde{H}P) = R(P)$ and $m(A\tilde{H}) < 1$ holds.*

Proof The sufficient part of the theorem is implied by Banach's fixed-point theorem. Now, concerning the necessary part, let us assume $T_P(A)$ to be invertible. Due to Devinatz and Shinbrot [31, Lemma 1], we have $A = \tilde{A}B$, where $\tilde{A}$ is unitary and B invertible with $R(BP) = R(P)$. Thus, $T_P(\tilde{A})$ is invertible, too, and we find $\tilde{A}(P\tilde{A}^*P + Q\tilde{A}^*Q) = I - \tilde{A}(P\tilde{A}^*Q + Q\tilde{A}^*P) = I - L$ where $\|L\| < 1$ owing to [31; Corollary 1]. This obviously tells us that we may take $B^{-1}(P\tilde{A}^*P + Q\tilde{A}^*Q)$ for $\tilde{H}$.

Remarks Concerning the sufficient part of the theorem, we may completely neglect the Hilbert space structure. It still works for arbitrary

Banach spaces $\mathfrak{X}$, e.g., $\mathfrak{X} = L^p(\mathbb{R}^n)(1 \leq p \leq \infty)$, and bounded linear projectors P on $\mathfrak{X}$ with $P^2 = P$. Thus, we may also treat systems on $(L^p(\mathbb{R}^n))^m$.

Problem *Under which conditions does the necessary part of the theorem hold for a given Banach space $\mathfrak{X}$ and a special WHO of type* (5.1)?

Now we proceed to extend the results to the case of multiple-part composite WHOs.

Theorem 19 *Let A_j be invertible operators and P_j, $j = 1, \ldots, N$, orthogonal projectors such that $P_jP_k = P_j\delta_{j,k}$ and $\sum_{j=1}^N P_j = I$ holds on a Hilbert space $\mathcal{H}$. Then the WHO*

$$W_N := \sum_{j=1}^{N} A_j P_j \tag{5.11}$$

is invertible if

$$\sum_{j=1}^{N} [m(A_j)]^2 < 1. \tag{5.12}$$

Proof With constants $\mu_j \neq 0$, write

$$\begin{aligned} W_N &= \left(I - \sum_{j=1}^{N} \mu_j^{-1}(\mu_j I - A_j)P_j\right)\left(\sum_{k=1}^{N} \mu_k P_k\right) \\ &= (I - U)T; \end{aligned} \tag{5.13}$$

then it is obvious that T is invertible by

$$T^{-1} = \sum_{k=1}^{N} \mu_k^{-1} P_k \tag{5.14}$$

and the following estimate holds:

$$\begin{aligned} \|U\|_{\mathcal{L}} &= \sup_{\|u\|=1} \sum_{j=1}^{N} \|\mu_j^{-1}(\mu_j I - A_j)\|_{\mathcal{L}} \, \|P_j u\|_{\mathcal{H}} \\ &\leq \sup_{\|u\|=1} \left(\sum_{j=1}^{N} \|\mu_j^{-1}(\mu_j I - A_j)\|_{\mathcal{L}}^2 \sum_{k=1}^{N} \|P_k u\|_{\mathcal{H}}^2\right)^{1/2} \\ &< 1, \end{aligned} \tag{5.15}$$

since one may choose the μ_j appropriately such that

$$\sum_{j=1}^{N} \|\mu_j^{-1}(\mu_j I - A_j)\|_{\mathcal{L}}^2 = \sum_{j=1}^{N} [m(A_j)]^2 + \varepsilon < 1.$$

Remarks (1) If condition (5.12) is violated but there exists an invertible operator $B \in \mathcal{L}(\mathfrak{X})$ such that (5.12) holds with $B_j := BA_j$, then W_N is still invertible.

(2) For the Banach spaces $\mathfrak{X} = L^p(\mathbb{R}^n)$, $1<p<\infty$, condition (5.12) is to be replaced by

$$\sum_{j=1}^{N} [m(A_j)]^{p'} < 1$$

and guarantees the invertibility of the WHO W_N (5.11), if the projectors P_j are multiplications by characteristic functions.

Problem *Is condition* (5.12), *apart from a pre-multiplication by an invertible $B \in \mathcal{L}(\mathcal{H})$, a necessary one or how is it to be modified?*

Corollary 1 *Let*

$$W_N = \sum_{j=1}^{N} A_j P_j$$

be an N-part composite WHO, where the P_j are orthogonal projectors as above and the $A_j \in \mathcal{L}(\mathcal{H})$ are self-adjoint having numerical ranges $W(A_j) \subset [\delta, M]$ for $j = 1, \ldots, N$ with $0 < \delta < M$. If the inequality

$$\frac{\sqrt{N}-1}{\sqrt{N}+1} < \frac{\delta}{M} =: \beta < 1 \tag{5.16}$$

holds, W_N is invertible on $\mathcal{H}$.

Proof Condition (5.16) implies

$$\sqrt{N}\frac{1-\beta}{1+\beta} < 1 \quad \text{or} \quad N\left(\frac{M-\delta}{M+\delta}\right)^2 < 1,$$

yielding

$$\sum_{j=1}^{N} m(A_j)^2 < 1.$$

Remark For $N = 4$ we obtain $\frac{1}{3} < \delta/M = \beta < 1$, resulting, for example, in $W(A_j) \subset (\frac{1}{2}, \frac{3}{2})$ as a sufficient condition.

Let us briefly discuss an application to the problem described in section 2.4 by applying the following:

Corollary 2 *Let $W_N = \sum_{j=1}^{N} \mathbf{A}_j P_j$, where $\mathbf{A}_j := F^{-1}\mathbf{\Phi}_j \cdot F$ are operating on $\mathcal{H} = (L^2(\mathbb{R}^n))^m$, the $\mathbf{\Phi}_j$ being $m \times m$ symbol matrices from $(L^\infty(\mathbb{R}^n))^m$ and* *the P_j, $j=1,\ldots,N$, being orthogonal projectors on $\mathcal{H}$ as above. If the condition*

$$\|\det(\mathbf{I}-\mathbf{\Phi}_j(\xi))\|_{L^\infty(\mathbb{R}^n)}<\frac{1}{\sqrt{N}} \tag{5.17}$$

is fulfilled, then W_N is invertible.

In section 2, we were led to the symbol matrices

$$\mathbf{D}_j(\xi):=\begin{pmatrix} \mu_j\,|\xi|^2-\kappa_j & 0 & 0 \\ 0 & \mu_j\,|\xi|^2-\kappa_j & 0 \\ 0 & 0 & (\lambda_j+2\mu_j)\,|\xi|^2-\kappa_j \end{pmatrix}. \tag{5.18}$$

By pre-multiplication by

$$\Psi(\xi):=\begin{pmatrix} \mu\,|\xi|^2-\kappa & 0 & 0 \\ 0 & \mu\,|\xi|^2-\kappa & 0 \\ 0 & 0 & (\lambda+2\mu)\,|\xi|^2-\kappa \end{pmatrix} \tag{5.19}$$

with suitable constants μ, κ independent of j making $\sup_j|\mu-\mu_j|$ a minimum, we obtain

$\mathbf{\Phi}_j(\xi):=I-\mathbf{\Psi}\cdot\mathbf{D}_j=$

$$\begin{pmatrix} \dfrac{(\mu-\mu_j)\,|\xi|^2-(\kappa-\kappa_j)}{\mu\,|\xi|^2-\kappa} & 0 & 0 \\ 0 & \dfrac{(\mu-\mu_j)\,|\xi|^2-(\kappa-\kappa_j)}{\mu\,|\xi|^2-\kappa} & 0 \\ 0 & 0 & \dfrac{(\lambda-\lambda_j+2\mu-2\mu_j)\,|\xi|^2-(\kappa-\kappa_j)}{(\lambda+2\mu)\,|\xi|^2-\kappa} \end{pmatrix} \tag{5.20}$$

Corollary 2 then gives sufficient inequalities for $|\mu-\mu_j|$ and/or $|\kappa-\kappa_j|$, $j=1,\ldots,N$, for W_N to be invertible. By analogous estimates, one may give sufficient conditions, or improve former ones, for the unique solubility of the problems posed in section 2.

5.2 Some generalizations of Simonenko's theorem on multiple-part composite WHOs

The outline of the proof of Simonenko's theorem on convolution operators on cones (cf., Theorem 11 in section 4) showed that for

$W_N=\sum_{j=1}^N A_jP_j$ with $P_j=\chi_{\Gamma_j}$ and cones $\Gamma_j\subset\mathbb{R}^n$, the following properties should hold:

(i) smoothness of $\partial\Gamma_j\cap S^{n-1}$,
(ii) the boundaries of only two cones are allowed to intersect (apart from the common apex $x=0$).

We now want to relax these conditions.

(a) For finite points $x_0\in\mathbb{R}^n$, the operators $A_j:=\lambda_jI+k_j*\ (k_j\in L^1)$ are locally equivalent to λ_jI and $\sum_{j=1}^N\lambda_j\chi_{\Gamma_j}$ is locally Fredholmian iff $\lambda_j\neq 0$ for all j such that $\operatorname{mes}(\Gamma_j\cap K_\rho(x_0))>0$ for all $\rho>0$. Therefore, deformations, confined to compact parts of $\mathbb{R}^n$, of the cones Γ_j to measurable disjoint sets E_j decomposing $\mathbb{R}^n$ do not affect the statements of the theorem.
(b) At 'infinity', the sets Γ_j may behave only asymptotically like cones, i.e., they are allowed to be decomposable into $\Gamma_j=\Gamma_j'\cup E_j$, where the Γ_j' are smooth cones as before and $\operatorname{mes}(E_j\cap K_1(x_0))\to 0$ as $|x_0|\to\infty$ (cf., Speck [62]).
(c) In the case of an infinitely distant point x_0 lying on the boundaries of more than two cones, Theorem 19 gives a sufficient criterion for W_N to be locally Fredholmian at x_0.
(d) The authors want to publish a condition for the Fredholm property for $T_P(A)$, A invertible and $P=\chi_E\cdot$ for $E:=\{x\in\mathbb{R}^3: x_1>0, x_2>0, x_3\in\mathbb{R}\}$.

Due to homotopy arguments, all composite WHOs of these more general types are Fredholmian with an index equal to zero.

Arriving at this point of the discussion, the *algebraic method* proves its strength, since it starts with commutator relations dispensing completely with problems of compactification of $\mathbb{R}^n$. A very far-reaching result in this field goes back to Cordes [63]:

Theorem 20 *Let $a,b\in C(\mathbb{R}^n)\cap L^\infty(\mathbb{R}^n)$ such that*

$$cm_x(a):=\sup_{|y|\leq 1}|a(x-y)-a(x)|\to 0 \quad as \quad |x|\to\infty \tag{5.21}$$

and the same holding for b. Then the operator

$$A:=a(\cdot)\cdot F^{-1}b(\cdot)\cdot F$$

is compact on $L^2(\mathbb{R}^n)$.

Remarks (1) Simonenko [39, 44] claimed $a\in C(\dot{\mathbb{R}}^n)$ and $b\in C(\tilde{\mathbb{R}}^n-\{0\})$ or the roles of a and b interchanged.

(2) In this context, we are naturally led to the following

Problem *Characterize all $\Phi\in L^\infty(\mathbb{R}^n)$ such that $A:=F^{-1}\Phi\cdot F$ is of local type with respect to $\dot{\mathbb{R}}^n$, i.e., $[\omega\cdot, A]\sim 0$ for all $\omega\in C(\dot{\mathbb{R}}^n)$.*

The equivalence of the Fredholm criterion for an arbitrary generalized convolution operator of this type (cf., section 4!), with the regularity of a residual class of presymbol-functions, has been established by Speck [64].

5.3 Bisingular integral equations and multiple-part composite WHOs

Since the Fredholm property of the bisingular operator given by

$$Lu := (a_0 I + a_1 S_1 + a_2 S_2 + a_{12} S_{12})u \tag{5.22}$$

may not be inferred directly by Simonenko's method, owing to the fact that the *partial Cauchy operators* S_j acting on $L^p(\mathbb{R}^n)(n \geq 2)$ are not of local type, we proceed once again to the discussion of L for the case of $\Gamma_1 = \Gamma_2 = \mathbb{R}$. The theory of bisingular integral operators as developed by Pilidi [54] and Dudučava [48, 49] applies here also:

Theorem 21 *Let $a_\nu \in C((\dot{\mathbb{R}})^2)$ be given functions; then, the following statements are equivalent.*

(i) *L is Fredholmian on $L^2(\mathbb{R}^2)$.*

(ii) $[a_0(z_1, .) + a_1(z_1, .) \operatorname{sgn} \xi_1]I + [a_2(z_1, .) + a_{12}(z_1, .) \operatorname{sgn} \xi_1]S_2$ *for all* $z_1 \in \dot{\mathbb{R}}$ *fixed and* $\xi_1 = \pm 1$ *and*

$$[a_0(., z_2) + a_2(., z_2) \operatorname{sgn} \xi_2]I + [a_1(., z_2) + a_{12}(., z_2) \operatorname{sgn} \xi_2]S_1$$

for all $z_2 \in \dot{\mathbb{R}}$ *fixed and* $\xi_2 = \pm 1$ *are invertible one-dimensional singular integral operators with respect to* x_2 *and* x_1*, respectively.*

(iii)

$$a_0(z_1, z_2) + a_1(z_1, z_2) \operatorname{sgn} \xi_1 + a_2(z_1, z_2) \operatorname{sgn} \xi_2 \\ + a_{12}(z_1, z_2) \operatorname{sgn} \xi_1 \operatorname{sgn} \xi_2 =: \Phi \neq 0 \tag{5.23}$$

for all $z_1, z_2 \in \dot{\mathbb{R}}$ *and* $\xi_1, \xi_2 = \pm 1$ *and, additionally,*

$$[\arg \Phi]_{z_2=-\infty}^{\infty} = 0 \quad \text{for all} \quad z_1 \in \dot{\mathbb{R}} \quad \text{and} \quad \xi_1, \xi_2 = \pm 1,$$
$$[\arg \Phi]_{z_1=-\infty}^{\infty} = 0 \quad \text{for all} \quad z_2 \in \dot{\mathbb{R}} \quad \text{and} \quad \xi_1, \xi_2 = \pm 1.$$

Remark If the coefficients are even in $C(\dot{\mathbb{R}}^2)$, the last two additional conditions may be dropped since they hold automatically. This is the case in

Theorem 22 *Let a_ν be elements of the two-dimensional Wiener algebra. Then the following statements are equivalent.*

(i) *L is Fredholmian on $L^2(\mathbb{R}^2)$.*

(ii) $F^{-1}LF = \sum_{j=1}^{4} A_j P_j$ *is Fredholmian where the* $P_j = \chi_{\Gamma_j} \cdot$ *denote the four* quadrant-projectors *and* $A_j := F^{-1}M_j \cdot F$, *where*

$$\left.\begin{aligned} M_1(x_1, x_2) &:= (a_0 + a_1 + a_2 + a_{12})(x_1, x_2), \\ M_2(x_1, x_2) &:= (a_0 - a_1 + a_2 - a_{12})(x_1, x_2), \\ M_3(x_1, x_2) &:= (a_0 + a_1 - a_2 - a_{12})(x_1, x_2), \\ M_4(x_1, x_2) &:= (a_0 - a_1 - a_2 + a_{12})(x_1, x_2). \end{aligned}\right\} \tag{5.24}$$

(iii) *Condition* (5.23) *holds.*

In order to imply invertibility we combine this with Theorem 19 and obtain

Theorem 23 *Let the functions* M_j, $j = 1, 2, 3, 4$, *given by* (5.24) *and be elements of* $L^\infty(\mathbb{R}^2)$ *such that*

$$\sum_{j=1}^{4} [m(M_j)]^2 < 1. \tag{5.25}$$

Then L, *defined by* (5.22), *is invertible on* $L^2(\mathbb{R}^2)$.

Remark Apart from pre-multiplication by a $L^\infty(\mathbb{R}^2)$-function M with $\operatorname{ess\,inf} |M| > 0$ on $\mathbb{R}^2$, the last condition is fulfilled in the case of

$$W(M_j) \subset K_{1/2}(1) \subset \mathbb{C}.$$

References

1 Noble, B., *Methods Based on the Wiener–Hopf Technique for the Solution of Partial Differential Equations,* Pergamon Press, London, 1958.

2 Wiener, N. and Hopf, E., Über eine Klasse singulärer Intergralgleichungen, *S. B. Preuss. Akad. Wiss., Phys.-Math. Kl.,* **30–32,** 696–706, 1931.

3 Kreĭn, M. G., Integral equations on a half-line with a kernel depending upon the difference of the arguments. *AMS Transl.,* **22,** 163–288, 1962.

4 Gohberg, I. Z. and Kreĭn, M. G., Systems of integral equations on a half-line with kernels depending on the difference of arguments. *AMS Transl.,* **14,** 217–287, 1960.

5 Gohberg, I. Z. and Feldman, I. A., *Faltungsgleichungen und Projektionsverfahren zu ihrer Lösung,* Birkhäuser Verlag, Stuttgart, 1974.

6 Gerlach, E., Zur Theorie einer Klasse von Integrodifferentialgleichungen, Dissertation, TU Berlin, 1969.

7 Kremer, M. Über eine Klasse singulärer Integralgleichungen vom Faltungstyp, Dissertation, TU Berlin, 1969.

8 Dudučava, R. V., Wiener–Hopf integral operators with discontinuous symbols, *Sov. Math. Dokl.,* **14,** 1001–1005, 1973.

9 Dudučava, R. V., On convolution integral operators with discontinuous coefficients, *Sov. Math. Dokl.,* **15,** 1302–1306, 1974.

10 Talenti, G., Sulle equazioni integrali di Wiener–Hopf., *Boll. Un. Mat. Ital.* **7,** Suppl. fasc. 1, 18–118, 1973.

11 Višik, M. I. and Èskin, G. I., General boundary value problems with discontinuous conditions at the boundary, *Sov. Math. Dokl.* **5,** 1154–1157, 1964.

12 Višik, M. I. and Èskin, G. I., Equations in convolutions in a bounded region. *Russ. Math. Surv.*, **20,** 85–151, 1965.

13 Višik, M. I. and Èskin, G. I., Elliptic equations in convolution in a bounded domain and their applications, *Russ. Math. Surv.*, **22,** 13–75, 1967.

14 Dikanskiĭ, A. S., Problems adjoint to elliptic pseudodifferential boundary value problems, *Sov. Math. Dokl.*, **12,** 1520–1525, 1971.

15 Dikanskiĭ, A. S., Conjugate problems of elliptic differential and pseudodifferential boundary value problems in a bounded domain, *Math. USSR Sb.*, **20,** 67–83, 1973.

16 Radlow, J., Diffraction by a quarter plane, *Arch. Rat. Mech. Anal.*, **8,** 139–158, 1961.

17 Radlow, J., Note on the diffraction at a corner, *Arch. Rat. Mech. Anal.*, **19,** 62–70, 1965.

18 Hörmander, L., Estimates for translation invariant operators in L^p spaces, *Acta Math.*, **104,** 93–140, 1960.

19 Latz, N., Untersuchungen über ein System von simultanen Integralgleichungen aus der Theorie der Beugung elektromagnetischer Wellen an rechtwinkligen dielektrischen Keilen, Dipl.-arbeit, Univ. Saarbrücken, 1963.

20 Radlow, J., Diffraction by a right-angled dielectric wedge, *Int. J. Eng. Sci.*, **3,** 429–439, 1965.

21 Kuo, N. H. and Plonus, M. A., A systematic technique in the solution of diffraction by a right-angled dielectric wedge, *J. Math. Phys.*, **46,** 394–407, 1967.

22 Latz, N., Untersuchungen über ein skalares Übergangswertproblem aus der Theorie der Beugung elektromagnetischer Wellen an dielektrischen Keilen, Dissertation, Univ. Saarbrücken, 1968.

23 Kraut, E. A. and Lehman, G. W., Diffraction of electromagnetic waves by a right-angle dielectric wedge, *J. Math. Phys.* **10,** 1340–1348, 1969.

24 Rawlins, A. D., Diffraction by a dielectric wedge, *J. Inst. Math. Applic.*, **19,** 261–279, 1977.

25 Meister, E. and Latz, N., Ein System singulärer Integralgleichungen aus der Theorie der Beugung elektromagnetischer Wellen an dielektrischen Keilen, *Z. Angew. Math. Mech.*, **44,** T47–T49, 1964.

26 Latz, N., Wiener–Hopf-Gleichungen zu speziellen Ausbreitungsproblemen elektromagnetischer Schwingungen. Habilitationsschrift, TU Berlin, 1974.

27 Shinbrot, M., On singular integral operators, *J. Math. Mech.*, **13,** 395–406, 1964.

28 Devinatz, A., Toeplitz operators on H^2 spaces, *Trans. AMS,* **112,** 304–317, 1964.

29 Shinbrot, M., On the range of general Wiener–Hopf operators, *J. Math. Mech.*, **18,** 587–601, 1969.

30 Shinbrot, M., The solution of some integral equations of Wiener–Hopf type, *Quart. Appl. Math.*, **28,** 15–36, 1970.

31 Devinatz, A. and Shinbrot, M., General Wiener–Hopf operators, *Trans. AMS*, **145,** 467–494, 1969.

32 Shinbrot, M., An inversion formula for certain general Wiener–Hopf operators, *Arch. Rat. Mech. Anal.*, **37,** 342–362, 1970.

33 Shinbrot, M., An inversion formula for analytic families of general Wiener–Hopf operators, *Indiana Univ. Math. J.*, **20,** 945–948, 1971.

34 Devinatz, A., On Wiener–Hopf operators, in *Functional Analysis: Proceedings of a Conference at the University of California at Irvine*, Academic Press, New York and London, 1967, pp. 81–118.

35 Reeder, J., On the invertibility of general Wiener–Hopf operators, *Proc. AMS*, **27,** 72–76, 1971.

36 Reeder, J., General Wiener–Hopf operators and complete biorthogonal systems, *Indiana Univ. Math. J.*, **23,** 107–119, 1973.

37 Pellegrini, V., Unbounded general Wiener–Hopf operators, *Indiana Univ. Math. J.*, **21,** 85–90, 1971.

38 Pellegrini, V., General Wiener–Hopf operators and the numerical range of an operator, *Proc. AMS*, **38,** 141–146, 1973.

39 Simonenko, I. B., Operators of convolution type in cones, *Math. USSR Sb.*, **3,** 279–293, 1967.

40 Goldenstein, L. S. and Gohberg, I. Z., On a multidimensional integral equation on a half-space whose kernel is a function of the difference of the arguments, and on a discrete analogue of this equation, *Sov. Math. Dokl.*, **1,** 173–176, 1960.

41 Breuer, M. and Cordes, H. O., On Banach algebras with σ-symbol, *J. Math. Mech.*, **13,** 313–323, 1964.

42 Breuer, M., Banachalgebren mit Anwendungen auf Fredholmoperatoren und singuläre Integralgleichungen, *Bonner Math. Schr.* No. 24, 1965.

43 Cordes, H. O. and Herman, E. A., Gelfand theory of pseudodifferential operators, *Amer. J. Math.*, **90,** 681–717, 1968.

44 Simonenko, I. B., A new general method of investigating linear operator equations of the type of singular integral equations, *Sov. Math. Dokl.*, **5,** 1323–1326, 1964.

45 Speck, F.-O., Über verallgemeinerte Faltungsoperatoren und eine Klasse von Integrodifferentialgleichungen, Dissertation, TH Darmstadt, 1974.

46 Strang, G., Toeplitz operators in a quarter-plane, *Bull. AMS*, **76,** 1303–1307, 1970.

47 Douglas, R. G. and Howe, R., On the C^*-algebra of Toeplitz operators on the quarter-plane, *Trans. AMS*, **158,** 203–217, 1971.

48 Dudučava, R. V., On bisingular integral operators and convolution operators on a quadrant, *Sov. Math. Dokl.*, **16,** 330–334, 1975.

49 Dudučava, R. V., Bisingular integral operators and boundary value problems of the theory of analytic functions in spaces of generalized functions, *Sov. Math. Dokl.*, **16,** 1324–1328, 1975.

50 Kraut, E., On equations of the Wiener–Hopf type in several complex variables, *Proc. AMS*, **23,** 24–26, 1969.

51 Latz, N., Über eine Integralgleichung vom Faltungstyp, *Meth. Verf. Math. Phys.*, **3,** 73–83, 1970.

52 Kremer, M., Über eine Algebra "nicht normaler" Wiener–Hopf-Operatoren, I and II, *Math. Ann.*, **220,** 77–86, 87–95, 1976.

53 Simonenko, I. B., Multidimensional discrete convolutions, *Mat. Issled.*, **3,** 108–122, 1968 (in Russian).

54 Pilidi, V. S., On multidimensional bisingular operators, *Sov. Math. Dokl.*, **12,** 1723–1726, 1971.

55 Pilidi, V. S. and Sazonov, L. I., *A priori* estimates for characteristic bisingular integral operators, *Sov. Math. Dokl.*, **15,** 1064–1067, 1974.

56 Pilidi, V. S., Index computation for a bisingular operator, *Funct. Anal. Appl.*, **7,** 337–338, 1973.

57 Kakičev, V. A., Boundary value problems of linear conjugation for functions holomorphic in bicylindrical regions, *Sov. Math. Dokl.*, **9,** 222–226, 1968.

58 Przeworska-Rolewicz, D., *Equations with Transformed Argument*, Elsevier, Amsterdam, 1973.

59 Busenberg, S. N., Iterative solution of a Wiener–Hopf problem in several variables, *Proc. AMS*, **29,** 39–46, 1971.

60 Gerlach, E. and Latz, N., Zur Spektraltheorie bei Faltungsoperatoren, *Z. Angew. Math. Mech.*, **57,** T231–T232, 1977.

61 Halmos, P. R., *A Hilbert-space Problem Book*, Springer, Berlin, Heidelberg, New York, 1974.

62 Speck, F.-O., Eine Erweiterung des Satzes von Rakovčik und ihre Anwendung in der Simonenko-Theorie, Math. Ann., **228,** 93–100, 1977.

63 Cordes, H. O., On compactness of commutators of multiplications and convolutions, and boundedness of pseudodifferential operators, *J. Funct. Anal.*, **18,** 115–131, 1975.

64 Speck, F.-O., Über verallgemeinerte Faltungsoperatoren und ihre Symbole, in *Function-Theoretic Methods for Partial Differential Equations. Proceedings of an International Symposium, Darmstadt*, Lecture Notes in Mathematics No. 561, Springer Verlag, Berlin, New York, Heidelberg, 1976, pp. 459–471.

Professor Dr E. Meister and Dr F.-O. Speck,
Schlossgartenstrasse 7,
Fachbereich Mathematik,
Technische Hochschule,
D-6100 Darmstadt,
Bundesrepublik Deutschland

J. J. Moreau

Application of convex analysis to some problems of dry friction

1 Introduction

The core of what is meant today by 'convex analysis' consists in studying convex subsets of linear spaces, convex real functions defined on such spaces, the extremal problems involving them and their minimax counterpart. This subject has received considerable attention during recent decades, stimulated by the frequent occurrence of convexity assumptions in optimization, economics and the related numerical analysis. Such modern developments as variational inequalities or monotone operators are closely interrelated with convex analysis.

To the author's historical knowledge, mechanics was the first domain of science to make a precise use of the concept of a convex set (17th century): the equilibrium positions of a solid body lying on a horizontal plane and subject to gravity are characterized by the condition that the vertical line drawn through its centre of mass meets the convex hull of the points of support. Investigating the statics or the dynamics of systems with unilateral constraints in this spirit constituted the author's primary motivation for taking some part in the recent development of convex analysis (*see*, for example, Moreau [1, 2]).

On the other hand, the importance of convexity assumptions regarding the potential function of a force law has been repeatedly stressed, mainly after Hill [3], in relation to stability, sometimes with hints at thermodynamics.

The concepts of the subdifferential of a convex function, now of general use in many domains, and of the superpotential of a force law, were defined by the author in order to include classical force laws and perfect (possibly unilateral) constraints in a unified treatment (*see*, for example, Moreau [4–6]).

The duality of linear spaces plays a prominent role in modern convex analysis. Regarding this concept also, mechanists acted as forerunners; in fact, the mathematical structure of a pair of linear spaces, placed in duality by a bilinear form, constitutes the essence of the traditional method of virtual work or virtual power.

The present paper, attemptedly self-consistent, develops an example of the application of modern convex analysis to dry friction and ends with two general theorems. This theme was first presented at a small symposium on convexity in 1970 (Moreau [7]); subsequently, modern convex analysis has been widely applied to resistance laws of various sorts (Moreau [8]), mainly to *plasticity theory* (Moreau [9–11], Nayroles [12–15], Debordes and Nayroles [16]) possibly with strain hardening (Nguyen and Halphen [17], Nguyen [18]). The concrete and elementary situation taken as an example in what follows may be studied as an introduction to these more elaborate topics.

2 The classical formulation of Coulomb's law

Let $\mathscr{S}_0$ denote a perfectly rigid body, assumed to be fixed; another perfectly rigid body $\mathscr{S}_1$ moves in contact with $\mathscr{S}_0$. This means that the respective boundary surfaces Σ_0 and Σ_1, supposed geometrically smooth, remain tangent at a point M, *a priori* moving in both of them. All the following is relative to some definite instant; in the corresponding configuration, let $\boldsymbol{\nu}$ denote the unit normal vector to Σ_0 and Σ_1 at the point M. By definition, the *sliding velocity* $\mathbf{V}$ of $\mathscr{S}_1$ on $\mathscr{S}_0$ is the velocity vector relative to $\mathscr{S}_0$ of the element M_1 of $\mathscr{S}_1$ which happens to be in M at the instant under consideration. Under the usual geometrical and kinematical smoothness assumptions, it is elementarily proved that $\boldsymbol{\nu} \cdot \mathbf{V} = 0$, i.e., $\mathbf{V}$ belongs to the two-dimensional linear spaces Π consisting of the vectors tangent at M to Σ_0 and Σ_1.

The contact forces exerted by $\mathscr{S}_0$ on $\mathscr{S}_1$ are supposed to reduce to a single force $\mathbf{R}$ acting on the element M_1; let us decompose this vector into

$$\mathbf{R} = \mathbf{F} + N\boldsymbol{\nu} \quad \text{with} \quad \mathbf{F} \in \Pi.$$

The Coulomb law of dry friction, *when the normal component* $N \geqslant 0$ *is treated as known*, states a relation between $\mathbf{F}$ and $\mathbf{V}$ traditionally formulated as follows.

There exists $f \geqslant 0$, *the friction coefficient, such that*

$$\textit{if} \quad \mathbf{V} = 0 : |\mathbf{F}| \leqslant fN, \tag{1}$$

$$\textit{if} \quad \mathbf{V} \neq 0 : |\mathbf{F}| = fN \tag{2}$$

and the vectors $\mathbf{F}$ *and* $\mathbf{V}$ *are parallel with opposite directions.* (*Here* $|\ |$ *represents the Euclidean norm.*)

Such a juxtaposition of two apparently heterogeneous statements concerning the events $\mathbf{V} = 0$ and $\mathbf{V} \neq 0$ might look purely empirical. Actually, the use of some elementary concepts of convex analysis will emphasize their strong consistency.

3 Generalization

In the same situation as above, let us consider the closed disc

$$D=\{\mathbf{\Phi}\in\Pi:|\mathbf{\Phi}|\leqslant fN\}.$$

The Coulomb law is equivalently expressed by

$$\left.\begin{array}{r}\mathbf{F}\in D,\\ \forall\mathbf{\Phi}\in D:\mathbf{V}\cdot(\mathbf{\Phi}-\mathbf{F})\geqslant 0.\end{array}\right\}\tag{3}$$

In fact, when $\mathbf{V}\neq 0$, this means that the set

$$\{\mathbf{\Phi}\in\Pi:\mathbf{V}\cdot(\mathbf{\Phi}-\mathbf{F})\leqslant 0\},$$

i.e., the closed half-plane having $\mathbf{F}$ as a boundary point and $\mathbf{V}$ as an outward normal vector is a *supporting half-plane* of the set D at the point $\mathbf{F}$ (i.e., this half-plane has only boundary points in common with D and $\mathbf{F}$ is one of them). By the elementary properties of the circle, this is equivalent to (2). In the case $\mathbf{V}=0$, the equivalence of (3) to (1) is trivial.

From this stage it is quite natural to generalize the formulation into a law of *anisotropic friction*, as it may physically result from the directional structure of the material surfaces in contact: the disc D will be replaced by some subset C of Π, containing the origin.

On the other hand, one aim of this paper is to emphasize the consideration of many-dimensional pairs of linear spaces. Generally speaking, a linear space of *velocities* $\mathscr{V}$ and a linear space of *forces* $\mathscr{F}$ will be introduced. These spaces are *placed in duality* by the bilinear form 'power': for $\mathbf{V}\in\mathscr{V}$ and $\mathbf{F}\in\mathscr{F}$, we shall denote by $\langle\mathbf{V},\mathbf{F}\rangle$ the power of the force $\mathbf{F}$ if the motion has the velocity $\mathbf{V}$. In the preceding example, $\mathscr{V}$ and $\mathscr{F}$ were two copies of the same two-dimensional Euclidean space Π, placed in duality with itself by the Euclidean scalar product.

Similarly to (3), let us define a *friction law* as the relation between $\mathbf{V}\in\mathscr{V}$ and $\mathbf{F}\in\mathscr{F}$ formulated as follows:

$$\left.\begin{array}{r}\mathbf{F}\in C,\\ \forall\mathbf{\Phi}\in C:\langle\mathbf{V},\mathbf{\Phi}-\mathbf{F}\rangle\geqslant 0,\end{array}\right\}\tag{4}$$

where C is a given subset of $\mathscr{F}$. *This subset is assumed to contain the origin of* $\mathscr{F}$, i.e., zero is a possible value of $\mathbf{F}$, compatible in particular with the value zero of $\mathbf{V}$. Then, by setting $\mathbf{\Phi}=0$, it turns out that for every pair $\mathbf{F}$, $\mathbf{V}$ satisfying (4), the power $\langle\mathbf{V},\mathbf{F}\rangle$ is non-positive: friction, as described by relation (4), is a *dissipative* phenomenon.

4 The principle of maximum dissipation

Relation (4) is obviously equivalent to the following statement: *the set of the elements* $\mathbf{F} \in \mathscr{F}$ *which the relation associates with a given* $\mathbf{V} \in \mathscr{V}$ *is identical with the set of the points of* C *where the function* $\mathbf{\Phi} \mapsto \langle \mathbf{V}, \mathbf{\Phi} \rangle$ *attains its infimum relative to* C.

In most of the mechanical situations which a relation of the form (4) is meant to describe, it is required that every value of $\mathbf{V}$ be feasible, i.e., for every $\mathbf{V} \in \mathscr{V}$ the above set is non-empty. Here is the most usual mathematical assumption ensuring that: there exists a topology on the space $\mathscr{F}$ relative to which the set C is compact, while the real function $\Phi \mapsto \langle \mathbf{V}, \mathbf{\Phi} \rangle$ is continuous for every $\mathbf{V}$ in $\mathscr{V}$. In the case of finite-dimensional $\mathscr{V}$ and $\mathscr{F}$, this will naturally be the topology defined by the use of components in these linear spaces; then, the continuity of linear functions is automatic and we only have to make the assumption that C is *closed and bounded*. For infinite-dimensional cases, such as those arising in the mechanics of continua, it is necessary to specify some topology on $\mathscr{F}$ among those which are said to be *compatible with the duality defined by the bilinear form* $\langle .,. \rangle$ or *topologies of the dual pair* $(\mathscr{V}, \mathscr{F}, \langle .,. \rangle)$; in order that compactness involve the mildest restriction about C, this should be the coarsest of these topologies, i.e., the *weak topology* usually denoted by $\sigma(\mathscr{F}, \mathscr{V})$ (*see*, for example, Robertson and Robertson [23]).

In the proper friction phenomenon, the non-negative expression $-\langle \mathbf{V}, \mathbf{F} \rangle$ is equal to the power transformed into heat and is called the *dissipated power*: thus, relation (4) may be entitled 'the principle of maximum dissipation'.

5 Indicator functions and subdifferentials

Every subset C of $\mathscr{F}$ may be described by giving its *indicator function* ψ_C

$$\psi_C(\mathbf{\Phi}) = \begin{cases} 0 & \text{if } \mathbf{\Phi} \in C, \\ +\infty & \text{if } \mathbf{\Phi} \notin C. \end{cases}$$

Using this, one writes relation (4) equivalently in the form

$$\forall \mathbf{\Phi} \in \mathscr{F} : < -\mathbf{V}, \mathbf{\Phi} - \mathbf{F} > + \psi_C(\mathbf{F}) \leqslant \psi_C(\mathbf{\Phi})$$

which states that $\psi_C(\mathbf{F})$ is finite, that the *affine function*

$$\mathbf{\Phi} \mapsto < -\mathbf{V}, \mathbf{\Phi} - \mathbf{F} > + \psi_C(F)$$

is a *minorant* of the function ψ_C and that this minorant is *exact* at the point $\mathbf{F}$, i.e., it takes the same value at this point as ψ_C (namely zero). The element $-\mathbf{V}$ of $\mathcal{V}$ constitutes the *slope* or *gradient* of the considered affine function in the sense of the duality $\langle .\, , . \rangle$. According to the terminology introduced by the author [4] and now usual in the whole field of convex analysis, the gradient of an affine minorant of a function $g : \mathcal{F} \to]-\infty, +\infty]$; if this minorant is exact at the point $\mathbf{F}$, is called a *subgradient* of g at the point $\mathbf{F}$. The set, denoted by $\partial g(\mathbf{F})$, of the subgradients of g at the point $\mathbf{F}$ is a (possibly empty) convex subset of $\mathcal{V}$ called the *subdifferential* of g at $\mathbf{F}$.

In this notation, the relation (4) is equivalently written as

$$-\mathbf{V} \in \partial \psi_C(\mathbf{F}). \tag{5}$$

Usually, C is a closed convex set (*see* section 7 below); (5) means that $\mathbf{F} \in C$ and that $\mathbf{V}$ is, in a classical generalized sense, a *normal inward vector* to this set at the point $\mathbf{F}$ (in particular, $\mathbf{V} = 0$ if $\mathbf{F}$ is internal to C).

6 Dissipation function

Let g be a function with values in $]-\infty, +\infty]$, defined, for instance, on the member $\mathcal{F}$ of the considered dual pair of linear spaces. The handling of the affine minorants of g induces us to construct the function g^*, with values in $]-\infty, +\infty]$, defined on $\mathcal{V}$ by

$$g^*(\mathbf{W}) = \sup_{\mathbf{\Phi} \in \mathcal{F}} [\langle \mathbf{W}, \mathbf{\Phi} \rangle - g(\mathbf{\Phi})]. \tag{6}$$

It is called the *conjugate* or *polar function* of g. An affine function $\Phi \to \langle \mathbf{W}, \mathbf{\Phi} \rangle - \alpha$ is a minorant of g if and only if the real number α satisfies $\alpha \geqslant g^*(\mathbf{W})$.

In the special case, $g = \psi_C$, the expression $\langle \mathbf{W}, \mathbf{\Phi} \rangle - \psi_C(\mathbf{\Phi})$ takes the value $-\infty$ when $\mathbf{\Phi} \notin C$; therefore,

$$\psi_C^*(\mathbf{W}) = \sup_{\mathbf{\Phi} \in C} \langle \mathbf{W}, \mathbf{\Phi} \rangle.$$

This function is classically known under the (rather improper) name of the *support function* of the set C, relative to the considered duality. It is evidently *sublinear,* i.e., convex and positively homogeneous of degree 1.

In the present situation, it will prove more convenient to introduce the function φ

$$\varphi(\mathbf{W}) = \psi_C^*(-W) = -\inf_{\mathbf{\Phi} \in C} \langle \mathbf{W}, \mathbf{\Phi} \rangle, \tag{7}$$

i.e., the support function of the set $-C$. Using φ yields an equivalent formulation of the relations (4) or (5)

$$\left.\begin{array}{r}\mathbf{F}\in C,\\ -\langle\mathbf{V},\mathbf{F}\rangle=\varphi(\mathbf{V}).\end{array}\right\}\tag{8}$$

In other words, the values of $\mathbf{F}\in\mathscr{F}$ that the relation associates with a given $\mathbf{V}\in\mathscr{V}$ are the elements of C such that the dissipated power $-\langle\mathbf{V},\mathbf{F}\rangle$ is equal to $\varphi(\mathbf{V})$.

Hence the name of *the dissipation function*, given to φ.

For instance, in the case of Coulomb's law (3),

$$\varphi(\mathbf{V})=fN\,|\mathbf{V}|.$$

7 The convexity of C

Returning to the definition (6) of g^*, one immediately finds that the relation $\mathbf{W}\in\partial g(\mathbf{F})$ is equivalent to

$$g^*(\mathbf{W})+g(\mathbf{F})-\langle\mathbf{W},\mathbf{F}\rangle=0\tag{9}$$

where the = sign may be replaced by $\leqslant$ because the left-hand side is essentially non-negative.

The above does not, in general, involve the symmetry between the spaces $\mathscr{V}$ and $\mathscr{F}$. In fact, a polar function is, by construction, the supremum of a collection of continuous affine functions; therefore, it is convex and lower semi-continuous (l.s.c.) (relative to every topology compatible with the considered duality). As we started with an arbitrary $g:\mathscr{F}\to]-\infty,+\infty]$, it cannot be expected, in general, that g would in turn be the polar function of g^*. However, standard separation arguments (i.e., the Hahn–Banach theorem) may be used to prove that g is equal to g^* if and only if g is convex and lower semi-continuous (for some of the topologies of the dual pair $(\mathscr{V},\mathscr{F},\langle.\,,.\rangle)$, consequently for all of them). *If such is the case, the symmetry of* (9) *implies that* $\mathbf{W}\in\partial g(\mathbf{F})$ *is equivalent to* $\mathbf{F}\in\partial g^*(\mathbf{W})$.

In the following, we shall deal with the special case $g=\psi_C$; this function is convex and l.s.c. if and only if *the subset C of $\mathscr{F}$ is convex and closed.* If such is the case, the friction law, as expressed equivalently by (4), (5) or (8), is also equivalent to

$$\mathbf{F}\in\partial\psi_C^*(-\mathbf{V}),$$

i.e., in view of the definition (7) of φ,

$$-\mathbf{F}\in\partial\varphi(\mathbf{V}).\tag{10}$$

Remark 1 A relation of this form between some velocity $\mathbf{V}$ and some force $\mathbf{F}$ may be called a *resistance law*, admitting the (convex and l.s.c.)

function φ as *superpotential* or *pseudopotential.* The more general case, where φ is not necessarily sublinear, was studied in Moreau [8], where the connection of the superpotential with the dissipated power was also investigated.

Remark 2 In all the following, the set C is assumed to be convex. As far as the contact friction is concerned, the contrary would seem unrealistic. In fact, one must keep in mind that the point contact between two bodies is only a schematic representation of some contact which takes place on a very small area $\mathscr{A}$. We may imagine, this area to be arbitrarily divided into two others, $\mathscr{A}_1$ and $\mathscr{A}_2$, in which the sliding velocity has the same value, namely zero in what follows. Let $\mathbf{R}_1$ and $\mathbf{R}_2$ be the resultant forces experienced by $\mathscr{S}_1$ through $\mathscr{A}_1$ and $\mathscr{A}_2$, respectively. Then

$$\mathbf{R}=\mathbf{R}_1+\mathbf{R}_2 \tag{11}$$

and, concerning the normal components, one has $N=N_1+N_2$. The values of N_1 and N_2 in the last equation depend on some 'micro-information' about the distribution of pressure in $\mathscr{A}$. In order to obtain a law which does not depend on the microscopic pressure distribution in $\mathscr{A}$ for the global reaction $\mathbf{R}$, one must admit the following as the law of friction in every subarea such as $\mathscr{A}_1$ (or $\mathscr{A}_2$, or $\mathscr{A}$ itself): for zero sliding velocity and an arbitrary non-negative pressure component, the set of the possible values of $\mathbf{R}_1$ (or $\mathbf{R}_2$ or $\mathbf{R}$) is a conic subset Γ, with vertex at the origin in the space of the three-dimensional vectors. And (11) entails the inclusion $\Gamma+\Gamma\subset\Gamma$, which means that Γ is convex. Returning to the formulation (4), one finds that Γ is the cone generated in the space of three-dimensional vectors by the set $C+N\boldsymbol{\nu}$; hence, the convexity of C.

8 Product spaces

Let $(\mathscr{V}_1, \mathscr{F}_1, \langle.,.\rangle_1)$ and $(\mathscr{V}_2, \mathscr{F}_2, \langle.,.\rangle_2)$ be two dual pairs of linear spaces. The two product spaces $\mathscr{V}=\mathscr{V}_1\times\mathscr{V}_2$ and $\mathscr{F}=\mathscr{F}_1\times\mathscr{F}_2$ are placed in duality by the bilinear form $\langle.,.\rangle$, defined as follows: for $v=(v_1, v_2)\in\mathscr{V}_1\times\mathscr{V}_2$ and $f=(f_1, f_2)\in\mathscr{F}_1\times\mathscr{F}_2$, set

$$\langle v, f\rangle=\langle v_1, f_1\rangle_1+\langle v_2, f_2\rangle_2. \tag{12}$$

For instance, if v_1, v_2 are some independent velocity parameters of a mechanical system and f_1, f_2 are the associated force parameters such that the terms on the right in (12) represent their respective powers, the bilinear form $\langle v, f\rangle$ represents the power of the whole.

In this framework, the following is easily established. Let $g_1:\mathscr{F}_1\to]-\infty,+\infty]$ and $g_2:\mathscr{F}_2\to]-\infty,+\infty]$ and let g be the function defined for every $f=(f_1, f_2)$ in $\mathscr{F}$ by

$$g(f)=g_1(f_1)+g_2(f_2).$$

If g_1^* and g_2^* are the respective polar functions of g_1 and g_2, the polar function g^* in the sense of the duality (12) is defined for every $v=(v_1, v_2)$ in $\mathscr{V}$ by

$$g^*(v)=g_1^*(v_1)+g_2^*(v_2).$$

Concerning the subdifferential sets, on the other hand, one has, in the sense of the three respective dualities, the equivalence

$$v\in\partial g(f)\Leftrightarrow v_1\in\partial g_1(f_1) \quad\text{and}\quad v_2\in\partial g_2(f_2). \tag{13}$$

9 An example of composite friction law

Let $\mathscr{S}_1$ be one of the wheels, with radius a, by which some vehicle $\mathscr{S}_2$ is supported, possibly with skidding, upon the horizontal ground $\mathscr{S}_0$. We shall treat this wheel as a perfectly rigid body presenting a single point of contact M with the plane surface of the ground. Let us describe the friction at this point by means of the notations of section 2. The reaction $\mathbf{R}$ exerted by the ground on the wheel is written as

$$\mathbf{R}=\mathbf{F}+N\boldsymbol{\nu}$$

and, in view of (5), the Coulomb law takes the form

$$-\mathbf{V}\in\partial\psi_D(\mathbf{F}), \tag{14}$$

where D denotes the closed disc with radius fN, with centre at the origin in the two-dimensional Euclidean linear space Π; recall that $\mathbf{V}$ is the sliding velocity of $\mathscr{S}_1$ relative to $\mathscr{S}_0$.

On the other hand a *brake* is supposed to act on the wheel. Let $\mathbf{i}$ denote a unit vector of the wheel axis, assumed to be parallel to the ground. Let h be the moment, relative to this oriented axis, of the forces that the wheel $\mathscr{S}_1$ experiences from the vehicle body $\mathscr{S}_2$. Neither the driving torque nor the friction in the bearings are taken into account, so that h is, in fact, the braking torque. Let ω be the angular velocity of $\mathscr{S}_1$ relative to $\mathscr{S}_2$. We assume that the operation of the brake involves a dry friction with a given normal component. This is expressed by a relation between the real numbers h and ω, namely the one-dimensional case of the general formalism presented in the foregoing. Here the space of velocities and the space of forces are two copies of the real line $\mathbb{R}$ and the bilinear form 'power' reduces to the ordinary product. In the sense of this duality, the brake law is written as

$$-\omega\in\partial\psi_I(h) \tag{15}$$

where I denotes a given interval $[-b, +b]$; this summarizes the familiar relations: $h = -b \operatorname{sgn} \omega$ if $\omega \neq 0$ and h arbitrary in $[-b, +b]$ if $\omega = 0$.

In view of the preceding section, taking into account (14) and (15), we note that, equivalently, we have the relation

$$-(\omega, \mathbf{V}) \in \partial\psi_{I\times D}(h, \mathbf{F}) \tag{16}$$

in the sense of the duality $(\mathscr{V}, \mathscr{F}, \langle . , . \rangle)$. Here, $\mathscr{V}$ and $\mathscr{F}$ are two copies of the three-dimensional linear space $\mathbb{R}\times \Pi$; by definition,

$$\langle(\omega, \mathbf{V}), (h, \mathbf{F})\rangle = \omega h + \mathbf{V} \cdot \mathbf{F} \tag{17}$$

is the total power of the torque h and of the ground reaction $\mathbf{R} = \mathbf{F} + N\boldsymbol{\nu}$ acting simultaneously on the wheel.

Here is the first problem we are to deal with in the following. The wheel load, i.e., the normal component N of $\mathbf{R}$, will be treated as known. Let $\mathbf{G} \in \Pi$ denote the horizontal component of the resultant force experienced by the vehicle $\mathscr{S}_2$ from the wheel $\mathscr{S}_1$. Let $\mathbf{W} \in \Pi$ denote the velocity of the wheel centre, which is also the velocity of the corresponding point of $\mathscr{S}_2$ or the velocity of M, the 'geometrical' point of contact with the ground. Under the assumption that the wheel is sufficiently light and the motion sufficiently slow for the inertia of the wheel to be negligible, we are to summarize the combination of possible skidding on the ground and of possible brake action into a simple relation between $\mathbf{W}$ and $\mathbf{G}$. It will turn out, under the above simplifying assumptions, that the wheel may be forgotten and the interaction between the vehicle and the ground be described as the anisotropic friction related to a certain convex set C.

This consists in the *elimination* of the variables ω, $\mathbf{V}$, h, $\mathbf{F}$ from the following set of relations.

(i) The kinematical relation

$$\mathbf{W} = \mathbf{V} - \omega a \mathbf{j} \tag{18}$$

expressing the fact that the wheel is a rigid body; here $\mathbf{j}$ is the unit vector $\boldsymbol{\nu} \times \mathbf{i}$.

(ii) The quasi-equilibrium equations of the wheel:

$$\mathbf{G} - \mathbf{F} = 0, \tag{19}$$

$$h + a\mathbf{j} \cdot \mathbf{F} = 0. \tag{20}$$

(iii) The composite friction law (16).

An adequate use of various rules of the 'subdifferential calculus' would do the job, but it will be more instructive to place the reasoning in a general setting.

10 Subdifferentials and linear mappings

Let $(\mathscr{V}, \mathscr{F})$ and $(\mathscr{V}', \mathscr{F}')$ be two dual pairs of linear spaces; both corresponding bilinear forms will be denoted by $\langle .\,, .\rangle$. Let $L:\mathscr{F}'\to\mathscr{F}$ be a linear mapping; in infinite-dimensional cases, it will be assumed that L is continuous in the weak topologies of the dual pairs. Let $g:\mathscr{F}\to]-\infty, +\infty]$ be convex and l.s.c.; then, the composite function $g' = g\circ L:\mathscr{F}'\to]-\infty, +\infty]$ is convex and l.s.c. A classical rule of the 'subdifferential calculus' is the following (Rockafellar [24]):

If there exists a point in the range of L, where the function g is finite and continuous (in some topology of the dual pair $(\mathscr{V}, \mathscr{F})$), one has, for every F' in $\mathscr{F}'$,

$$\partial(g\circ L)(F') = L^*(\partial g(L(F'))), \tag{21}$$

where L^* denotes the transpose of L.

Application Returning to our mechanical problem, let us set $\mathscr{V} = \mathscr{F} = \mathbb{R}\times \Pi$ with the duality defined in (17). Moreover, let $\mathscr{V}' = \mathscr{F}' = \Pi$ with the duality defined by the Euclidean scalar product and define $L:\mathscr{F}'\to\mathscr{F}$ by

$$L(\mathbf{G}) = (-a\mathbf{j}\cdot\mathbf{G}, \mathbf{G}).$$

Elementary computation yields the transpose $L^*:\mathscr{V}\to\mathscr{V}'$, namely

$$L^*(\omega, V) = \mathbf{V} - \omega a\mathbf{j}.$$

Then, (18) amounts to

$$\mathbf{W} = L^*(\omega, \mathbf{V}), \tag{22}$$

while (19) and (20) are condensed into

$$(h, \mathbf{F}) = L(\mathbf{G}). \tag{23}$$

Take $g = \psi_{I\times D}$, so that (16) takes the form

$$-(\omega, \mathbf{V}) \in \partial g(h, \mathbf{F}). \tag{24}$$

The elimination of ω, $\mathbf{V}$, h, $\mathbf{F}$ from (22), (23), (24) yields

$$-\mathbf{W} \in L^*(\partial g(L(\mathbf{G}))). \tag{25}$$

This, indeed, is the necessary and sufficient condition to be satisfied by $\mathbf{W}$ and $\mathbf{G}$ in order that there exist ω, $\mathbf{V}$, h, $\mathbf{F}$ which also satisfy the above conditions.

Suppose now that the interval I and the disc D do not degenerate into single points; then the zero of $\mathscr{F}$ constitutes a point in the range of L where g is continuous; thus (21) holds, making (25) equivalent to

$$-\mathbf{W} \in \partial(g\circ L)(\mathbf{G}).$$

Here is the expression for the function $g \circ L$:

$$(g \circ L)(\mathbf{G}) = \psi_{I \times D}(-a\mathbf{j} \cdot \mathbf{G}, \mathbf{G})$$
$$= \begin{cases} 0 & \text{if } -a\mathbf{j} \cdot \mathbf{G} \in I \quad \text{and} \quad \mathbf{G} \in D, \\ +\infty & \text{otherwise.} \end{cases}$$

In other words $g \circ L$ is the indicator function of the closed convex subset $C = B \cap D$ of Π, where B denotes the strip

$$B = \left\{ \mathbf{G} \in \Pi : -\frac{b}{a} \leqslant \mathbf{j} \cdot \mathbf{G} \leqslant \frac{b}{a} \right\}.$$

Hence, the final form of (25),

$$-\mathbf{W} \in \partial \psi_C(\mathbf{G}), \tag{26}$$

which constitutes a *friction law* in the general sense of section 3. As announced, this presents the interaction between the vehicle and the ground by forgetting the wheel. Recall that N, the load supported by the wheel, is treated as known.

Let us now discuss the various cases.

(i) If $b \geqslant afN$, the width of the strip B is greater than or equal to the diameter of the disc D; hence, $C = D$. This means that the brake is so tightly applied that the wheel stays always locked; thus, the interaction between the vehicle and the ground amounts to the simple Coulomb friction.

(ii) If $b < afN$ the shape of C is shown in Fig. 1. Recall that (26) expresses that $\mathbf{G} \in C$ and that, in the classical generalized sense, $\mathbf{W}$ is an *inward normal vector* to C at the point $\mathbf{G}$ (in particular, $\mathbf{W}$ is necessarily zero if $\mathbf{G}$ is internal to C). The presence of the rectilinear parts in the boundary of C implies that a value of $\mathbf{W}$ parallel to $\mathbf{j}$ (i.e., normal to the wheel plane) corresponds to an infinity of possible values of the force $\mathbf{G}$. The presence of corner points in the boundary implies that these corner

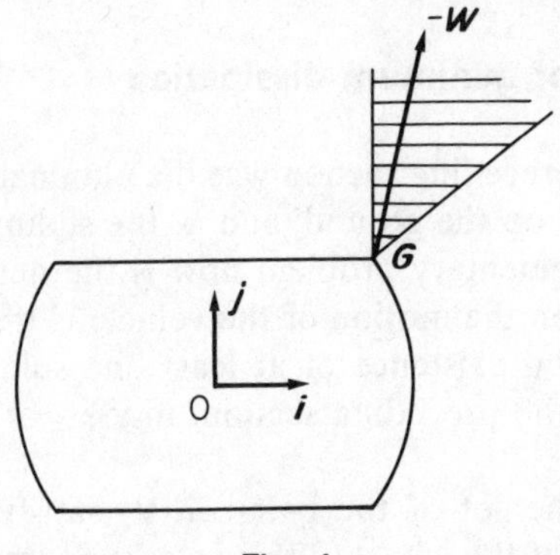

Fig. 1

values of **G** correspond to an infinity of values for **W**, the set of them form a closed convex angular region in Π.

Note that some more refined arguments of convex analysis (*see*, for example, Rockafellar [20], Theorem 23.8) allow us to remove the assumption $b \neq 0$ made in the foregoing. The case $b = 0$ is that where no brake is applied; then, C reduces to a line segment and (26) describes an extreme case of anisotropic friction. Such a side-slipping free wheel is the key device of the Amsler planimeter and of some other ancient integrating instruments.

Remark The above computation involving a pair of mutually transpose linear mappings L and L^* is more than an occasional mathematical trick. It is based on the fact that the rigidity of the wheel constitutes a *perfect* mechanical constraint; in fact, the external forces applied to the wheel may be summarized as:

(i) the force **F** applied to the contact point (as the normal component $N\boldsymbol{\nu}$ does not matter here);
(ii) the force $-\mathbf{G}$ applied to the wheel centre;
(iii) the axial torque h.

By the definition of a perfect constraint (*see* more developments and examples in Moreau [6, 10]) the (quasi-)equilibrium condition of the wheel may be expressed by the fact that the above system of forces should yield a zero power for every set of values of the velocity parameters **V**, **W**, ω satisfying the kinematical equations of constraint, namely (22). In other words,

$$-\mathbf{G} \cdot L^*(\boldsymbol{\omega}, \mathbf{V}) + \mathbf{F} \cdot \mathbf{V} + h\omega = 0$$

must hold for every $(\omega, \mathbf{V})$; by the definition of transpose mappings, this is precisely (23).

11 The principle of minimum dissipation

The purpose of the preceding section was the elimination of **V**, the sliding velocity of the wheel on the ground, and ω, the sliding angular velocity in the brake. A complementary problem now is the determination of these sliding velocities when the motion of the vehicle is treated as known, i.e., when **W** is given. The existence of at least one solution ω, **V** for every $\mathbf{W} \in \Pi$ results from the preceding section, in view of the compactness of C.

For a given **W**, the set of the pairs $(\omega, \mathbf{V})$ satisfying the kinematical condition of rigidity of the wheel (22) is an affine submanifold $\mathscr{E}$ of $\mathbb{R} \times \Pi$, namely

$$\mathscr{E} = \{(\omega, \mathbf{V}) \in \mathbb{R} \times \Pi : \mathbf{V} - \omega a \mathbf{j} = \mathbf{W}\}.$$

Introducing its indicator function $\psi_{\mathscr{E}}$, one observes that the subdifferential set $\partial\psi_{\mathscr{E}}(\omega, \mathbf{V})$, empty if $(\omega, \mathbf{V}) \notin \mathscr{E}$, consists otherwise in the subspace of $\mathbb{R} \times \Pi$ orthogonal to $\mathscr{E}$; this is precisely the set of the pairs $(h, \mathbf{F})$ satisfying the quasi-equilibrium condition (23) for some $\mathbf{G}$. Thus, (22) and (23) are equivalently condensed into

$$(h, \mathbf{F}) \in \partial\psi_{\mathscr{E}}(\omega, \mathbf{V}). \tag{27}$$

(Regarding such an interpretation of perfect constraints as 'resistance laws' with pseudopotentials, *see* Moreau [8, 10].)

On the other hand, using section 6, the composite friction law (16) is equivalently written as

$$(-h, -\mathbf{F}) \in \partial\varphi(\omega, \mathbf{V}), \tag{28}$$

where $\varphi : \mathbb{R} \times \Pi \to \mathbb{R}$ denotes the total dissipation function

$$\begin{aligned} \varphi(\omega, \mathbf{V}) = \psi^*_{I \times D}(-\omega, -\mathbf{V}) &= \psi^*_I(-\omega) + \psi^*_D(-\mathbf{V}) \\ &= b\,|\omega| + fN\,|\mathbf{V}|. \end{aligned} \tag{29}$$

The elimination of h and $\mathbf{F}$ from (27) and (28) yields

$$(0, 0) \in \partial\psi_{\mathscr{E}}(\omega, \mathbf{V}) + \partial\varphi(\omega, \mathbf{V}). \tag{30}$$

Since the function φ is continuous, the *addition rule for subdifferentials* (cf., Moreau [4] or the books [19–22]) may be applied so that (30) exactly expresses that $(\omega, \mathbf{V})$ is a point of $\mathbb{R} \times \Pi$, where the function $\psi_{\mathscr{E}} + \varphi$ attains its minimum. Recalling that the function $\psi_{\mathscr{E}}$ takes the value zero on $\mathscr{E}$ and $+\infty$ elsewhere, one concludes as follows.

If $\mathbf{W}$ *is given, the values of* $\omega, \mathbf{V}$ *solving the problem minimize the dissipation function* (29) *under the kinematical condition* (18).

Consequently, if one is only interested in the unknown ω, it turns out that the values of it which solve the problems are exactly the points of $\mathbb{R}$ where the function

$$\omega \mapsto fN\,|\mathbf{W} + \omega a \mathbf{j}| + b\,|\omega|$$

attains its minimum.

Remark The preceding expresses a minimization 'principle' for the dissipation function restricted to the set of the 'kinematically admissible' velocities. The same is a classical feature in plasticity theory. (Concerning the use of convex analysis in treating the variational properties of elastoplastic systems, *see* Moreau [25].) Another interesting example, involving a continuous system, is that of a heavy perfectly flexible inextensible rope lying with dry friction on a horizontal plane. Here, the velocity distribution entailed in quasistatic evolution by some imposed motion of the rope extremities, is characterized by minimizing the dissipation function on the set of the velocity distributions agreeing with these end conditions and with inextensibility.

12 Quotient spaces

We present now an abstract structure in which the reader will recognize a generalization of the foregoing.

Let us consider a mechanical system in a given configuration. Let $\mathscr{V}$ denote a linear space, the elements of which constitute, in some general sense, the possible values of the *velocity* of the system if it passes through the considered configuration. Let $\mathscr{F}$ denote a linear space the elements of which are, in a general sense, the possible values of the various forces the system may experience in this configuration. These two spaces are placed in duality by the bilinear-form 'power' noted $\langle .,. \rangle$.

The spaces $\mathscr{V}$ and $\mathscr{F}$ may have very diverse functional realizations, namely spaces of vector fields, of tensor fields, etc. and the considered mathematical procedure can usually be applied to a given mechanical situation in several different ways (*see*, for example Moreau [10], Nayroles [14].)

Let us suppose that the system is subject to a friction force F obeying the law

$$-V \in \partial\psi_C(F), \tag{31}$$

where C is a given convex subset of $\mathscr{F}$, closed in the topologies of the dual pair $(\mathscr{V}, \mathscr{F}, \langle .,. \rangle)$.

In addition, the system experiences a *moving constraint* or 'driving' which implies that its velocity V belongs to a certain affine submanifold $\mathscr{E}$ of $\mathscr{V}$, a translate of some given linear subspace $\mathscr{W}$ of $\mathscr{V}$, closed in the topologies of the dual pair $(\mathscr{V}, \mathscr{F}, \langle .,. \rangle)$. Specifying $\mathscr{E}$ among the various translates of $\mathscr{W}$ amounts to prescribing an element W of the quotient space $\mathscr{V}/\mathscr{W}$: we shall refer to this element as the *driving velocity* (mathematically, $\mathscr{E}$ and W are the same thing but using two notations seems clearer in mechanical applications).

The moving constraint is assumed to be perfect, i.e., the 'force' $R \in \mathscr{F}$ exerted on the system by the driving device is orthogonal to the affine manifold $\mathscr{E}$, in the sense of the duality $(\mathscr{V}, \mathscr{F}, \langle .,. \rangle)$

$$R \in \mathscr{W}^{\perp}. \tag{32}$$

The constraint is also assumed to be *firm*, i.e., the driving device is strong enough to provide any value of R satisfying (32). (Concerning the concept of the firmness of a mechanical constraint, *see* Moreau [26].) The opposite, $G=-R$, may be interpreted as the *resistance of the system to the driving.*

Inertia is neglected, so that the motion is characterized by the quasi-static equation

$$G=F. \tag{33}$$

The duality $(\mathscr{V},\mathscr{F},\langle .\,,.\rangle)$ classically induces a duality between the pair $\mathscr{V}/\mathscr{W}$ and $\mathscr{W}^\perp$. Our purpose is the elimination of V and F, yielding a relation between the elements W and G of the latter dual pair.

Proposition 1 *If the above conditions are satisfied, the elements W and G satisfy*

$$-W\in\partial\psi_D(G) \tag{34}$$

in the sense of the dual pair of spaces $\mathscr{V}/\mathscr{W}$ *and* $\mathscr{W}^\perp$, *with* $D=C\cap\mathscr{W}^\perp$.

Conversely, if $\mathscr{W}^\perp$ *meets the interior of C (in some topology of the dual pair* $\mathscr{V}$, $\mathscr{F}$), *every* (W,G) *satisfying* (34) *corresponds to at least one pair* (V,F) *agreeing with above conditions.*

Under the same topological assumption, the dissipation function γ *of the friction law* (34) *is defined on* $\mathscr{V}/\mathscr{W}$ *by*

$$\gamma(W)=\min_{U\in\mathscr{C}}\varphi(U), \tag{35}$$

where φ *denotes the dissipation function of the friction law* (31).

For the proof, we may call L the natural injection of $\mathscr{W}^\perp$ into $\mathscr{F}$ and L^* its transpose, namely the natural surjection of $\mathscr{V}$ onto $\mathscr{V}/\mathscr{W}$. This allows us to write the conditions of the problem in the form

$$F=L(G)\qquad W=L^*(V)$$

by which the elimination of V and F from (31) leads to

$$-W\in L^*(\partial\psi_C(L(G))). \tag{36}$$

Since $\psi_C\circ L=\psi_D$, the computation rule of section 10 yields the equivalence of (34) to (36), because the assumption that $\mathscr{W}^\perp$ meets the interior of C means the existence of a point of the range of L where ψ_C is finite and continuous. Without this assumption, however, (36) is easily proved to entail (34); this is the first statement of the proposition.

Finally, (35) results from the computation rule for $(\psi_C\circ L)^*$ (*see* Rockafellar [24], Theorem 3).

The above proposition, involving the quotient space $\mathscr{V}/\mathscr{W}$, may be described as a way of processing some *partial information* about the considered mechanical system. Actually, science is always dealing with partial information about nature; thus, it could be said that similar constructions of quotient spaces implicitly underlie every scientific act.

As in Section 11, let us turn now to the complementary question of characterizing the value of V corresponding to a given W, which, incidentally, will throw some light on the expression (35) of γ.

Proposition 2 *In the affine manifold $\mathscr{E}$, the set of the elements V satisfying the conditions of the problem, if not empty, is equal to the set of the points where the restriction $\varphi|_{\mathscr{E}}$ of the dissipation function φ attains a finite minimum.*

If moreover, there exists a point in $\mathscr{E}$ where φ is finite and continuous (in some topology of the dual pair $\mathscr{V}$, $\mathscr{F}$) the two above sets are equal, even if the first one is empty.

In fact, the conditions $V\in\mathscr{E}$ and $F\in\mathscr{W}^{\perp}$ are condensed into

$$F\in\partial\psi_{\mathscr{E}}(V),$$

while (31) is, by section 7, equivalent to

$$-F\in\partial\varphi(V).$$

Therefore, the values of V satisfying the conditions of the problem are characterized by

$$0\in\partial\psi_{\mathscr{E}}(V)+\partial\varphi(V). \tag{37}$$

In view of the trivial inclusion

$$\partial\psi_{\mathscr{E}}(V)+\partial\varphi(V)\subset\partial(\psi_{\mathscr{E}}+\varphi)(V), \tag{38}$$

this property implies that the function $\psi_{\mathscr{E}}+\varphi$ (equal to φ on $\mathscr{E}$ and taking the value $+\infty$ elsewhere) achieves a finite minimum at the point V. Suppose the existence of at least one pair V, F satisfying the conditions of the problem; then, in view of (8),

$$\varphi(V)=-\langle V,F\rangle.$$

Let V' denote another point where the restriction of φ to $\mathscr{E}$ attains its minimum; then, $\varphi(V)=\varphi(V')$. As $V-V'\in\mathscr{W}$ and $F\in\mathscr{W}^{\perp}$, one has

$$\langle V,F\rangle=\langle V',F\rangle;$$

thus,

$$\varphi(V')=-\langle V',F\rangle$$

which proves that V' satisfies, with the same F, the conditions of the problem. This proves the first part of the proposition.

Finally the existence of a point in $\mathscr{E}$ where φ is finite and continuous implies that the inclusion (38) is actually an equality of sets (*see*, for example, Moreau [4]; then, the fact that $\varphi|_{\mathscr{C}}$ attains a finite minimum at the point V is equivalent to (37).

Remark All the preceding could be adapted to the case where the system experiences, in addition, some given constant *load*, namely $G \in \mathscr{F}$. This amounts to replacing C by its translate $C+G$. If this set meets $\mathscr{W}^{\perp}$, there may exist, for some given driving velocity $W \in \mathscr{V}/\mathscr{W}$, a quasi-static evolution of the system.

References

1 Moreau, J. J., Quadratic programming in mechanics: Dynamics of one-sided constraints, *SIAM J. Control,* **4,** 153–158, 1966.

2 Moreau, J. J., Principes extrémaux pour le problème de la naissance de la cavitation, *J. Mécanique,* **5,** 439–470, 1966.

3 Hill, R., New horizons in the mechanics of solids, *J. Mech. Phys. Solids,* **5,** 66–74, 1956.

4 Moreau, J. J., Fonctionnelles sous-différentiables, *C. R. Acad. Sci. Paris, Sér. A,* **258,** 4117–4119, 1963.

5 Moreau, J. J., La notion de sur-potentiel et les liaisons unilatérales en élastostatique, *C. R. Acad. Sci. Paris, Sér. A,* **267,** 954–957, 1968.

6 Moreau, J. J., La convexité en statique, in *Analyse Convexe et ses Applications,* Lecture Notes in Economics and Mathematical Systems no. 102 (J. P. Aubin, ed.) Springer Verlag, Berlin, Heidelberg and New York, 1974, pp. 141–167.

7 Moreau, J. J., Convexité et frottement, Université de Montréal, Département d'Informatique, Publication no. 32 (multigraph 30 pp), 1970.

8 Moreau, J. J., Sur les lois de frottement, de plasticité et de viscosité, *C. R. Acad. Sci. Paris, Sér. A,* **271,** 608–611, 1970.

9 Moreau, J. J., Sur l'évolution d'un système élasto-visco-plastique, *C.R. Acad. Sci. Paris, Sér. A,* **273,** 118–121, 1971.

10 Moreau, J. J., On unilateral constraints, friction and plasticity, in *New Variational Techniques in Mathematical Physics* (G. Capriz and G. Stampacchia, eds), CIME, II ciclo 1973, Edizioni Cremonese, Roma, 1974, pp. 175–322.

11 Moreau, J. J., Application of convex analysis to the treatment of elastoplastic systems, in *Applications of Functional Analysis to Problems of Mechanics,* Lecture Notes in Mathematics (P. Germain and B. Nayroles, eds), Springer Verlag, Berlin, Heidelberg and New York, 1976, pp. 56–89.

12 Nayroles, B., Quelques applications variationnelles de la théorie des fonctions duales à la mécanique des solides, *J. Mécanique,* **10,** 263–289, 1971.

13 Nayroles, B., Opérations algébriques en mécanique des structures, *C.R. Acad. Sci. Paris, Sér. A,* **273,** 1075–1078, 1971.

14 Nayroles, B., Point de vue algébrique, convexité et intégrandes convexes en mécanique des solides, in *New Variational Techniques in Mathematical Physics* (G. Capriz and G. Stampacchia, eds), CIME, II ciclo 1973, Edizioni Cremonese, Roma, 1974, pp. 323–404.

15 Nayroles, B., Deux théorèmes de minimum pour certains systèmes dissipatifs, *C.R. Acad. Sci. Paris, Sér. A*, **282,** 1035–1038, 1976.

16 Debordes, O. and Nayroles, B., Sur la théorie et le calcul à l'adaptation des structures élastoplastiques, *J. Mécanique*, **15,** 1–53, 1976.

17 Nguyen, Q. S. and Halphen, B., Sur les lois de comportement élasto-viscoplastiques à potentiel généralisé, *C.R. Acad. Sci. Paris, Sér. A*, **277,** 319–322, 1973.

18 Nguyen, Q. S., Matériaux élasto-visco-plastiques et élastoplastiques à potentiel généralisé, *C.R. Acad. Sci. Paris, Sér. A*, **277,** 915–918, 1973.

19 Moreau, J. J., Fonctionnelles convexes, Séminaire sur les équations aux dérivées partielles, Collège de France, Paris, 1966–67 (multigraph 108pp.).

20 Rockafellar, R. T., *Convex Analysis*, Princeton University Press, Princeton, 1970.

21 Laurent, P. J., *Approximation et Optimisation*, Hermann, Paris, 1972.

22 Ekeland, I. and Teman, R., *Analyse Convexe et Problèmes Variationnels*, Dunod-Gauthier-Villars, Paris, 1974.

23 Robertson, A. P. and Robertson, W. J., *Topological Vector Spaces*, Cambridge University Press, London, 1966.

24 Rockafellar, R. T., Integrals which are convex functionals II, *Pacific J. Math.* **39,** 439–469, 1971.

25 Moreau, J. J., Systèmes élastoplastiques de liberté finie, Séminaire d'Analyse convexe, Montpellier, **3,** exposé no. 12 (30pp), 1973.

26 Moreau, J. J., *Mécanique Classique*, Masson, Paris, 1971, Vol. II.

Professor J. J. Moreau,
Institut de Mathématiques,
Université des Sciences et Techniques du Languedoc,
Place Eugéne Bataillon,
34060 Montpellier-Cedex,
France

I. Müller

Entropy in non-equilibrium – a challenge to mathematicians

Introduction

This paper consists of three parts. The first part illustrates the thermostatic theory of Carathéodory for homogeneous processes in a thermoelastic body. The second part gives a brief account of rational thermodynamics of a thermoelastic body. And the third part presents a synoptic view of Carathéodory's work and of rational thermodynamics. Similarities in objective and procedure between those theories are exhibited in this synopsis and it culminates in a challenge to mathematicians to discover the conditions under which entropy exists in non-equilibrium.

1 The thermostatics of Carathéodory

Carathéodory's ideas on thermostatics are illustrated here for homogeneous processes in thermoelastic bodies.

Adiabatic processes

In the case under consideration, the objective of thermostatics is the determination of the empirical temperature $\vartheta(t)$ for a given deformation gradient $F_{iA}(t)$. To reach that objective, thermostatics relies on the equation of balance of internal energy, viz.,

$$\left(\int_V \rho\varepsilon \, \mathrm{d}V\right)^{\cdot} + \oint_{S(V)} q_i \, \mathrm{d}a_i - \int_V t_{ij} F^{-1}_{Ai} \dot{F}_{jA} \, \mathrm{d}V = 0,$$

where $\rho\varepsilon$ and q_i are density and flux of internal energy, respectively, and $t_{ij} F^{-1}_{Ai} \dot{F}_{jA}$ is the density of production of internal energy due to the power of the stress t_{ij}. Carathéodory considers only adiabatically isolated bodies for which $\oint_{S(V)} q_i \, \mathrm{d}a_i$ is zero; and, since we are only interested here in homogeneous processes, the balance of internal energy assumes the form

$$(\rho\varepsilon V)^{\cdot} - (t_{ij}F_{Ai}^{-1}V)\dot{F}_{jA} = 0 \quad \text{or} \quad \dot{E} - \gamma_\alpha \dot{F}_\alpha = 0, \tag{1.1}$$

where for brevity we define $E \equiv \rho\varepsilon V$, $\gamma_\alpha \equiv t_{ij}F_{Ai}^{-1}V$ and $F_\alpha \equiv F_{iA}$.

However, the balance of internal energy is not, in its present form, an equation for $\vartheta(t)$, because ϑ does not even occur in the balance and, instead, new quantities have appeared, namely E and γ_α. In this situation, we rely on experience, which tells us that E and γ_α are related to the history of the functions $\vartheta(t)$ and $F_\alpha(t)$ in a materially dependent manner by the constitutive equations

$$E(t) = \mathop{\mathcal{N}}_{s=0}^{\infty}[\vartheta(t-s), F_{iA}(t-s)], \qquad \gamma_\alpha = \mathop{\mathcal{Y}_\alpha}_{s=0}^{\infty}[\vartheta(t-s), F_{iA}(t-s)]. \tag{1.2}$$

$\mathcal{N}$ and $\mathcal{Y}$ are constitutive functionals. We assume that the body has fading memory in the sense that $E(t)$ and $\gamma_\alpha(t)$ are influenced more by values of ϑ and F_α in the recent past that in the distant past. Rather obviously, this implies that for very slowly varying functions $\vartheta(t)$ and $F_\alpha(t)$, the values of E and γ_α at time t depend only on the values of ϑ and F_α at that same time, i.e., for slowly varying functions $\vartheta(t)$, $F_\alpha(t)$ the constitutive functionals reduce to constitutive functions and we write

$$E(t) = E(\vartheta(t), F_\alpha(t)), \qquad \vartheta_\alpha(t) = \gamma_\alpha(\vartheta(t), F_\alpha(t)). \tag{1.3}$$

When we insert the constitutive relations (1.2) into the balance of internal energy $(1.1)_2$ we obtain a functional equation for $\vartheta(t)$, $F_\alpha(t)$ and each solution of that equation is called an *adiabatic process*. In particular, for slowly varying functions $\vartheta(t)$, $F_\alpha(t)$, we have the constitutive relations (1.3) and, if these are inserted into $(1.1)_2$, we obtain a differential equation for $\vartheta(t)$, $F_\alpha(t)$ and each solution of that equation is called a *quasistatic adiabatic process.*

Remark If we knew the constitutive functionals, or functions, it would be a purely mathematical problem to determine adiabatic processes. However, there is no material for which the constitutive relations are known and, therefore, thermodynamicists are looking for restrictions on the generality of the constitutive functionals. Such restrictions are based on physical principles in which we have confidence, and one of these principles is expressed in Carathéodory's axiom of inaccessibility, which is a statement about the solutions of the functional equation $\dot{E} - \gamma_\alpha \dot{F}_\alpha = 0$ for $\vartheta(t)$, $F_\alpha(t)$.

Carathéodory's axiom of inaccessibility and Carathéodory's lemma

The axiom reads as follows. In each neighbourhood of an arbitrary initial state there exist states which cannot be reached by adiabatic processes.

For motivation of this axiom, let us look at the special case of an adiabatic process that starts with ϑ^i, F^i_α and ends with ϑ^f, F^f_α such that $F^f_\alpha = F^i_\alpha$, i.e., the process carries the body back to its original form. What can we say about the final temperature ϑ^f? Of course, it has to satisfy the balance of internal energy

$$E(\vartheta^f, F^i_\alpha) - E(\vartheta^i, F^i_\alpha) = \int_{t^i}^{t^f} \gamma_\alpha \dot{F}_\alpha \, dt,$$

from which it is obvious that there will not only be one ϑ^f; rather ϑ^f will depend on how we conduct the process. But we may feel that, whatever value ϑ^f has, it cannot be smaller than ϑ^i, because of ubiquitous dissipation. Thus, in this case, all states $(\vartheta^f, F^i_\alpha)$ are inaccessible from $(\vartheta^i, F^i_\alpha)$ for which $\vartheta^f < \vartheta^i$. One will admit that this is plausible, and equally plausible predictions can be made about the outcome of other adiabatic processes; they all serve to motivate Carathéodory's axiom.

The axiom becomes relevant to thermostatics by Carathéodory's lemma which reads: let $\dot{x}_0 - \bar{X}_\alpha \dot{x}_\alpha = 0$ be a Pfaffian equation in which the $\bar{X}_\alpha$'s are finite, continuous and differentiable functions of x_0 and x_α. If one knows that in each neighbourhood of an arbitrary point in the space of the x_0, x_α there are points that cannot be reached along curves which satisfy the Pfaffian equation, the Pfaffian form $\dot{x}_0 - \bar{X}_\alpha \dot{x}_\alpha$ must have an integrating factor.†

It is obvious how this lemma applies to thermostatics, because the balance of internal energy (1.1) together with the constitutive relations (1.3) for quasistatic processes forms the Pfaffian equation $\dot{E}(\vartheta, F_\gamma) - \gamma_\alpha(\vartheta, F_\gamma)\dot{F}_\alpha = 0$ and Carathéodory's axiom ensures that 'in each neighbourhood of an arbitrary point in the space of the ϑ, F_α there are points that cannot be reached along curves which satisfy the Pfaffian equation'. Therefore, the Pfaffian form $\dot{E} - \gamma_\alpha \dot{F}_\alpha$ has integrating factors and we write

$$\dot{H}_\Lambda = \Lambda(\dot{E}(\vartheta, F_\gamma) - \gamma_\alpha(\vartheta, F_\gamma)\dot{F}_\alpha), \tag{1.4}$$

where H_Λ is the potential corresponding to the integrating factor Λ and that factor itself is a function of E, F_α or of ϑ, F_α.

Remark The existence of integrating factors as such is not useful for the purpose of finding restrictions on the constitutive relations. For one thing, if there is one integrating factor, there are many. But Carathéodory's argument continues and he shows that *one* of the integrating factors is a universal function of temperature alone. Thus he comes to the concepts of entropy and absolute temperature.

† For proof, *see* Carathéodory's paper.

Entropy and absolute temperature

Carathéodory considers two different bodies I and II, which we take both to be thermoelastic bodies, so that – according to the preceding argument – we may write

$$\dot{H}_{\Lambda^{\mathrm{I}}} = \Lambda^{\mathrm{I}}(\dot{E}^{\mathrm{I}} - \gamma^{\mathrm{I}}_{\alpha}\dot{F}^{\mathrm{I}}_{\alpha}) \quad \text{and} \quad \dot{H}_{\Lambda^{\mathrm{II}}} = \Lambda^{\mathrm{II}}(\dot{E}^{\mathrm{II}} - \gamma^{\mathrm{II}}_{\beta}\dot{F}^{\mathrm{II}}_{\beta}. \tag{1.5}$$

A compound body is formed by bringing the bodies I and II into contact along a wall which allows the exchange of internal energy. We assume that, as a consequence, in equilibrium, the two bodies will have the same temperature and the energy equation for the compound body in a quasistatic adiabatic process has the form

$$\dot{E}(\vartheta, F^{\mathrm{I}}_{\gamma}, F^{\mathrm{II}}_{\delta}) - \gamma^{\mathrm{I}}_{\alpha}(\vartheta, F^{\mathrm{I}}_{\gamma})\dot{F}^{\mathrm{I}}_{\alpha} - \gamma^{\mathrm{II}}_{\beta}(\vartheta, F^{\mathrm{II}}_{\delta})\dot{F}^{\mathrm{II}}_{\delta} = 0. \tag{1.6}$$

Carathéodory's axiom and lemma apply to the compound body just as they apply to single bodies and, therefore, we have integrating factors and potentials for the Pfaffian form $\dot{E} - \gamma^{\mathrm{I}}_{\alpha}\dot{F}^{\mathrm{I}}_{\alpha} - \gamma^{\mathrm{II}}_{\beta}\dot{F}^{\mathrm{II}}_{\beta}$ and we may write

$$\dot{H}_{\Lambda} = \Lambda(\dot{E}(\vartheta, F^{\mathrm{I}}_{\gamma}F^{\mathrm{II}}_{\delta}) - \gamma^{\mathrm{I}}_{\alpha}(\vartheta, F^{\mathrm{I}}_{\gamma})\dot{F}^{\mathrm{I}}_{\alpha} - \gamma^{\mathrm{II}}_{\beta}(\vartheta, F^{\mathrm{II}}_{\delta})\dot{F}^{\mathrm{II}}_{\beta}), \tag{1.7}$$

where Λ may be a function of $E, F^{\mathrm{I}}_{\alpha}, F^{\mathrm{II}}_{\beta}$.

In the appendix, it is shown how we can make use of the fact that $E = E^{\mathrm{I}} + E^{\mathrm{II}}$ to show that the integrating factors of the partial bodies are functions of the following type:

$$\Lambda^{\mathrm{I}} = \frac{\alpha^{\mathrm{I}}(H_{\Lambda^{\mathrm{I}}})}{T(\vartheta)} \quad \text{and} \quad \Lambda^{\mathrm{II}} = \frac{\alpha^{\mathrm{II}}(H_{\Lambda^{\mathrm{II}}})}{T(\vartheta)}. \tag{1.8}$$

We conclude that the temperature dependence of the integrating factors of these entirely different bodies is governed by the same function $T(\vartheta)$ and we express this by saying that $T(\vartheta)$ is a universal function of ϑ.

Insertion of (1.8) into (1.5) leads to an equation of the form

$$\dot{H}_{\Lambda} = \frac{\alpha(H_{\Lambda})}{T(\vartheta)}(\dot{E} - \gamma_{\alpha}\dot{F}_{\alpha}) \quad \text{or} \quad \frac{1}{\alpha(H_{\Lambda})}\dot{H}_{\Lambda} = \frac{1}{T(\vartheta)}(\dot{E} - \gamma_{\alpha}\dot{F}_{\alpha})$$

for both partial bodies, so that with

$$H \equiv \int_{H^0_{\Lambda}}^{H_{\Lambda}} \frac{\mathrm{d}x}{\alpha(x)}$$

(where H^0_{Λ} is some constant) we may write

$$\dot{H} = \frac{1}{T(\vartheta)}(\dot{E} - \gamma_{\alpha}\dot{F}_{\alpha}). \tag{1.9}$$

Thus, we see that among all integrating factors there is *one* which is a function of temperature alone and that function is universal. The corresponding potential, which has been denoted by H, is called the *entropy*.

Equation (1.9) may now be used to derive restrictions on the constitutive functions in (1.3), because it implies the integrability conditions

$$\frac{\mathrm{d}\ln T}{\mathrm{d}\vartheta}=\frac{\dfrac{\partial\gamma^{\alpha}}{\partial\vartheta}}{\gamma_{\alpha}-\dfrac{\partial E}{\partial F_{\alpha}}} \tag{1.10}$$

which are fairly restrictive indeed: not only must the combination $(\partial\gamma^{\alpha}/\partial\vartheta)(\gamma_{\alpha}-(\partial E/\partial F_{\alpha}))^{-1}$ be only a function of ϑ, but it must be the *same* function in all bodies, since $T(\vartheta)$ is a universal function.

The function $T(\vartheta)$ can be determined – to within a multiplicative constant – by integration of (1.10), if the right-hand side of that equation has been determined experimentally as a function of ϑ for just *one* body. When this is done (and if the multiplicative constant of integration is chosen to be positive) it turns out that $T(\vartheta)$ is a positive-valued, monotonically-increasing function of temperature. Therefore $T(\vartheta)$ may be used as a measure for temperature and it is called the *absolute temperature.* With a proper choice of the multiplicative constant of integration, the absolute temperature measures temperatures in kelvins.

The equations (1.10) imply the restrictions which we were expecting to get from Carathéodory's axiom and they are restrictions on the constitutive functions which determine the material properties of a body in quasistatic processes. This is the reason why we say that Carathéodory's ideas apply to *thermostatics* rather than thermodynamics.

Remark Apart from giving the desired restrictions on constitutive equations, the above arguments have led to the concepts of entropy and absolute temperature. In particular, entropy has interesting properties and we proceed to discuss two of those.

Additivity and growth of entropy

We rewrite Eqn (1.7) for the compound body with $\Lambda=1/T(\vartheta)$ as the integrating factor so that the corresponding potential H_{Λ} is the entropy H:

$$\dot{H}=\frac{1}{T}(\dot{E}-\gamma_{\alpha}^{\mathrm{I}}\dot{F}_{\alpha}^{\mathrm{I}}-\gamma_{\beta}^{\mathrm{II}}\dot{F}_{\beta}^{\mathrm{II}})$$

or, with $E=E_{\mathrm{I}}+E_{\mathrm{II}}$,

$$\dot{H}=\frac{1}{T}((\dot{E}_{\mathrm{I}}-\gamma_{\alpha}^{\mathrm{I}}\dot{F}_{\alpha}^{\mathrm{I}})-(\dot{E}_{\mathrm{II}}-\gamma_{\beta}^{\mathrm{II}}\dot{F}_{\beta}^{\mathrm{II}})).$$

Hence, with (1.9), which holds for the partial bodies I and II, it follows that

$$\dot{H} = (H_{\mathrm{I}} + H_{\mathrm{II}}) \,. \tag{1.11}$$

We conclude that the entropy of the compound body is the sum of the entropies of the partial bodies. One can easily see that, among all potentials, only entropy has this additivity property. Carathéodory makes this comment: the additivity of entropy has induced many physicists to consider entropy as a physical quantity which, similar to the mass, belongs to each body and depends on the state of this body. In other words, entropy is an additive constitutive quantity.

It is obvious from (1.9) that $\dot{H} = 0$ in a quasistatic adiabatic process, because in such a process $\dot{E}(\vartheta, F_\gamma) - \gamma_\alpha(\vartheta, F_\gamma)\dot{F}_\alpha = 0$. In an arbitrary adiabatic process, we have

$$\dot{H} > 0, \tag{1.12}$$

i.e., entropy grows.

For proof of (1.12), we consider all possible adiabatic processes which start at ϑ^i, F^i_α and end at ϑ^f, F^f_α. According to the balance of internal energy, $E(\vartheta^f, F^f_\alpha)$ is given by the equation

$$E(\vartheta^f, F^f_\alpha) = E(\vartheta^i, F^i_\alpha) + \int_{t^i}^{t^f} \gamma_\alpha \dot{F}_\alpha \, \mathrm{d}t$$

and it will differ for different processes since the power term $\int_{t^i}^{t^f} \gamma_\alpha \dot{F}_\alpha \, \mathrm{d}t$ is different. We assume that all possible values of the power term lie in a connected interval. Therefore, the possible values $E(\vartheta^f, F^f_\alpha)$, ϑ^f and $H(\vartheta^f, F^f_\alpha)$ also each lie in a connected interval. Now we ask whether the interval of the $E(\vartheta^f, F^f_\alpha)$ contains $E(\vartheta^i, F^i_\alpha)$, and the answer is that it may or it may not, depending on the values of F^f_α. The same is true for the interval of the ϑ^f: it may or may not contain ϑ^i. But the interval of the $H(\vartheta^f, F^f_\alpha)$ will certainly contain $H(\vartheta^i, F^i_\alpha)$, because all possible adiabatic processes include the quasistatic adiabatic process in which H does not change. Now the question arises as to where $H(\vartheta^i, F^i_\alpha)$ lies within the interval of the $H(\vartheta^f, F^f_\alpha)$, and one can easily see that $H(\vartheta^i, F^i_\alpha)$ cannot lie at an interior point. In fact, were $H(\vartheta^i, F^i_\alpha)$ at an interior point, we could – by a suitable adiabatic process – first reach any value H in the neighbourhood of $H(\vartheta^i, F^i_\alpha)$ and then – by a quasistatic adiabatic process – go over to any desired set of values F^i_α.Thus we should be able to reach any point in the neighbourhood of ϑ^i, F^i_α by adiabatic processes which contradicts Carathéodory's axiom. We conclude that $H(\vartheta^i, F^i_\alpha)$ must lie at one on the endpoints of the interval of the $H(\vartheta^f, F^f_\alpha)$ so that H can only either grow or decrease in a non-quasistatic adiabatic process. *One* experiment with any body ensures that H in fact grows. This proves (1.12).

2 Rational thermodynamics of a thermoelastic body

Thermodynamic processes

The objective of thermodynamics is the determination of the following five fields

$$\left.\begin{aligned}&\rho(\bar{X}_A, t), &&\text{density;}\\&x_i(\bar{X}_A, t), &&\text{motion;}\\&\vartheta(\bar{X}_A, t), &&\text{empirical temperature;}\end{aligned}\right\}\tag{2.1}$$

for all particles $\bar{X}_A$ of a body and for all times.

To determine these fields, we need field equations, and these are derived from the equations of balance of mechanics and thermodynamics, namely the equations of balance of

$$\begin{aligned}&\text{mass,} &&\frac{\partial\rho}{\partial t}+\frac{\partial}{\partial x_j}(\rho v_j)=0;\\&\text{momentum,} &&\frac{\partial\rho v_i}{\partial t}+\frac{\partial}{\partial x_j}(\rho v_i v_j - t_{ij})=0;\\&\text{energy,} &&\frac{\partial\rho(\varepsilon+\frac{1}{2}v^2)}{\partial t}+\frac{\partial}{\partial x_j}(\rho(\varepsilon+\tfrac{1}{2}v^2)v_j - t_{ij}v_i + q_j)=0;\end{aligned}\tag{2.2}$$

where v_i is the velocity, t_{ij} is the stress and ε and q_i are the specific internal energy and flux of internal energy respectively.

While these are five equations, they are not suitable to serve as field equations for the five fields (2.1). In fact, the temperature ϑ does not even occur in (2.2) and, instead, the new quantities ε, q_i and t_{ij} have appeared. Experience shows that these quantities are related to the fields (2.1) in a materially dependent manner by constitutive equations. In particular, we speak of a simple thermoelastic body, if the constitutive relations have the forms

$$\left.\begin{aligned}t_{ij}&=t_{ij}\left(F_{iA},\vartheta,\frac{\partial\vartheta}{\partial x_j}\right),\\q_i&=q_i\left(F_{iA},\vartheta,\frac{\partial\vartheta}{\partial x_j}\right),\\\varepsilon&=\varepsilon\left(F_{iA},\vartheta,\frac{\partial\vartheta}{\partial x_j}\right).\end{aligned}\right\}\tag{2.3}$$

When these constitutive equations are inserted into the equations of balance, we obtain five field equations for $\rho(\bar{X}_A, t)$, $x_i(\bar{X}_A, t)$ and $\vartheta(\bar{X}_A, t)$ and every solution of these field equations is called a *thermodynamic process*.

Remark If the constitutive functions were known, we should now face the mathematical problem of solving this system of field equations for certain boundary- and initial-value problems. But since there is no material for which the constitutive functions are known, thermodynamicists search for restrictions on the generality of these functions and, in order to obtain such restrictions, they make use of physical principles and, above all, of the entropy principles.

Entropy principle

The entropy postulate of thermodynamics is motivated by the ideas of Clausius and Carathéodory on thermostatics and by considerations of statistical mechanics as well as by the formulae of the kinetic theory of gases. It may be expressed in four parts.

1. Entropy is an additive quantity, so that we may write an equation of balance of entropy:

$$\frac{\partial \rho\eta}{\partial t}+\frac{\partial}{\partial x_j}(\rho\eta v_j+\Phi_j)=\sigma.$$

2. The specific entropy and the entropy flux are constitutive quantities. In particular, in a simple thermoelastic body, we have

$$\begin{aligned}\eta &= \eta\left(F_{iA},\vartheta,\frac{\partial\vartheta}{\partial x_i}\right),\\ \Phi_i &= \Phi_i\left(F_{iA},\vartheta,\frac{\partial\vartheta}{\partial x_i}\right).\end{aligned} \tag{2.4}$$

3. Entropy production is non-negative for all thermodynamic processes, so that we have the entropy inequality $\sigma \geqslant 0$ or

$$\frac{\partial \rho\eta}{\partial t}+\frac{\partial}{\partial x_j}(\rho\eta v_j+\Phi_j)\geqslant 0 \qquad \forall \text{ thermodynamic processes.} \tag{2.5}$$

4. The normal component of the entropy flux is continuous at a wall where the temperature is continuous, i.e.,

$$[\Phi_i e_i]=0 \quad \text{if} \quad [\vartheta]=0, \tag{2.6}$$

where e_i is the unit normal to the wall and the brackets indicate the difference of the bracketed quantity at the right and the left of the wall.

The key to the evaluation of the entropy principle is Statement 3, according to which the entropy inequality need not hold for arbitrary fields of density, motion and temperature but only for thermodynamic processes, i.e., solutions of the field equations. We may express this by saying that the fields that satisfy the inequality are 'constrained' by the requirement that they be solutions of the field equations. Liu has shown that one can get rid of these constraints by the use of Lagrange multipliers. He showed that the requirement of the entropy postulate (Statement 3) is equivalent to the requirement that the inequality

$$\frac{\partial \rho\eta}{\partial t}+\frac{\partial}{\partial x_j}(\rho\eta v_j+\Phi_j)-\Lambda^{\rho}\left(\frac{\partial \rho}{\partial t}+\frac{\partial}{\partial x_i}(\rho v_j)\right)-\Lambda^{v_i}\left(\frac{\partial \rho v_i}{\partial t}+\frac{\partial}{\partial x_j}(\rho v_i v_j-t_{ij})\right)$$
$$-\Lambda^{\varepsilon}\left(\frac{\partial \rho(\varepsilon+\frac{1}{2}v^2)}{\partial t}+\frac{\partial}{\partial x_j}(\rho(\varepsilon+\tfrac{1}{2}v^2)v_j-t_{ij}v_i+q_j)\right)\geqslant 0 \quad (2.7)$$

hold for arbitrary analytic fields $\rho(\bar{X}_A, t)$, $x_i(\bar{X}_A, t)$, $\vartheta(\bar{X}_A, t)$. The Λ's in (2.7) are called Lagrange multipliers and they may be functions of the variables F_{iA}, ϑ, $(\partial\vartheta/\partial x_i)$.

When the constitutive relations (2.3) and (2.4) are inserted into the inequality (2.7), and when all indicated differentiations are carried out, it turns out that the left-hand side of that inequality is linear in the derivatives

$$\dot{F},\ \dot{\vartheta},\ \frac{\partial F_{iA}}{\partial x_j},\ \frac{\partial^2\vartheta}{\partial x_i\partial x_j},\ \frac{\partial^2\vartheta}{\partial t\partial x_j}. \quad (2.8)$$

Now, since the inequality must hold for arbitrary fields ρ, x_i and ϑ, it must, in particular, hold for arbitrary values of the derivatives (2.8). Thus, we should be able to violate the inequality unless the coefficients of these derivatives are zero. This argument leads to the following intermediate results:

$$\left.\begin{aligned} &\mathrm{d}\eta=\Lambda^{\varepsilon}(\vartheta)\left(\mathrm{d}\varepsilon-\frac{1}{\rho}t_{ij}F^{-1}_{Aj}\,\mathrm{d}F_{iA}\right)\\ &\Phi_i=\Lambda^{\varepsilon}(\vartheta)q_i,\\ &\frac{\mathrm{d}\Lambda^{\varepsilon}}{\mathrm{d}\vartheta}q_i\frac{\partial\vartheta}{\partial x_i}\geqslant 0\end{aligned}\right\} \quad (2.9)$$

and ε and η are only functions of F_{iA} and ϑ, independent of $\partial\vartheta/\partial x_i$.

Remark These results as such are not useful, since they still contain the Lagrange multiplier Λ^{ε} which has entered the theory as a mere auxiliary quantity. It is true that the inequality has restricted Λ^{ε} to be a function of ϑ alone, but we need to know the form of that function for explicit results. We proceed to show that Λ^{ε} is closely related to the absolute temperature.

Absolute temperature

For the determination of $\Lambda^\varepsilon(\vartheta)$, we rely upon the continuity condition for the entropy flux which is Statement 4 of the entropy postulate. We consider a wall between two simple thermoelastic solids I and II which cannot maintain a temperature difference between its sides, so that $[\vartheta]=0$. By (2.6), we thus have $[\Phi_i e_i]=0$ and by (2.9)

$$[\Lambda^\varepsilon(\vartheta)q_i e_i]=0.$$

Now, the normal component of the flux of internal energy is continuous at any wall under rather general conditions. Therefore $q_i e_i$ may be taken out of the bracket and, in fact, out of the whole equation. We are left with $[\Lambda^\varepsilon(\vartheta)]$ or, explicitly,

$$\Lambda^\varepsilon_{\mathrm{I}}(\vartheta)=\Lambda^\varepsilon_{\mathrm{II}}(\vartheta) \tag{2.10}$$

and we conclude that $\Lambda^\varepsilon(\vartheta)$ is a *universal* function of temperature.

Equation $(2.9)_1$ implies an integrability condition for η which reads

$$\frac{\mathrm{d}\ln\Lambda^\varepsilon}{\mathrm{d}\vartheta}=-\frac{\dfrac{1}{\rho}\dfrac{\partial t_{ij}}{\partial\vartheta}F^{-1}_{Aj}}{\dfrac{1}{\rho}t_{ij}F^{-1}_{Aj}-\dfrac{\partial\varepsilon}{\partial F_{iA}}}. \tag{2.11}$$

Here we see a rather strong restriction on the constitutive equations for ε and t_{ij}. These equations must be such that the term on the right-hand side of (2.11) is a function of ϑ only and, moreover, a universal function. Just as in the case of Carathéodory's argument, we may now calculate $\Lambda^\varepsilon(\vartheta)$ by integration of (2.11), if the right-hand side of that equation has been measured for only *one* material as a function of ϑ. In this manner, it turns out that Λ^ε is a positive-valued, monotonically-decreasing function of ϑ, if the constant of integration is given a positive value. We write

$$T(\vartheta)=\frac{1}{\Lambda^\varepsilon(\vartheta)} \tag{2.12}$$

and call T the absolute temperature. This is the way in which rational thermodynamics introduces the absolute temperature.

This definition of the absolute temperature is consistent with the previous definition in Carathéodory's theory. Indeed, elimination of Λ^ε between $(2.9)_1$ and (2.12) leads to

$$\mathrm{d}\eta=\frac{1}{T(\vartheta)}\left(\mathrm{d}\varepsilon-\frac{1}{\rho}t_{ij}F^{-1}_{iA}\,\mathrm{d}F_{jA}\right), \tag{2.13}$$

and this is in complete agreement with the previous equation (1.9). In the comparison, we need only to realize that (1.9) refers to homogeneous processes.

Apart from (2.13), we derive from (2.9) and (2.12) the results

$$\Phi_i = \frac{1}{T} q_i \quad \text{and} \quad -\frac{1}{T^2}\frac{\mathrm{d}T}{\mathrm{d}\vartheta} q_i \frac{\partial \vartheta}{\partial x_i} \geqslant 0. \tag{2.14}$$

By use of $(2.14)_1$, we may write the entropy inequality for a body of volume V in the form

$$\left(\int_V \rho\eta \, \mathrm{d}V\right)^{\cdot} \geqslant \oint_{\partial(V)} \frac{q_i \, \mathrm{d}a_i}{T}. \tag{2.15}$$

This is the extension of Carathéodory's result (1.12) and it reduces to that result for adiabatic processes, where q_i vanishes on the surface of the body. Equation $(2.14)_2$ shows that q_i points in the opposite direction to $\partial\vartheta/\partial x_i$.

3 Synopsis

The careful reader will have noticed that the arguments of Carathéodory and the arguments of rational thermodynamics run parallel to a certain extent and the purpose of this chapter is to exhibit this parallism in a synoptic manner. This will help us to find out which arguments of Carathéodory's have a counterpart in rational thermodynamics and which do not.

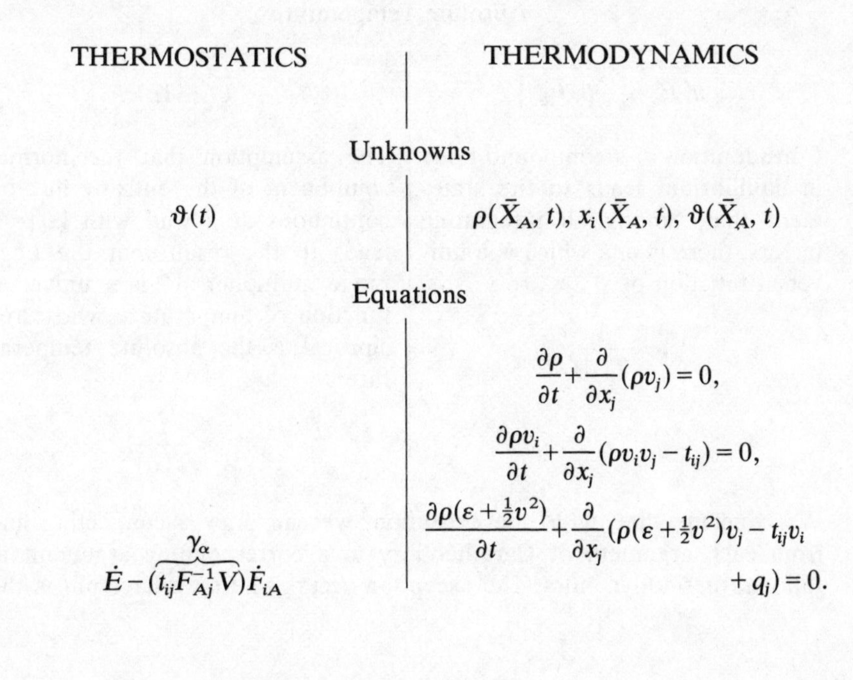

THERMOSTATICS	THERMODYNAMICS
Unknowns	
$\vartheta(t)$	$\rho(\bar{X}_A, t)$, $x_i(\bar{X}_A, t)$, $\vartheta(\bar{X}_A, t)$
Equations	
	$\frac{\partial\rho}{\partial t} + \frac{\partial}{\partial x_j}(\rho v_j) = 0,$
	$\frac{\partial\rho v_i}{\partial t} + \frac{\partial}{\partial x_j}(\rho v_i v_j - t_{ij}) = 0,$
$\dot{E} - \overbrace{(t_{ij}F^{-1}_{Aj}V)}^{\gamma_\alpha}\dot{F}_{iA}$	$\frac{\partial\rho(\varepsilon + \frac{1}{2}v^2)}{\partial t} + \frac{\partial}{\partial x_j}(\rho(\varepsilon + \frac{1}{2}v^2)v_j - t_{ij}v_i + q_j) = 0.$

THERMOSTATICS	THERMODYNAMICS

Constitutive Quantities

E, γ_α	ε, t_{ij}, q_i

Restrictive Principle

Carathéodory's axiom of inaccessibility	???
↓	↓
$\dot{H}_\Lambda - \Lambda(\dot{E} - \gamma_\alpha \dot{F}_\alpha) = 0$	$\frac{\partial \rho\eta}{\partial t} + \frac{\partial}{\partial x_j}(\rho\eta v_j + \Phi_j) - \Lambda^\rho\left(\frac{\partial \rho}{\partial t} - \frac{\partial}{\partial x_j}(\rho v_j)\right) - \Lambda^{v_i}\left(\frac{\partial \rho v_i}{\partial t} + \frac{\partial}{\partial x_j}(\rho v_i v_j - t_{ij})\right) - \Lambda^\varepsilon\left(\frac{\partial \rho(\varepsilon + \frac{1}{2}v^2)}{\partial t} + \frac{\partial}{\partial x_j}(\rho(\varepsilon + \frac{1}{2}v^2)v_j - t_{ij}v_i + q_j)\right) \geqslant 0$

Absolute Temperature

$\vartheta, F^{\mathrm{I}}_\alpha$ | $\vartheta, F^{\mathrm{II}}_\beta$	I ⇑ II
Consideration of a compound body in equilibrium leads to the statement that, among all integrating factors, there is *one* which is a universal function of ϑ: $\Lambda = \frac{1}{T(\vartheta)}$.	The assumption that the normal component of the entropy flux be continuous at a wall with $[\vartheta] = 0$ leads to the result that the Lagrange multiplier Λ^ε is a universal function of temperature whose reciprocal is the absolute temperature: $\Lambda^\varepsilon = \frac{1}{T(\vartheta)}$.

We conclude that, with one exception, we can draw a connecting line from each argument of Carathéodory to a corresponding argument in rational thermodynamics. The exception occurs where we have put in the question marks. In rational thermodynamics there is no counterpart to Carathéodory's axiom. I suggest that the mathematicians consider it as a challenge to fill this gap.

The nature of this challenge becomes clearer, when we introduce the following abbreviations

$$v_k = \{\rho, x_i, \vartheta\}, \qquad k = 1, \ldots, 5$$
$$x_A = \{x_a, t\}, \qquad A = 1, \ldots, 4$$
$$N_{kA} = \begin{pmatrix} \rho & \rho v_i \\ \rho v_i & \rho v_i v_j - t_{ij} \\ \rho(\varepsilon + \frac{1}{2}v^2) & \rho(\varepsilon + \frac{1}{2}v^2)v_j - t_{ij}v_i + q_j \end{pmatrix},$$
$$S_A = \{\rho\eta v_j + \Phi_j, \rho\gamma\},$$
$$\Lambda^k = \{\Lambda^\rho, \Lambda^{v_i}, \Lambda^\varepsilon\}.$$

In this notation, the balance equations assume the form $\partial N_{kA}/\partial x^A = 0$ and the entropy inequality reads $\partial S_A/\partial x^A \geqslant 0$.

Here then is the challenge to mathematicians, set again in juxtaposition to Carathéodory's theory.

Carathéodory's axiom is a statement on the solution of the equation

$$\dot{E} = \gamma_\alpha \dot{F}_\alpha = 0,$$

which ensures the existence of the integrating factor Λ and of the potential H_Λ such that

$$\dot{H}_\Lambda - \Lambda(\dot{E} - \gamma_\alpha \dot{F}_\alpha) = 0$$

holds.

Find the statement on the solution of the set of the five field equations

$$\frac{\partial N_{kA}}{\partial x^A} = 0$$

which ensures the existence of the multipliers Λ^k and of the four-vector S_A such that the inequality

$$\frac{\partial S_A}{\partial x^A} - \Lambda^k \frac{\partial N_{kA}}{\partial x_A} \geqslant 0$$

holds!

Once this challenge is answered, we shall know whether and under which conditions it is legitimate to carry the concept of entropy from thermostatics over to non-equilibrium thermodynamics.

Bibliography

Carathéodory, C., Untersuchungen über die Grundlagen der Thermodynamik, *Math. Annalen*, **67**, 355, 1909.

Müller, I., *Thermodynamik, Grundlagen der Materialtheorie*, Bertelsmann Universitätsverlag, Düsseldorf, 1973.

Liu, I.-S., Method of Lagrange multipliers for exploitation of the entropy inequality, *Arch. Rat. Mech. Anal.*, **46,** 131, 1972.

I. Müller
Fachbereich 6 – Theoretische Physik,
Gesamthochschule Paderborn,
Postfach 1621,
479 Paderborn,
West Germany.

Appendix

Proof of (1.8)†

From (1.5), we conclude that

$$\left.\begin{matrix}\Lambda^{\mathrm{I}}\\ \Lambda^{\mathrm{II}}\end{matrix}\right\} \text{ is a function of } \begin{cases}\vartheta, F^{\mathrm{I}}_{\alpha}\ (\alpha=1,2,\dots n)\\ \vartheta, F^{\mathrm{II}}_{\beta}\ (\beta=1,2,\dots m)\end{cases} \text{ or}$$

$$\text{of } \left.\begin{matrix}\vartheta, H_{\Lambda^{\mathrm{I}}}=F^{\mathrm{I}}_{\alpha}\ (\alpha=2,\dots n)\\ \vartheta, H_{\Lambda^{\mathrm{II}}}, F^{\mathrm{II}}_{\beta}\ (\beta=2,\dots m)\end{matrix}\right\} \quad \text{(A.1)}$$

and, from (1.7), it follows that

$$\Lambda \text{ is a function of } \vartheta, F^{\mathrm{I}}_{\alpha}, F^{\mathrm{II}}_{\beta}\begin{pmatrix}\alpha=1,2\dots n\\ \beta=1,2\dots m\end{pmatrix} \text{ or}$$

$$\text{of } \vartheta, H_{\Lambda^{\mathrm{I}}}, H_{\Lambda^{\mathrm{II}}}, F^{\mathrm{I}}_{\alpha}, F^{\mathrm{II}}_{\beta}\begin{pmatrix}\alpha=2,\dots n\\ \beta=2,\dots m\end{pmatrix}. \quad \text{(A.2)}$$

By use of $E=E_{\mathrm{I}}+E_{\mathrm{II}}$, we may rewrite (1.7) in the form

$$\dot{H}_{\Lambda}=\Lambda((\dot{E}_{\mathrm{I}}-\gamma^{\mathrm{I}}_{\alpha}F^{\mathrm{I}}_{\alpha})+(\dot{E}_{\mathrm{II}}-\gamma^{\mathrm{II}}_{\beta}F^{\mathrm{II}}_{\beta}))$$

or, by (1.5)

$$\dot{H}_{\Lambda}=\frac{\Lambda}{\Lambda^{\mathrm{I}}}\dot{H}_{\Lambda^{\mathrm{I}}}+\frac{\Lambda}{\Lambda^{\mathrm{II}}}\dot{H}_{\Lambda^{\mathrm{II}}},$$

whence we conclude that

$$\left.\begin{matrix}\Lambda/\Lambda^{\mathrm{I}}\\ \Lambda/\Lambda^{\mathrm{II}}\end{matrix}\right\} \text{ are functions of } H_{\Lambda^{\mathrm{I}}}, H_{\Lambda^{\mathrm{II}}}. \quad \text{(A.3)}$$

It follows from $(\mathrm{A.3})_1$ and $(\mathrm{A.1})_1$ that

$$\Lambda \text{ is independent of } F^{\mathrm{II}}_{\beta}(\beta=2,\dots m)$$

† In this treatment, we assume invertibility of the constitutive functions as required by the argument.

from $(A.3)_2$ and $(A.1)_2$ that

Λ is independent of $F^{\mathrm{I}}_{\alpha}(\alpha=2,\ldots n);\rightarrow \Lambda=\Lambda(\vartheta, H_{\Lambda^{\mathrm{I}}}, H_{\Lambda^{\mathrm{II}}})$;

from $(A.3)_1$ that

Λ^{I} is a function of ϑ, $H_{\Lambda^{\mathrm{I}}}$, $H_{\Lambda^{\mathrm{II}}}$ or, in fact, only of ϑ, $H_{\Lambda^{\mathrm{I}}}$, since body I must be unaffected by $H_{\Lambda^{\mathrm{II}}}(\vartheta, F^{\mathrm{II}}_{\beta})$,

from $(A.3)_2$ that

Λ^{II} is a function of ϑ, $H_{\Lambda^{\mathrm{I}}}$, $H_{\Lambda^{\mathrm{II}}}$ or, in fact, only of ϑ, $H_{\Lambda^{\mathrm{II}}}$, since body II must be unaffected by $H_{\Lambda^{\mathrm{I}}}(\vartheta, F^{\mathrm{I}}_{\alpha})$.

Thus, $\Lambda=\Lambda(\vartheta, H_{\Lambda^{\mathrm{I}}}, H_{\Lambda^{\mathrm{II}}})$ and $\Lambda^{\mathrm{I}}=\Lambda^{\mathrm{I}}(\vartheta, H_{\Lambda^{\mathrm{I}}})$ with $\Lambda/\Lambda^{\mathrm{I}}$ independent of ϑ and $\Lambda^{\mathrm{II}}=\Lambda^{\mathrm{II}}(\vartheta, H_{\Lambda^{\mathrm{II}}})$ with $\Lambda/\Lambda^{\mathrm{II}}$ independent of ϑ; whence it follows that Λ, Λ^{I}, Λ^{II} must have the forms

$$\Lambda=\frac{\psi(H_{\Lambda^{\mathrm{I}}}, H_{\Lambda^{\mathrm{II}}})}{T(\vartheta)}, \qquad \Lambda^{\mathrm{I}}=\frac{\alpha^{\mathrm{I}}(H_{\Lambda^{\mathrm{I}}})}{T(\vartheta)}, \qquad \Lambda^{\mathrm{II}}=\frac{\alpha^{\mathrm{II}}(H_{\Lambda^{\mathrm{II}}})}{T(\vartheta)}$$

where ψ, α^{I}, α^{II} and $T(\vartheta)$ are arbitrary functions.

A. Pliś

On dynamical models with retardation

In relativistic mechanics, retardations occur because interactions are propagated with finite velocities. In the non-relativistic case, the usual theory contains no retardation. In this paper, we shall present a retarded differential equation for a non-relativistic particle interacting with movable objects.

The retardation is caused by the force from the objects acting on the particle, because the force depends on the localization of the objects which, in turn, depends on the past localizations of the particle.

The retarded differential equation is of the form

$$m\frac{d^2x}{dt^2}=F[x_t], \tag{1}$$

where m denotes the mass of the particle, F is a given functional and x_t denotes the function defined and equal to $x(s)$ on the interval $0\leqslant s\leqslant t$. Equation (1) has the meaning that the second derivative of x at t depends in a known way on the value of x at t and on the values of x before t. The following initial conditions are considered:

$$x=u, \qquad \frac{dx}{dt}=v \quad \text{at} \quad t=0. \tag{2}$$

If the objects are non-relativistic particles, there exists a retarded differential equation consistent with the Newtonian equations. In that case, the retarded differential equation is obtained by a certain elimination of other particles from the Newtonian equations.

The elimination can be carried out in the following manner. Consider the Newtonian equations

$$m\frac{d^2x}{dt^2}=f(t,x,x_1,\dots,x_n) \qquad m_1\frac{d^2x_1}{dt^2}=f_1(t,x,x_1,\dots,x_n)$$

$$\vdots$$

$$m_n\frac{d^2x_n}{dt^2}=f_n(t,x,x_1,\dots,x_n),$$

with the initial conditions

$$x(0) = u, \qquad x_1(0) = u_1, \ldots, x_n(0) = u_n$$
$$x'(0) = v, \qquad x_1'(0) = v_1, \ldots, x_n'(0) = v_n.$$

Clearly these equations are without retardation. To pass to a model with retardation, let us consider the case in which we are interested only in the particle, with trajectory $x(t)$.

For a while, let us treat the function $x(t) = y(t)$ as known. Then we can find the functions $x_1, \ldots, x_n$ from the differential equations

$$m_1 \frac{d^2 x_1}{dt^2} = f_1(t, y(t), x_1, \ldots, x_n),$$
$$m_n \frac{d^2 x_n}{dt^2} = f_n(t, y(t), x_1, \ldots, x_n),$$

with the initial conditions

$$x_1(0) = u_1, \ldots, x_n(0) = u_n, \qquad x_1'(0) = v_1, \ldots, x_n'(0) = v_n$$

provided the solution of the considered Cauchy problem is unique. Then we have $x_1(t) = x_1[y_t], \ldots, x_n(t) = x_n[y_t]$, where $x_1[\ ], \ldots, x_n[\ ]$ denote functionals.

Hence, the function $x = x(t)$ satisfies the equation

$$m \frac{d^2 x}{dt^2} = f(t, x, x_1[x_t], \ldots, x_n[x_n]) = F[x_t]. \tag{4}$$

Under the assumption of uniqueness of solutions, Eqn (4) is equivalent to the Newtonian equations. We have uniqueness if, for example, functions $f, f_1, \ldots, f_n$, and hence functional $F[\ ]$, defined by (4), satisfy the Lipschitz condition.

Example Consider the following system:

$$\frac{d^2 x}{dt^2} = h(t, x) + x - x_1, \qquad \frac{d^2 x_1}{dt^2} = x_1 - x,$$

where $h(t, x)$ is a given function. The solution of the equation

$$\frac{d^2 x_1}{dt^2} = x_1 - y(t)$$

can be given explicitly as a functional $\tilde{x}_1[y_t]$. Functional $\tilde{x}_1$, defined by non-homogeneous ordinary differential equation of the second order with constant coefficients, is expressible by integrals and exponential functions.

Equation (1) has now the form

$$\frac{d^2x}{dt^2} = h(t, x) + x - \tilde{x}_1[x_t],$$

where $\tilde{x}_1$ is given by integrals and exponential functions.

Remark The exact elimination of additional particles presented in this paper is not possible in quantum mechanics.

Andrzej Pliś
Instytut Matematyczny PAN,
Oddział Kraków,
ul. L. Solskiego 30,
Kraków,
Poland

M. Roseau

Reflexions sur le problème de la diffraction par un demi-plan; condition de radiation de Sommerfeld et non-unicité

1

Comme il est bien connu, la théorie de la diffraction se propose d'étudier l'effet sur une grandeur physique, en état de propagation par ondes dans un milieu non borné, de l'interaction avec un obstacle ayant des caractéristiques définies; son objet principal est le calcul de l'onde secondaire, c'est à dire l'onde refléchie et l'onde diffractée, éventuellement l'onde de surface, lorsque l'onde primaire ou incidente est connue.

Les solutions exactes sont, on le sait, peu nombreuses et relatives à des geométries d'obstacles simples, cylindre, sphère, demi-plan; cependant leur intérêt est multiple, qu'il soit de preciser les conditions d'un problème bien posé ou de fournir les éléments de comparaison nécessaires dans l'application de certaines méthodes approchées.

Nous proposons dans ce texte d'appliquer à plusieurs problèmes de diffraction d'ondes planes par un demi-plan, une méthode d'analyse introduite pour l'étude d'ondes liquides de gravité en profondeur variable [1].

1.1 Problème du dock

L'axe des y est porté par l'horizontale suivant le dock et la surface libre, l'axe des x a direction et sens de la verticale descendante. Le potentiel $\varphi(x, y)$ satisfait à $\Delta\varphi - k^2\varphi = 0$ dans le demi-plan $x > 0$, et aux conditions aux limites

$$\frac{\partial\varphi}{\partial x} + \varphi = 0 \quad x = 0, y > 0 \quad \text{surface libre,}$$

$$\frac{\partial\varphi}{\partial x} = 0 \qquad x = 0, y < 0 \quad \text{dock.}$$

1.2 Diffusion du bruit aerodynamique par une plaque semi-infinie de grand flexibilité

Le potentiel $\varphi(x, y)$ vérifie $\Delta\varphi + k^2\varphi = 0$ avec $x + iy = \rho e^{i\theta}$, $\rho > 0$, $-\pi \leqslant \theta \leqslant \pi$ et les conditions aux limites

$$\eta \left[\frac{\partial\varphi}{\partial y}(\rho e^{i\pi}) + \frac{\partial\varphi}{\partial y}(\rho e^{-i\pi})\right] = \varphi(\rho e^{i\pi}) - \varphi(\rho e^{-i\pi}), \qquad \rho > 0,$$

$$\frac{\partial\varphi}{\partial y}(\rho e^{i\pi}) = \frac{\partial\varphi}{\partial y}(\rho e^{-i\pi}), \qquad \rho > 0.$$

1.3 Diffraction par une paroi semi-infinite d'ondes liquides en bassin tournant

Le potentiel $\varphi(x, y)$ est solution des équations

$$\Delta\varphi + k^2\varphi = 0 \qquad \forall x + iy = \rho e^{i\theta}, \qquad -\pi \leqslant \theta \leqslant \pi, \qquad \rho > 0 \tag{1}$$

$$2\Omega \frac{\partial\varphi}{\partial x} + i\mu \frac{\partial\varphi}{\partial y} = 0, \qquad \theta = \pm\pi, \qquad \rho > 0. \tag{2}$$

Ω et μ respectivement mesure de la rotation du bassin et de la pulsation du mouvement d'ondes.

Ces problèmes ont été discutés en [2–4] successivement, par réduction à une équation intégrale de Wiener–Hopf, la solution étant au départ représentée avec l'aide d'une fonction de Green appropriée. Cependant la technique utilisée dans [3, 4] rend nécessaire d'attribuer au paramètre k une partie imaginaire non-nulle, hypothèse qui ne correspond pas à la réalité physique. Notre approche tout à fait différente est affranchie de cette condition et fait apparaitre en outre dans le cas du problème 1.3 deux types de solution aux caractères distincts.

2

On se contenera de décrire dans ce qui suit les solutions du problème 1.3, en rappelant que les paramètres k, Ω, μ sont réels et que l'on considère le cas des ondes de première classe, c'est à dire $(\mu + 2\Omega)/(\mu - 2\Omega) = \sigma^2 > 0$; on peut supposer $k > 0$ et aussi $\sigma > 1$, car les équations du problème sont invariantes par la transformation $(x, y, \Omega, \mu) \rightarrow (x, -y, -\Omega, \mu)$. Des conditions de régularité à l'origine (φ bornée, $\partial\varphi/\partial x$, $\partial\varphi/\partial y$ de carré intégrable au voisinage de l'origine, c'est à dire énergie localement finie), et de comportement à l'infini (onde plane incidente donnée ou tendance vers 0) sont, en outre, à considérer.

On postule pour $\varphi(x, y)$ une représentation ayant la forme

$$\varphi(x, y) = \int_C \exp\left\{\frac{ik}{2}\left[x\left(z+\frac{1}{z}\right)+iy\left(z-\frac{1}{z}\right)\right]\right\}\cdot g(z)\,dz + $$
$$+\int_\Gamma \exp\left\{\frac{ik}{2}\left[x\left(z+\frac{1}{z}\right)-iy\left(z-\frac{1}{z}\right)\right]\right\}\cdot h(z)\,dz, \quad (3)$$

où C, Γ sont des contours tracés sur le feuillet

$$-\pi-\frac{\pi}{4}<\arg z<\pi+\frac{\pi}{4}$$

du plan complexe z; ceux ci dépendent du secteur $\pi/4 \leqslant \theta \leqslant \pi$, $-\pi/3 \leqslant \theta \leqslant \pi/3$, $-\pi \leqslant \theta \leqslant -\pi/4$ auquel appartient le point $x+iy=\rho e^{i\theta}$, le prolongement analytique de la solution étant assuré quand on passe de la représentation valable dans l'un des secteurs à celle valable dans le secteur voisin (Fig. 1). Les singularités z_1, z_2 que doit contourner dans certains cas le chemin Γ sont associées au comportement de la solution á l'infini; elles sont données telles que

$$|z_1|=|z_2|, \qquad \theta_1=\arg z_1=-\arg z_2, \qquad 0<\arg z_1<\pi.$$

Les fonctions $h(z)$, $g(z)$ sont définie par

$$h(z)=\frac{1}{z}\left[c+\frac{az_1^{1/2}}{z^{1/2}-z_1^{1/2}}+\frac{bz_2^{1/2}}{z^{1/2}-z_2^{1/2}}\right], \quad (4)$$

$$g(z)=\frac{1}{z}\left[c-\frac{az_1^{1/2}}{z^{1/2}+z_1^{1/2}}-\frac{bz_2^{1/2}}{z^{1/2}+z_2^{1/2}}\right]\frac{\sigma^2z^2-1}{z^2-\sigma^2}, \quad (5)$$

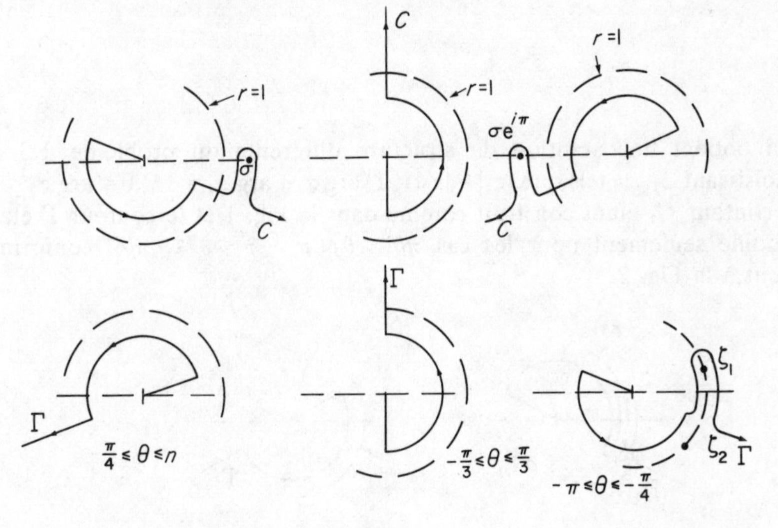

Fig. 1

a, b, c coefficients tels que

$$c=\frac{a+b}{1+\sigma^2}, \tag{6}$$

$$\left(\sigma^2 z_1-\frac{1}{z_1}\right) a+\left(\sigma^2 z_2-\frac{1}{z_2}\right) b=0. \tag{7}$$

La solution $\varphi(x, y)$ correspondant à ces données et calculée par (3), (4), (5) est régulière à l'origine (énergie localement finie) et, normalisée en choisissant $a=i/4\pi$, se comporte asymptotiquement quand $\rho\to\infty$ comme

$$\varphi\sim 0, \qquad \theta_1<\theta<\pi \quad \text{(zone d'ombre)}$$
$$\varphi\sim\exp\left[ik(x\cos\theta_1+y\sin\theta_1)\right], \qquad -\theta_1<\theta<\theta_1,$$
$$\varphi\sim\exp\left[ik(x\cos\theta_1+y\sin\theta_1)\right] - 4\pi ib\exp\left[ik(x\cos\theta_1-y\sin\theta_1)\right], \qquad -\pi<\theta<-\theta_1.$$

En outre il existe des ondes de bord sur les faces de telle sorte que

$$\varphi\sim i\frac{(\sigma-1)(\sigma^4-1)}{2\sigma^{3/2}(\sigma^2\,e^{-i\theta_1}-e^{i\theta_1})}\sin\frac{\theta_1}{2}\exp\left[-\frac{ik}{2}(\sigma+\sigma^{-1})\rho\right], \qquad \theta=\pi,\ \rho\to\infty,$$

$$\varphi\sim-\frac{(\sigma+1)(\sigma^4-1)}{2\sigma^{3/2}(\sigma^2\,e^{-i\theta_1}-e^{i\theta_1})}\sin\frac{\theta_1}{2}\exp\left[\frac{ik}{2}(\sigma+\sigma^{-1})\rho\right]+ +(1-4\pi ib)\exp(-ik\rho\cos\theta_1), \qquad \theta=-\pi,\ \rho\to\infty.$$

En fin l'on peut montrer que la différence entre $\varphi(x, y)$ et la représentation asymptotique que l'on vient de décrire satisfait á la condition de radiation de Sommerfeld á l'infini.

3

On obtient une solution de structure différente du problème 1.3 en choisissant z_1, z_2 tels que $|z_1|=|z_2|=1$, $\arg z_1+\arg z_2=2\pi$, $0<\arg z_1<\pi$, le contour C_1 étant construit comme dans la Fig. 1 et le contour Γ étant modifié seulement pour les cas $\pi/4\leqslant\theta\leqslant\pi$, $-\pi\leqslant\theta\leqslant-\pi/4$, conformément à la Fig. 2.

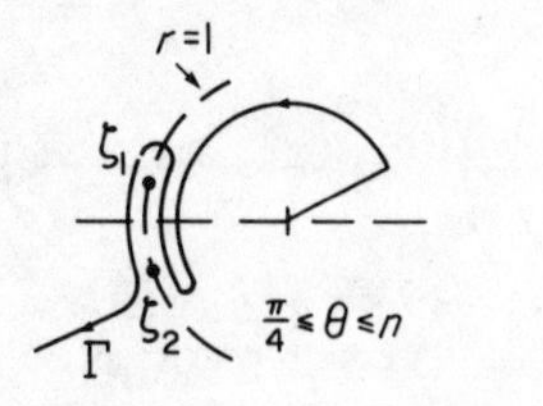

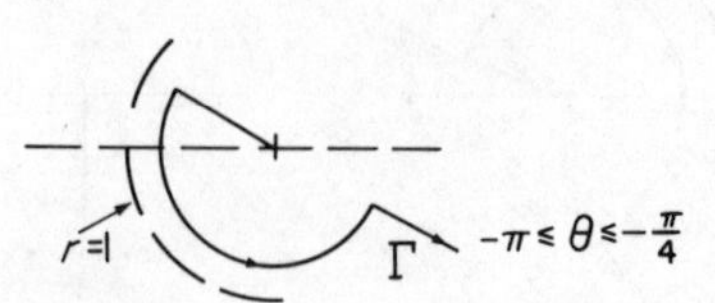

Fig. 2

La solution est encore représentée à l'aide de (3), (4), (5), (6), (7), cependant on prend

$$a=\frac{z_1^2-\sigma^2}{(\sigma^4-1)(1+z_1)z_1^{1/2}},\qquad b=\frac{\sigma^2 z_1^2-1}{(\sigma^4-1)(1+z_1)z_1^{1/2}},$$

$$c=\frac{(z_1-1)z_1^{-1/2}}{\sigma^4-1} \tag{8}$$

vérifiant (6), (7).

L'étude asymptotique de la solution φ, obtenue par (3), (4), (5) sous les conditions qui précèdent revèle que

$$\varphi\sim\sqrt{\left(\frac{2\pi}{k\rho}\right)}\,\mathrm{e}^{i(k\rho+\pi/4)}\frac{(1+\mathrm{e}^{i\theta})\,\mathrm{e}^{i\theta/2}}{1-\sigma^2\,\mathrm{e}^{2i\theta}},\qquad \theta\in(-\pi,\pi),\qquad \rho\sim\infty,$$

$$\varphi\sim-\frac{i\pi(\sigma+1)}{\sigma^{3/2}}\exp\left[-i\frac{k}{2}(\sigma+\sigma^{-1})\rho\right],\qquad \theta=\pi,\qquad \rho\sim\infty,$$

$$\varphi\sim\frac{\pi(\sigma-1)}{\sigma^{3/2}}\exp\left[i\frac{k}{2}(\sigma+\sigma^{-1})\rho\right],\qquad \theta=-\pi,\qquad \rho\sim\infty.$$

On peut établir que $h(z)$, $g(z)$ vérifient la relation

$$h(z)\cdot z+g(z)\frac{1}{z}=\frac{(1+z)z^{1/2}}{1-\sigma^2 z^2},\qquad \arg\frac{1}{z}=-\arg z,$$

indépendante de z_1, d'où l'on déduit facilement que la solution construite dans ce paragraphe est indépendante de z_1. En fait on peut la représenter par

$$\varphi(x,y)=\int_\Gamma \exp\left\{\frac{ik}{2}\left[x\left(z+\frac{1}{z}\right)-iy\left(z-\frac{1}{z}\right)\right]\right\}\frac{1+z}{z^{1/2}(1-\sigma^2 z^2)}\,\mathrm{d}z$$

où le contour Γ doit etre choisi conformément à la Fig. 3.

On a ainsi obtenu une solution regulière à l'origine, evanescente à l'infini dans le plan, hormis sur les bords superieur et inférieur du demi axe négatif Ox', qui sont porteurs d'ondes, et satisfaisant à la condition de radiation de Sommerfeld.

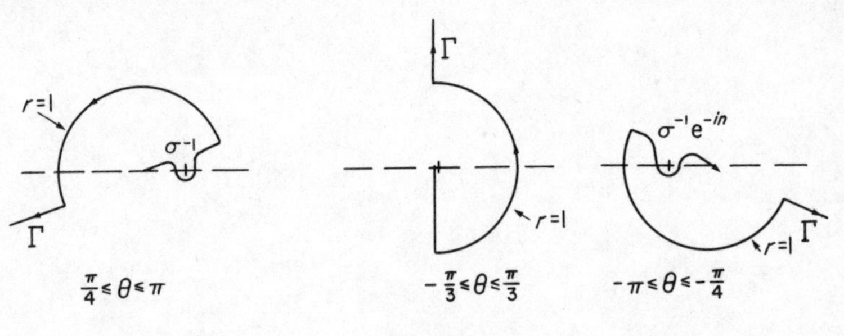

Fig. 3

References

1 Roseau, M., *Asymptotic Wave Theory*, North Holland, Amsterdam, 1976.
2 Heins, A. E., Water waves over a channel of finite depth with a dock, *Amer. J. Math.*, **70,** 730–748, 1948.
3 Crighton, D. C., and Leppington, F. G., Scattering of aerodynamic noise by a semi-infinite compliant plate, *J. Fluid Mech.*, **43,** 721–736, 1970.
4 Manton, M. J., Mysak, L. A., and Gorman, R. E., The diffraction of internal waves by a semi-infinite barrier, *J. Fluid Mech.*, **43,** 165–176, 1970.

Professor M. Roseau,
Universite Pierre et Marie Curie,
Mécanique Theoretique,
Tour 66,
4 Place Jussieu,
75230 Paris,
France.

I. N. Sneddon

Recent applications of integral transforms in the linear theory of elasticity†

Abstract

This paper is a survey of recent work on the theory of integral transforms and related topics and of their use in the analysis of boundary-value problems in the linear theory of elasticity. Recent publications in this field fall roughly into two categories. In the first place, there have been derived new results pertaining to long-established transforms such as Fourier, Hankel and Mellin transforms and to dual and triple integral equations involving them; an account is given of some of these developments. Some account is also given of results falling into the second category – those relating to new transforms devised in recent years. The most widely used of these in the theory of elasticity have been the finite Mellin transforms introduced by Naylor, so some indication is given of their use. Finally, mention is made of the differential transform.

1 Introduction

Integral transforms, particularly Fourier and Hankel transforms, have been used extensively in recent years in the analysis of boundary-value problems in the linear theory of elasticity. The application of these simple mathematical techniques is not confined to the classical theory (and indeed has been made to micropolar theory), but the older problems based on Navier's equations provide a simple context in which to demonstrate the power of the methods without involving undue physical complexity.

Descriptions of the earlier work on the use of integral transform techniques in elasticity are given in [30, 31, 58], and of more recent investigations in [4, 34]. Accounts of the application of transform

† This survey was written while the author was visiting Indiana University where he was supported by NSF grant MCS 75-038-94A2.

methods in particular branches of the theory of elasticity also form parts of books such as [7, 36, 35]. Recent treatments of integral transform techniques and related topics are to be found in [33, 26].

In this paper an account is given of some of the more recent work on the theory of integral transforms and related topics such as dual (and triple) integral equations. Recent work in this area falls roughly into two categories.

In the first place, new results pertaining to long-established transforms have been derived recently. For example, Sneddon [29] has derived a simple result concerning the double Fourier transforms of functions of the form $f(\rho)$, where $\rho^2 = x^2/a^2 + y^2/b^2$, which can be used effectively in the calculation of stress intensity factors for a plane crack with elliptical boundary and of corresponding physical quantities arising in the Boussinesq problem for a flat-ended elliptical punch; this is described in section 2.1. The use of integral transform methods in the solution of mixed boundary-value problems usually leads to dual or triple integral equations. A general procedure for deriving the solutions of a class of dual integral equations is described in section 2.2. The analysis in the vast majority of the published papers on dual integral equations is avowedly purely formal so it is a pleasure to end that section with an account of a paper [64] which gives a rigorous discussion of the uniqueness problem for dual integral equations of Titchmarsh type. In section 2.3, an account is given of the method of solving sets of simultaneous dual and triple integral equations by reducing the problem to that of solving a singular integral equation for a complex-valued function of a real variable.

Until recently, the most frequently used transforms in elasticity theory were those of Fourier and Hankel. In the last few years, Tweed and others have derived interesting properties of the Mellin transform and exploited the use of such results in the analysis of boundary-value problems concerned with finite elastic discs and cylinders. The salient features of this work are described in section 3.

The second category into which recent investigations in this area fall, is that concerned with the devising of new transforms. Of those, the most widely used in the theory of elasticity have been the Mellin type transforms introduced by Naylor [19]. In section 4, an account is given of these transforms and some indication is given of the application, by Tweed and others, of these results to the solution of problems involving circular discs of elastic material and thin elastic plates with circular holes.

The transform mentioned very briefly in section 5 is not a new one. Indeed, it is of some antiquity, being associated with the name of Weber, but it is only recently that it has been applied to elasticity by Srivastav and Olesiak.

Finally, in section 6, a very brief account is given of Ungar's differential transform, which, though not strictly an integral transform, is closely related to the Laplace transform and shows some promise of being of use in elastodynamics.

2 Fourier and Hankel transforms

We begin by considering some recent results concerning Fourier and Hankel transforms using the notation of [33], i.e. the operators $\mathscr{F}$, $\mathscr{F}^*$ are defined by

$$\mathscr{F}f(\xi)=(2\pi)^{-1/2}\int_{\mathbb{R}} f(x)\,e^{i\xi x}\,dx,$$

$$\mathscr{F}^*F(x)=(2\pi)^{-1/2}\int_{\mathbb{R}} F(\xi)e^{-i\xi x}\,d\xi,$$

so that, by Fourier's integral theorem, $\mathscr{F}^{-1}=\mathscr{F}^*$. Also $\mathscr{F}_c$ and $\mathscr{F}_s$ are defined by

$$\mathscr{F}_cf(\xi)=(2/\pi)^{1/2}\int_0^\infty f(x)\cos(\xi x)\,dx$$

$$\mathscr{F}_sf(\xi)=(2/\pi)^{1/2}\int_0^\infty f(x)\sin(\xi x)\,dx$$

with $\mathscr{F}_c^{-1}=\mathscr{F}_c$, $\mathscr{F}_s^{-1}=\mathscr{F}_s$ and $\mathscr{H}_\nu$, the operator of the Hankel transform of order ν, is defined by

$$\mathscr{H}_\nu f(\xi)=\int_0^\infty xf(x)J_\nu(\xi x)\,dx.$$

The Hankel inversion theorem states that $\mathscr{H}_\nu^{-1}=\mathscr{H}_\nu$.

2.1 A result concerning double Fourier transforms

We shall denote the double Fourier transform by $\mathscr{F}_2$ and its inverse by $\mathscr{F}_2^*$ so that

$$\mathscr{F}_2[f(x,y);(\xi,\eta)]=\frac{1}{2\pi}\int_{\mathbb{R}^2} f(x,y)\,e^{i(\xi x+\eta y)}\,dx\,dy$$

and

$$\mathscr{F}_2^*[F(\xi,\eta);(x,y)]=\frac{1}{2\pi}\int_{\mathbb{R}^2} F(\xi,\eta)\,e^{-i(\xi x+\eta y)}\,d\xi\,dn.$$

It is then easily shown (cf., [29]) that, if $\rho=\sqrt{(x^2/a^2+y^2/b^2)}$ $(a>b)$, then

$$\mathscr{F}_2[f(\rho);(\xi,\eta)]=ab\mathscr{H}_0[f(\rho);\lambda],\qquad \lambda=\sqrt{(a^2\xi^2+b^2\eta^2)}. \tag{2.1}$$

Another result which is very useful is

$$\mathcal{F}_2^*[(\xi^2+\eta^2)^{-1/2}\mathcal{H}_0 f(\lambda); (x, y)] = (2\pi)^{-1/2}a^{-1}\int_0^{\pi/2} [\hat{f}_2\{\rho \sin(t+\phi)\}$$

$$+\hat{f}_2\{\rho \sin(t-\varphi)\}](1-k^2\sin^2 t)^{-1/2}\,dt, \quad (2.2)$$

where $x = a\rho\cos\phi$, $y = b\rho\sin\phi$, $k^2 = (a^2-b^2)/a^2$ and where $\hat{f}_2 = \mathcal{F}_c\mathcal{H}_0 f$. It should also be noticed that if we denote the Abel transform of the second kind by $\mathcal{A}_2$ (cf. p. 321 of [33]), $\hat{f}_2(x) = \mathcal{A}_2[tf(t); x]$. Also

$$\mathcal{F}_2^*[(\xi^2+\eta^2)^{1/2}\mathcal{H}_0 f(\lambda); (x, y)] = (2\pi)^{-1/2}a^{-1}b^{-2}\int_0^{\pi/2} [\hat{f}_3\{\rho \sin(r+\phi)$$

$$+\hat{f}_3\{\rho \sin(t-\phi)\}](1-k^2\sin^2 t)^{1/2}\,dt \quad (2.3)$$

in which the function $\hat{f}_3$ is defined by the equation

$$\hat{f}_3(x) = -(d^2/dx^2)\hat{f}_2(x).$$

These equations have been applied to the solution of the Boussinesq problem for a flat-ended cylindrical punch with elliptical cross-section and to the calculation of the stress intensity factors of a crack in the shape of a flat elliptical disc.

2.2 Dual integral equations

The applications of integral transform methods to the solution of linear boundary-value problems in mathematical physics frequently lead to the solution of dual integral equations of the type

$$\int_0^\infty h_1(x, t)\phi(t)\,dt = f(x), \qquad 0 < x < 1 \quad (2.4)$$

$$\int_0^\infty h_2(x, t)\phi(t)\,dt = g(x) \qquad x > 1. \quad (2.5)$$

Many methods have been devised to solve such equations in the hundred years since the simplest of them was first considered by Weber. Recently a general technique for deriving solutions – as distinct from establishing them rigorously – has been outlined by Nasim and Sneddon [18].

The method consists in finding two functions m_1 and m_2 whose Mellin transforms $m_1^* = \mathcal{M}m_1$ and $m_2^* = \mathcal{M}m_2$ satisfy the functional equation

$m_1^*(s)h_1^*(s) = m_2^*(s)h_2^*(s)$ in some strip $\sigma_1 < \text{Re } s < \sigma_2$ of the complex s-plane. The unknown function ϕ is then a solution of the integral equation

$$\int_0^\infty \phi(u)k(ut)\,du = \psi(t) \tag{2.6}$$

whose kernel is defined by the equation $k(t) = \mathcal{M}^{-1}[m_1^*(s)h_1^*(s); t]$ and whose right hand side is defined by

$$\psi(t) = H(1-t)\int_0^t f(x)m_1(t/x)x^{-1}\,dx + H(t-1)\int_t^\infty g(x)m_2(t/x)x^{-1}\,dx. \tag{2.7}$$

For example, the classical dual relation of Titchmarsh type is that of finding the function ϕ satisfying the pair of equations (2.4) and (2.5) with $h_1(x) = 2^{2\alpha}x^{-2\alpha}J_\nu(x)$; $h_2(x) = 2^{2\beta}x^{-2\beta}J_\mu(x)$. In this case m_1 and m_2 turn out to be the kernels of the Erdélyi–Kober operators: (cf., pp. 38–55 of [32]).

The analysis contained in almost all published papers on dual integral equations is purely formal. An interesting exception is provided by the work of J. R. Walton. For example, in [64] Walton demonstrated rigorously that the behaviour of the solution of the dual integral equations of Titchmarsh type depends on the sign of $\lambda - \mu$ where λ is defined by $\lambda = \frac{1}{2}\mu + \frac{1}{2}\nu - \alpha + \beta$. Assuming that $\nu, \mu > -\frac{1}{2}$ and $\lambda > -1$, Walton proved that the solution of the equations (2.12) is unique if $\lambda > \mu$. He also showed that if $\lambda > \mu$ and $\lambda > \nu$ non-uniqueness can occur only in a pathological way, and produced counter-examples to uniqueness for the case $\mu < \lambda < \nu$.

The uniqueness theorems in the special case in which the Bessel function reduces to a sine or cosine were considered separately by Srivastav [39].

Notice should also be taken of recent work by R. Kalaba and others in which the theory of invariant imbedding has been used to transform a pair of dual integral equations into a Cauchy problem. *See*, for example, [2].

2.3 Simultaneous dual and triple integral equations

In the analysis of boundary-value problems concerning two dissimilar elastic half-spaces bonded, wholly or partly, along a plane, the use of integral transforms has led, in recent papers, to the solution of sets of simultaneous dual and triple integral equations. To illustrate the methods used to reduce such equations to a singular integral equation for a single complex-valued function we begin by considering the system of equations

$$\begin{aligned} \alpha\Phi_c(x)+\beta\Psi_c(x)&=C_1, && 0<x<1,\\ \beta\Phi_s(x)+\alpha\Psi_s(x)&=f_1(x), && 0<x<1,\\ \Psi_c(x)&=0 && x>1,\\ \Phi_s(x)&=0 && x>1, \end{aligned} \tag{2.8}$$

in which $\Phi_c=\mathcal{F}_c\phi_c$, $\Phi_s=\mathcal{F}_s\phi$, $\Psi_c=\mathcal{F}_c\psi$, $\Psi_s=\mathcal{F}_s\psi$, and C_1 is a constant. We let $\Psi_c(x)=r_1(x)H(1-x)$, $\Phi_s(x)=s_1(x)H(1-x)$, $x>0$ and introduce r, s where r is the even extension of r_1 to $(-1,1)$ and x is the odd extension of s_1 to the same interval; we find that the first two equations of the set are equivalent to a pair of simultaneous equations on $[0,1]$ and that if we write $\lambda=s+ir$, $g=f+iC_1$ with f defined to be the odd extension of f_1 to $(-1,1)$ the complex-valued function λ satisfies the singular integral equation

$$\beta\lambda(x)-\frac{\alpha}{\pi i}\int_{-1}^{1}\frac{\lambda(t)\,dt}{t-x}=g(x) \qquad -1<x<1 \tag{2.9}$$

whose solution is well-known.

This method has been applied in [16] to the problem of determining the displacement field at the rim of a penny-shaped crack situated at interface of two half-spaces of different elastic materials bonded together along their common plane boundary.

The closely related problem of determining the solution to pair of equations involving Abel transforms has been discussed by Lowengrub [15].

In [17] Lowndes follows Lowengrub [13] in using the Erdélyi–Kober operators to reduce a set of dual integral equations involving Hankel transforms of n unknown functions to a set of simultaneous integral equations.

Although not concerned with integral transforms the related investigations by Kelman and Feinerman on dual series relations are worthy of note; relevant references are [9], [5] and [6].

Similar methods apply to the solution of sets of simultaneous triple integral equations. For example, in determining the stress field in the vicinity of a pair of Griffith cracks located at the interface of two bonded dissimilar elastic half-planes, Lowengrub [14] was led to the problem of determining the solution of a set of simultaneous triple integral equations involving sine and cosine transforms. Proceeding as in the solution of the set (2.8) we are led to the solution of an integral equation of the type (2.9) but now with the range of integration of t replaced by $(-b,-a)\cup(a,b)$ with $0<a<b$. Lowengrub derives the solution of this equation by using the Plemelj formulae. It should be pointed out however that the

solution was discussed previously by Tricomi [43], Palócz [21] and Lewin [10].

3 Recent applications of Mellin transforms

In recent years there has been a great deal of activity, most of it generated by J. Tweed, concerned with the application of Mellin transforms to the solution of crack problems in the linear theory of elasticity. By formulating such problems as mixed boundary-value problems we are led to the discussion of dual and triple integral equations involving inverse Mellin transforms. In this section we shall look at equations of these types, but before proceeding to that we shall review some properties of the Mellin transform.

3.1 The Mellin transform

It is well known how the Mellin transform f^* of a function $f:\mathbb{R}^+\to\mathbb{R}$ defined by

$$f^*(s)=\mathcal{M}[f(r);s]=\int_0^\infty f(x)x^{s-1}\,dx \tag{3.1}$$

arises in a natural way in the solution of boundary-value problems concerning an infinite wedge $\{(r,\theta):\alpha<\theta<\beta,\ r>0\}$ where r and θ are plane polar coordinates.

The inversion theorem for the Mellin transform, i.e. the theorem which establishes a formula for the inverse $\mathcal{M}^{-1}$ of the operator is also familiar (*see*, for example, pp. 273–275 of [33]). The inverse $\mathcal{M}^{-1}$ is defined by the equation

$$\mathcal{M}^{-1}[f^*(s);x]=\frac{1}{2\pi i}\int_C x^{-s}f^*(s)\,ds. \tag{3.2}$$

In making use of the Mellin transform in applications, we have always to keep in mind that $\mathcal{M}^{-1}$ depends on the particular contour C which we employ. To illustrate this we consider the inverse transform $\mathcal{M}^{-1}[\cos(\pi s);x]$. If we interpret the integral as a Cauchy principal value we can easily show that if $0<\operatorname{Re}s<1$, $\mathcal{M}[(1-x)^{-1};s]=\pi\cot(\pi s)$, from which it follows immediately if $-n<\operatorname{Re}s<-n+1$, $\mathcal{M}[(1-x)^{-1}x^n;s]=\pi\cot(\pi s)$, with n an integer. Hence we see that if the contour C in the definition of $\mathcal{M}^{-1}$ lies in the strip $0<\operatorname{Re}s<1$, then $\mathcal{M}^{-1}[\cot(\pi s);x]=\pi^{-1}(1-x)^{-1}$, but if it lies in the strip $-n<\operatorname{Re}s<-n+1$, then $\mathcal{M}^{-1}[\cot(\pi s);x]=(1-x)^{-1}x^n$.

For this reason we should perhaps use a notation such as $\mathcal{M}_c^{-1}$ but, since in most cases no ambiguity is likely to arise, the simpler notation $\mathcal{M}^{-1}$ for the inverse of $\mathcal{M}$ is used.

We now consider the inverse Mellin transforms of functions of the form

$$A(s)=\int_\alpha^\beta p(t)t^{s+\lambda}\,\mathrm{d}t; \tag{3.3}$$

then

$$\mathcal{M}^{-1}[X(s)A(s);r]=\int_\alpha^\beta t^\lambda p(t)\mathcal{M}^{-1}[X(s);r/t]\,\mathrm{d}t. \tag{3.4}$$

For example, if $\sigma>0$ and we take C to lie in the strip $0<\operatorname{Re} s<n$

$$\begin{aligned}\mathcal{M}^{-1}[(s+\sigma)^{-1}A(s);r]&=r^\sigma\int_{\max(\alpha,r)}^\beta t^{\lambda-\sigma}p(t)\,\mathrm{d}t\cdot H(\beta-r)\\ \mathcal{M}^{-1}[\cot(\pi s/n)A(s);r]&=\frac{n}{\pi}\int_\alpha^\beta\frac{t^{n+\lambda}\rho(t)}{t^n-r^n}\,\mathrm{d}t\end{aligned} \tag{3.5}$$

where the integral on the right side of this equation is a Cauchy principal value but if, on the other hand, we take C to lie in the strip $-mn<\operatorname{Re} s<n-mn$, with $1-mn<-\sigma$, we may take

$$\begin{aligned}\mathcal{M}^{-1}[(s+\sigma)^{-1}A(s);r]&=-r^\sigma\int_\alpha^{\min(\beta,r)} t^{\lambda-\sigma}\rho(t)\,\mathrm{d}t\cdot H(r-\alpha)\\ \mathcal{M}^{-1}[\cot(\pi s/n)A(s);r]&=\frac{nr^{mn}}{\pi}\int_\alpha^\beta\frac{t^{\lambda-(m-1)n}p(t)\,\mathrm{d}t}{t^n-r^n}.\end{aligned} \tag{3.6}$$

3.2 Dual integral equations

We now illustrate how these simple results can be used in elucidating some difficult questions which arise in the discussion of crack problems in the linear theory of elasticity concerning thin circular discs and thin plates with circular holes.

We illustrate the methods used by considering the problem of determining the solution of the pair of dual integral equations

$$\begin{aligned}\mathcal{M}^{-1}[s^{-1}A(s);r]&=0 && 0<R<a\\ \mathcal{M}^{-1}[\cot(\pi s/n)A(s);r]&=r^nf(r) && a<r<b.\end{aligned} \tag{3.7}$$

If we write

$$A(s)=\int_a^b p(t^n)t^{s+n-1}\,\mathrm{d}t \tag{3.8}$$

then taking $\lambda=n-1$, $\sigma=0$, $m=1$, in equations (3.6) we find that (3.8) satisfies the dual integral equations (3.7) if p satisfies the integral equation

$$\frac{n}{\pi}\int_a^b \frac{t^{n-1}p(t^n)\,\mathrm{d}t}{t^n - r^n} = f(r).$$

Writing $\alpha = a^n$, $\beta = b^n$, $\phi(r) = f(r^{1/n})$, we see that this is equivalent to the integral equation

$$\frac{1}{\pi}\int_\alpha^\beta \frac{p(t)\,\mathrm{d}t}{t-x} = \phi(x) \qquad \alpha < x < \beta \tag{3.9}$$

whose solution has been given by Tricomi; this solution involves an arbitrary constant C. In any particular problem the constant C is determined by imposing a physical condition of some kind.

Tweed and Rooke [44] have made use of this method to solve the problem of determining the stress intensity factors and the crack energy of a symmetric array of n edge cracks in a circular cylinder under torsion. In this case, b is the radius of the cylinder and $b-a$ the length of each crack.

In a similar way Tweed and Rooke [54] solve the dual integral equations

$$\mathcal{M}^{-1}[(s+1)^{-1}A(s); r] = 0 \qquad 0<r<b$$
$$\mathcal{M}^{-1}[\cot(\pi s)A(s); r] = r^2\phi(r) \qquad b<r<1$$

by reducing them to a Tricomi-type equation. They use the solution to calculate the stress intensity factor and the crack energy of an edge crack in a finite elastic disc, whose radius is taken to be the unit of length.

In that paper they assume the crack to be opened up by constant pressure; in [25] the same authors consider the stress intensity factor of such a crack when the disc is rotating with constant angular velocity ω.

Using the same technique, Tweed and Rooke [55] derived an integral equation the solution of which is related to the stress intensity factor and the formation energy of a crack at the edge of a circular hole in an infinite elastic solid. Here the basic dual integral equations are

$$\mathcal{M}^{-1}[\cot(\pi s)(A(s); r] = r\phi(r) \qquad R<r<a+R$$
$$\mathcal{M}^{-1}[(1+s)^{-1}A(s); r] = 0 \qquad r>a+R$$

which they solve by a similar method.

Using the principle of superposition and the results of the papers cited above, Tweed and Rooke [57] also solved the problem of determining the stress field in an infinite solid containing a circular hole with a pair of radial edge cracks of different lengths.

The dual integral equations

$$\mathcal{M}^{-1}[\theta_n(s)A(s); r] = \psi(r) \qquad 1<r<c$$
$$\mathcal{M}^{-1}[(1+s)^{-1}A(s); r] = 0 \qquad c<r<\infty$$

$\theta_n(s) = \cot(\pi s/n) + \frac{1}{2}s[(ns/n - \cot(s+2)/n]$, arise in the analysis of the stress field in the vicinity of a crack in a symmetric array at the edge of a circular hole (whose radius is taken to be the unit of length) in an infinite elastic solid [25]. They can be solved also by a similar method.

8.3 Triple integral equations

A similar method can be devised for the solution of triple integral equations of the form

$$\begin{aligned} &\mathcal{M}^{-1}[(1+s)^{-1}A(s); r] = 0 \qquad && 0 < r < a \\ &\mathcal{M}^{-1}[\cot(\pi s/n)A(s); r] = r^n\psi(r) \qquad && a < r < b \\ &\mathcal{M}^{-1}[(1+s)^{-1}A(s); r] = 0 \qquad && b < t < \infty \end{aligned} \tag{3.10}$$

(cf. [45]). If the function $A(s)$ admits a representation of the form

$$A(s) = \int_a^b p(t^n)t^{s+n-1}\,\mathrm{d}t, \quad \int_a^b t^{n-2}p(t^n)\,\mathrm{d}t = 0 \tag{3.11}$$

then it follows from equations (3.3) and (3.6) with $\alpha = a$, $\beta = b$, $\sigma = 1$, $\lambda = n-1$, $m = 1$ that if we take the contour C in the definition of $\mathcal{M}^{-1}$ to lie in the strip

$$-n < \operatorname{Re} s < 1 - n,$$

then A as given by (3.11) will satisfy the equations (3.10) provided that p satisfies the integral equation

$$\frac{1}{\pi}\int_a^b \frac{nt^{n-1}}{t^n - r^n}\,p(t^n)\,\mathrm{d}t = \psi(r), \qquad a < r < b.$$

This, in turn, is easily reduced to the equation (3.11) with $\alpha = a^n$, $\beta = b^n$ and the function ϕ defined by the equation $\phi(x) = \psi(x^{1/n})$. The arbitrary constant C arising in the solution of this equation is determined from the second of the equations (3.11).

The special solution corresponding to $n = 2$ has been used by Tweed [45] to find the stress intensity factors of a pair of Griffith cracks defined by the relations $a < r < b$, $\theta = 0$ and $\theta = \pi$ and whose surfaces are loaded symmetrically.

By writing each of the components of the displacement field as a sum of an inverse Mellin transform involving A and a half-range Fourier series Tweed *et al.* [51] succeeded in reducing the problem of finding formulae for the stress intensity factor and crack energy of a radial crack in a disc of finite radius to that of solving the set of triple integral equations (3.10) with $n = 1$. The calculations for the case of a crack opened by the application of constant pressure to its faces are reported in detail in [51].

The case of the rotating disc has been treated in detail by Rooke and Tweed in [23] and that of the disc under point loading by the same authors in [24].

In discussing the stress field in the vicinity of a star-shaped array of cracks, Tweed and Rooke [56] reduced the boundary-value problem to that of solving the triple integral equations

$$\begin{aligned} \mathcal{M}^{-1}[(1+s)^{-1}A(s); r] &= \psi(r) \qquad 0<r<b \\ \mathcal{M}^{-1}[\theta_n(s)A(s); r] &= \psi(r) \qquad b<r<1 \\ \mathcal{M}^{-1}[(1+s)^{-1}A(s); r] &= 0 \qquad 1<r<\infty \end{aligned}$$

where the function θ_n is defined as before. These are solved by a similar method.

4 Applications of finite Mellin transforms

In this section we shall describe some recent applications to boundary-value problems of finite Mellin transforms introduced by Naylor [19].

4.1 The Naylor transforms

In the paper cited above, Naylor introduced the finite transforms $\mathcal{M}_R$ and $\mathcal{N}_R$ defined by the equations

$$\mathcal{M}_R[f(x); s] = \int_0^R [x^{s-1} + R^{2s}x^{-s-1}]f(x)\,dx \tag{4.1}$$

$$\mathcal{N}_R[f(x); s] = \int_0^R [x^{s-1} - R^{2s}x^{-s-1}]f(x)\,dx. \tag{4.2}$$

We can easily show from these definitions that

$$\mathcal{M}_R[H(x-t); x\to s] = s^{-1}(R^{2s}t^{-s} - t^s)H(R-t) \tag{4.3}$$

$$\mathcal{N}_R[H(x-t); x\to s] = s^{-1}t^{-s}(R^s - t^s)H(R-t) \tag{4.4}$$

$$\mathcal{M}_R\left[\frac{R^{2n}}{R^{2n}-x^nt^n} - \frac{t^n}{t^n-x^n}; x\to s\right] = \frac{\pi}{n}(R^{2s}t^{-s} - t^s)\cot(\pi s/n), |\text{Re } s|<n \tag{4.5}$$

$$\mathcal{N}_R\left[\frac{(Rx)^{1/2n}}{x^n-R^n}; x\to s\right] = \frac{\pi R^s}{n}\tan(\pi s/n); |\text{Re } s|<\tfrac{1}{2}n. \tag{4.6}$$

Since $\mathcal{M}_R[f(x); s] = \mathcal{M}[f_1(x); s]$ where $f_1(x) = f(x)H(R-x) + f(R^2/x)H(x-R)$, and $\mathcal{N}_R[f(x); s] = \mathcal{M}[f_2(x); s]$ where $f_2(x) = f(x) \times H(R-x) - f(R^2/x)H(x-R)$, the formulae for the inverse operators

$\mathcal{M}_R^{-1}$, $\mathcal{N}_R^{-1}$ may be deduced immediately from the inversion theorem for the Mellin transform.

Closely related to these transforms, though not itself a finite transform, is another transform introduced by Naylor in [19]. The transform $\mathcal{H}_R$ is defined by the equation

$$\mathcal{H}_R[f(x); s] = \int_R^\infty (x^{s-1} + R^{2s}x^{-s-1})f(x)\,dx.$$

If we write $f_3(x) = x^{-2}f(R^2x^{-1})H(R-x) + f(x)H(x-R)$ we see that

$$\mathcal{M}[f_3(x); s] = \mathcal{H}_R[f(x); s]$$

so that the formula for the inverse operator $\mathcal{H}_R^{-1}$ can be deduced from the inversion theorem for the Mellin transform.

4.2 Dual integral equations

We begin by considering the pair of dual integral equations

$$\begin{aligned} \mathcal{H}_R^{-1}[s^{-1}A(s); x] &= 0 \qquad & 0<x<a \\ \mathcal{H}_R^{-1}[\cot(\pi s/n)A(s); x] &= \psi(x) \qquad & a<x<R \end{aligned} \tag{4.7}$$

where the contour C in the definition of $\mathcal{H}_R^{-1}$ lies in the strip $|\text{Re } s| < n$ of the complex s-plane. If we replace $f(t)$ in equations (4.15) and (4.16) by $t^{n-1}p(t^n)$, we see that

$$A(s) = R^{-n}\int_a^R t^{n-1}(R^{2s}t^{-s} - t^s)p(t^n/R^n)\,dt \tag{4.8}$$

satisfies the dual equations (4.7) if the function p satisfies the integral equation

$$\frac{nR^{-n}}{\pi}\int_a^R t^{n-1}\left[\frac{R^{2n}}{R^{2n} - x^n t^n} - \frac{t^n}{t^n - x^n}\right]p(t^n/R^n)\,dt = \psi(x), \qquad a<x<R.$$

If we introduce functions h and g through the pairs of equations

$$h(\tau) = \begin{cases} -p(\tau), & (b<\tau<1) \\ \tau^{-2}p(\tau^{-1}), & (1<\tau<b^{-1}) \end{cases} \qquad g(\rho) = \begin{cases} \rho^{-1}\psi(R\rho^{1/n}), & (b<\rho<1) \\ \rho^{-1}\psi(R\rho^{-1/n}), & (1<\rho<b^{-1}) \end{cases}$$

where $b = (a/R)^n$ we obtain the singular integral equation

$$\frac{1}{\pi}\int_b^{b^{-1}} \frac{h(\tau)\,d\tau}{\tau - \rho} = g(\rho), \qquad (b<\rho<b^{-1}),$$

for the determination of h and hence of p.

This solution is due to Tweed [49] who has used it in [46] to derive a much simpler solution of the problem of determining the stress intensity

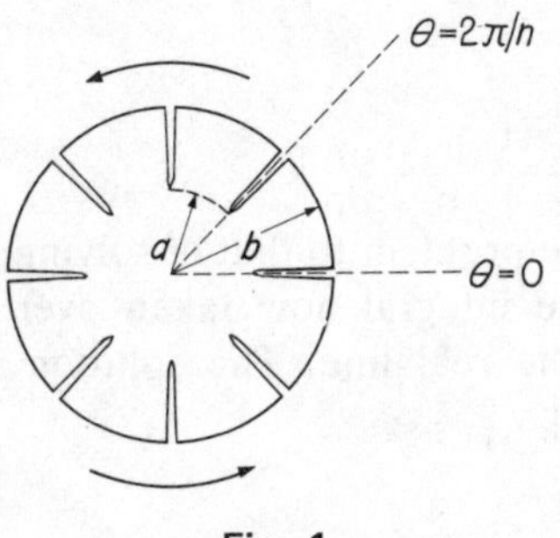

Fig. 1

factors and the crack formation energy of a radial system of edge cracks, of equal length, in a circular elastic cylinder under torsion (*see* Fig. 1 above). The method has been extended by Tweed and Longmuir [52] to cover the case where the cracks are not necessarily of equal length.

In a similar way we can derive the solution of the pair of dual integral equations

$$\begin{aligned} \mathscr{H}_R^{-1}[\cot(\pi s/n)A(s); x] &= \psi(x) \qquad R<x<a \\ \mathscr{H}_R^{-1}[s^{-1}A(s); x] &= 0 \qquad a<x<\infty \end{aligned} \tag{4.10}$$

the solution of this pair has been applied by Tweed [47] to determine the stress intensity factor of a crack $\{(r, \theta);\ R<r<Rb,\ \theta=0\}$ originating at the edge of a stress-free circular hole in an infinite elastic solid under a uniform longitudinal shear. In the problem in which there are two cracks $\{(r, \theta);\ R<r<Rb,\ \theta=0, \pi\}$ originating at the edge of a stress-free hole of radius R in an infinite elastic solid under a uniform longitudinal shear, the corresponding dual integrals are of the above type but with $\cot(\pi s/n)$ replaced by $\tan(\pi s/n)$. For details *see* [48].

4.3 Triple integral equations

Similar methods can be used to determine the solutions of triple equations.

We illustrate the method with the set of triple integral equations

$$\begin{aligned} \mathscr{H}_R^{-1}[s^{-1}A(s); x] &= 0 \qquad 0<x<a \\ \mathscr{H}_R^{-1}[\cot(\pi s/n)A(s); x] &= \psi(x) \qquad a<x<b \\ \mathscr{H}_R^{-1}[s^{-1}A(s); x] &= 0 \qquad b<x<R. \end{aligned} \tag{4.11}$$

Assuming a representation of the form

$$A(s) = \int_a^b t^{n-1}p(t^n/R^n)(R^{2s}t^{-s} - t^s)\,dt \tag{4.12}$$

with

$$\int_a^b t^{n-1}p(t^n/R^n)\,\mathrm{d}t=0 \tag{4.13}$$

we can again reduce the problem to that of solving an integral equation of the type (2.9) with the integral now taken over the union of the two disjoint intervals on the real line. The solution in the case $n=1$ was derived by Tweed [50].

5 Weber–Orr transforms

Brief mention should also be made of the fact that, in recent years, there have been several applications to elasticity of a transform introduced by Weber [66] and Orr [20] and whose inversion theorem was established rigorously by Watson (section 14.52 of [65]) and Titchmarsh [42]. Olesiak [4] has given an account of these transforms and of associated transforms due to him.

Dual integral equations involving such transforms have been discussed in [22], [37], [38], and [41].

Applications to elasticity are described in [22], [27], [28], [40] and [41].

6 Ungar's differential transform

We shall conclude our account of recent applications of integral transforms with some comments on the differential transform introduced by A. Ungar. Although the formal definition of this transform (cf., equation (6.2) below) is free of integrals, the transform is related to a formal double integral transform in which the integrals can be eliminated by the use of Cauchy's integral theorem and it retains many of the properties which make integral transforms so useful.

Before proceeding to the formal definition of the differential transform we consider the transform U defined by the equation

$$U_nf(\mathbf{x};\mu)=\frac{1}{2\pi i}\int_C \mathrm{d}\lambda\int_0^\infty s^n\mathrm{e}^{-s\mu^*(\mathbf{x};\lambda)+s\mu}f(\mathbf{x};\lambda)\,\mathrm{d}s \tag{6.1}$$

in which C is a closed contour in the complex λ-plane, to be specified later, and $\mu^*(\mathbf{x};\lambda)$ and $f(\mathbf{x};\lambda)$ are suitably differentiable functions of the real variables $\mathbf{x}=(x_1, x_2, \ldots, x_n)$ and the complex variable λ. Also, n denotes a positive integer (or zero). If we introduce a new variable of integration μ, defined by the equation $\mu=\mu^*(\mathbf{x};\lambda)$, we find that

$$U_nf(\mathbf{x};\mu)=\frac{1}{2\pi i}\int_C \mathrm{d}\bar{\mu}\int_0^\infty s^nf(\mathbf{x};\lambda^*)\lambda^*\mathrm{e}^{-s(\bar{\mu}-\mu)}\,\mathrm{d}s$$

where $\lambda^* = \lambda^*(\mathbf{x}; \mu)$ is the inverse of $\mu^*(\mathbf{x}; \lambda)$, i.e. $\mu^*(\mathbf{x}; \lambda^*(\mathbf{x}; \mu)) \equiv \mu$ and $\dot{\lambda}^* = \partial\lambda^*/\partial\mu$ is assumed to be different from zero in a neighbourhood of the point $\bar{\mu} = \mu$ in the complex $\bar{\mu}$-plane. The closed contour C in equation (6.1) is now specified to be such that its map C' in the $\bar{\mu}$-plane encircles the point $\bar{\mu} = \mu$. Evaluating the integral with respect to s, we find that the above double integral reduces to

$$\frac{n!}{2\pi i}\int_C \frac{f(\mathbf{x}; \lambda^*)\dot{\lambda}^*}{(\bar{\mu}-\mu)^{n+1}}\,\mathrm{d}\bar{\mu} = D_\mu^n\{f(\mathbf{x}; \lambda^*)\dot{\lambda}^*\}$$

where D_μ denotes the differential operator $\partial/\partial\mu$, and now $\dot{\lambda}^*$ denotes $\partial\lambda^*(x, \mu)/\partial\mu$.

In this way we are led to the formal definition

$$U_n f(\mathbf{x}; \mu) = D_\mu^n\{f(\mathbf{x}; \lambda^*)\dot{\lambda}^*\} \tag{6.2}$$

in which $\lambda^* = \lambda^*(\mathbf{x}; \mu)$ and $\dot{\lambda}^*$ has the interpretation given above.

The most important property of the differential transform is that it commutes with the operators of differentiation with respect to the variables x_j. If D_j denotes the differential operator $\partial/\partial x_j$, then it is easily shown that $U_n D_j = D_j U_n$ $(j = 1, 2, \ldots, N)$ (cf., [59]), and hence that $U_n \Delta_N = \Delta_N U_n$. Also we deduce immediately from the definition of U_n that $D_\mu U_n = U_{n+1}$.

To illustrate how to construct the differential transform of a given function we consider the following simple example communicated privately to the author by A. Ungar.

From Bateman's solution (p. 111 of [1]) of the two-dimensional wave equation we obtain the particular solution

$$\phi_0(x, y, t) = -\chi(\lambda)(x+t+\lambda)^{-1}\exp\{-s\mu^*(x, y; \lambda)\} \tag{6.3}$$

where λ is a parameter, the function χ is arbitrary and $\mu^* = \frac{1}{2}(x+t+\lambda)^{-1}\{\rho^2-(\lambda+t)^2\}$, with $\rho^2 = x^2+y^2$. Hence $\lambda^* = \rho^* - t - \bar{\mu}$, $\rho^{*2} = (x-\bar{\mu})^2 + y^2$, so that taking $\mu = 0$, $\lambda^* = \rho - t$, $\dot{\lambda}^* = -(x+\rho)/\rho$. Taking $n = 0$, we therefore obtain the solution

$$\phi(x, y, t) = \rho^{-1}(x+\rho)^{1/2}\chi(\rho-t), \tag{6.4}$$

(with χ arbitrary) of the two-dimensional wave equation.

The differential transform has been applied to problems in elastodynamics in the following way. Suppose that ϕ is a solution of a linear boundary-value problem and that it has a form appropriate for the application of the differential transform. Then, because of the relations (6.3), the differential transform $\psi = U_n\phi$ is another solution of the same differential equation. If the function ϕ is a regular solution, then, necessarily, its differential transform ψ is a solution with a singular point. In particular, solutions which describe the propagation of simple plane waves in layered media are transformed to solutions describing the propagation in layered media of waves emitted by point sources. For this reason it is useful to represent the latter such waves as differential transforms of simple plane waves.

Representations (as differential transforms of simple plane waves) of waves emitted by point sources have been given by Ungar and Robinson [63]. The displacement field due to a point source moving with uniform velocity in an elastic solid – originally derived in [3] – has been represented in a similar way in [60]. The latter representation was used in [60] and [61] to determine the motion of layered elastic media excited by uniformly moving point loads.

Further properties of the differential transform are described in the recent publication [62].

References

1 Bateman, H., *Partial Differential Equations*, Cambridge University Press, London, 1932.

2 Buell, J., Kagiwada, H., Kalaba, R., Ruspini, E. and Zagistin, E., Solution of a system of dual integral equations, *Int. J. Eng. Sci.*, **10**, 503, 1972.

3 Eason, G., Fulton, J. and Sneddon, I. N., Generation of waves in an infinite elastic solid by variable body forces., *Phil. Trans. Roy. Soc. London A*, **248**, 575, 1956.

4 Eason, G., Nowacki, W., Olesiak, Z. and Sneddon, I. N., *Integral Transform Methods in Elasticity*, Springer, Wien, 1977.

5 Feinerman, R. B. and Kelman, R. B., The convergence of least square approximations for dual orthogonal series, *Glasgow Math. J.*, **15,** 82, 1974.

6 Feinerman, R. B. and Kelman, R. B., Dual orthogonal series: an abstract approach., *Bull. Amer. Math. Soc.*, **81,** 733, 1975.

7 Галин, Л. А., Контактные Задачи Теории Упругости, Гтти, Москва, 1953; English translation, *Contact Problems in Elasticity* (Halina Moss, trans.), North Carolina State University Mathematics Group, Raleigh, N.C., 1961.

8 Gilbert, R. P., and Wienacht, R. J. (eds), *Function Theoretic Methods in Differential Equations*, Pitman Publishing, Co., London, 1977.

9 Kelman, R. B. and Feinerman, R. B., Dual orthogonal series, *SIAM J. Math. Anal.*, **5,** 489, 1974.

10 Lewin, L., The solution of singular integral equations over a multiple interval and applications to multiple diaphragms in rectangular waveguides, *SIAM J. Appl. Math.*, **16,** 417, 1968.

11 Longmuir, G. J. and Tweed, J., The longitudinal shear problem for an array of cracks at the edge of a circular hole in an infinite elastic solid, *J. Eng. Math.*, **10,** 305, 1976.

12 Longmuir, G. J. and Tweed, J., The elastic problem for a half-plane with a line crack in its interior, *Appl. Eng. Sci.*, **4,** 333, 1976.

13 Lowengrub, M., The solution of certain simultaneous pairs of dual integral equations, *Glasgow Math. J.*, **9,** 92, 1968.

14 Lowengrub, M., A pair of coplanar cracks at the interface of two bonded dissimilar elastic half-planes, *Int. J. Eng. Sci.*, **13,** 731, 1975.

15 Lowengrub, M., Systems of Abel-type integral equations, *see* pp. 277–295 of [8].

16 Lowengrub, M. and Sneddon, I. N., The effect of internal pressure on a penny-shaped crack at the interface of two bonded dissimilar elastic half-spaces, *Int. J. Eng. Sci.*, **12,** 387, 1974.

17 Lowndes, J. S., Simultaneous dual integral equations, *Glasgow Math. J.*, **14,** 73, 1973.

18 Nasim, C. and Sneddon, I. N., A general procedure for deriving solutions of dual integral equations, *J. Eng. Math.*, **12,** 115, 1978.

19 Naylor, D., On a Mellin type integral transform, *J. Math. Mech.*, **12,** 265, 1963.

20 Orr, W. McF., Extensions of Fourier's and the Bessel–Fourier theorems, *Proc. Roy. Irish Acad.*, **27,** 205, 1909; **29,** 10, 1911.

21 Palócz, I., The integral equation approach to currents and fields in plane parallel transmission lines, *J. Math. Mech.*, **15,** 541, 1966.

22 Parlas, S. C. and Michalopoulos, C. D., Axisymmetric contact problem for an elastic half-space with a cylindrical hole, *Int. J. Eng. Sci.*, **10,** 699, 1972.

23 Rooke, D. P., and Tweed, J., The stress intensity factors of a radial crack in a finite rotating disc, *Int. J. Eng. Sci.*, **10,** 709, 1972.

24 Rooke, D. P. and Tweed, J., The stress intensity factors in a point loaded disc, *Int. J. Eng. Sci.*, **11,** 285, 1973.

25 Rooke, D. P. and Tweed, J., The stress intensity factor of an edge crack in a finite rotating elastic disc, *Int. J. Eng. Sci.*, **11,** 279, 1973.

26 Ross, B. (ed), *Fractional Calculus and its Applications*, Lecture Notes in Mathematics No. 457, Springer, Berlin, 1975.

27 Rudnitskii, V. and Grilitski, V., Torsion of a double layer isotropic body by an annular punch, *Arch. Mech.*, **23,** 651, 1971.

28 Scott, R. A. and Miklowitz, J., Transient compressional waves in an infinite elastic space with a circular cylindrical cavity, *J. Appl. Mech.*, **31,** 627, 1964.

29 Sneddon, I. N., The stress intensity factor for an elliptical crack, *Int. J. Eng. Sci.*, to appear.

30 Sneddon, I. N., *Fourier Transforms*, McGraw-Hill, New York, 1951.

31 Sneddon, I. N., The use of transform methods in elasticity, Vol. I, North Carolina State University, Technical Report, AFOSR 64-1789, 1964; Vol. II, North Carolina State University Technical Report, AFOSR 65-0875, 1965.

32 Sneddon, I. N., The use in mathematical physics of Erdélyi–Kober operators and of some of their generalizations., pp. 37–79 of [26].

33 Sneddon, I. N., *The Use of Integral Transforms*, McGraw–Hill, New York, 1972.

34 Sneddon, I. N., *Metoda transformacji calkowych w mieszanych zagadnieniach brzegowych klasycznej teorii sprężystości*, Polsk. Akad. Nauk, Warszawa, 1974.

35 Sneddon, I. N., *The Linear Theory of Thermoelasticity*, Springer, Wien, 1974.

36 Sneddon, I. N. and Lowengrub, M., *Crack Problems in the Classical Theory of Elasticity*, Wiley, New York, 1969.

37 Srivastav, R. P., A pair of dual integral equations involving Bessel functions of the first and second kinds, *Proc. Edinburgh Math. Soc.*, **14,** 149, 1964.

38 Srivastav, R. P., An axisymmetric mixed boundary value problem for a half-space with a cylindrical cavity, *J. Math. Mech.*, **13,** 385, 1964.

39 Srivastav, R. P., Dual integral equations with trigonometric kernels and tempered distributions, *SIAM J. Math. Anal.*, **3,** 413, 1972.

40 Srivastav, R. P. and Lee, D., Axisymmetric crack problems for media with cylindrical cavities, *Int. J. Eng. Sci.*, **10,** 217, 1972.

41 Srivastav, R. P. and Narain, P., Stress distribution due to a pressurized exterior crack in an infinite isotropic elastic medium with a coaxial cylindrical cavity, *Int. J. Eng. Sci.*, **4,** 689, 1966.

42 Titchmarsh, E. C., Weber's integral equation, *Proc. London Math. Soc., ii*, **22,** 15, 1923.

43 Tricomi, F. G., The airfoil equation for a double interval, *ZAMP*, **2,** 402, 1951.

44 Tweed, J., and Rooke, D. P., The torsion of a circular cylinder containing a symmetric array of edge cracks, *Int. J. Eng. Sci.*, **10,** 801, 1972.

45 Tweed, J., The solution of certain triple integral equations involving inverse Mellin transforms, *Glasgow Math. J.*, **14,** 65, 1973.

46 Tweed, J., The solution of a torsion problem by finite Mellin transform techniques, *J. Eng. Math.*, **7,** 97, 1973.

47 Tweed, J., Some dual integral equations with an application in the theory of elasticity, *J. Elasticity*, **2,** 351, 1972.

48 Tweed, J., The solution of some dual equations with an application in the theory of elasticity, *J. Eng. Math.*, **7,** 273, 1973.

49 Tweed, J., Some dual integral equations involving inverse finite Mellin transforms, *Glasgow Math. J.*, **14,** 179, 1973.

50 Tweed, J., Some triple equations involving inverse finite Mellin transforms, *Proc. Edinburgh Math. Soc.*, **18,** 317, 1973.

51 Tweed, J., Das, S. C. and Rooke, D. P., The stress intensity factors of a radial crack in a finite disk, *Int. J. Eng. Sci.*, **10,** 323, 1972.

52 Tweed, J. and Longmuir, G. J., The torsion problem for a circular cylinder with radial edge cracks, *J. Eng. Math.*, **9,** 117, 1975.

53 Tweed, J. and Longmuir, G. J., The plane strain problem for an infinite annular solid with two pairs of racks at its inner boundary, *Letters Appl. Eng. Sci.*, **4,** 269, 1976.

54 Tweed, J. and Rooke, D. P., The stress intensity factor of an edge crack in a finite elastic disc, *Int. J. Eng. Sci.*, **11,** 65, 1973.

55 Tweed, J. and Rooke, D. P., The distribution of stream near the tip of a radial crack at the edge of a circular hole, *Int. J. Eng. Sci.*, **11,** 1185, 1973.

56 Tweed, J. and Rooke, D. P., The stress intensity factors of a star-shaped array of cracks in an infinite elastic solid, *Int. J. Eng. Sci.*, **12,** 423, 1974.

57 Tweed, J. and Rooke, D. P., The elastic problem for an infinite solid containing a circular hole with a pair of radial edge cracks of different lengths, *Int. J. Eng. Sci.*, **14,** 925, 1976.

58 Уфлянд, Я. С., Интегральные преобразования в задачах теории упругости, Издат. Акад. Наук СССР, Москва, 1963; English translation, Uflyand, Ya. S., *Integral Transforms and Problems in the Theory of Elasticity*

(W. J. A. Whyte, trans.), North Carolina State University Applied Mathematics Research Group, Raleigh, N.C., 1965.

59 Ungar, A., An operator related to the inverse Laplace transform, *SIAM J. Math. Anal.*, **5,** 367, 1974.

60 Ungar, A., Wave generation in an elastic half-space by a normal point load moving uniformly over the free surface, *Int. J. Eng. Sci.*, **14,** 935, 1976.

61 Ungar, A., The propagation of elastic waves from moving normal point loads in layered media, *Pure Appl. Geophysics*, **114,** 845, 1976.

62 Ungar, A., The differential transform technique for solving problems of wave propagation, in *IUTAM Symposium on Modern Problems in Elastic Wave Propagation*, Northwestern University, Ill., USA, 1977.

63 Ungar, A. and Robinson, N. I., The application of the differential transform technique to point source problems of elasticity, *Int. J. Eng. Sci.*, **15,** 157, 1977.

64 Walton, J. R., The question of uniqueness for dual integral equations of Titchmarsh, *Proc. Roy. Soc. Edinburgh A*, **76,** 267, 1977.

65 Watson, G. N., *A Treatise on the Theory of Bessel Functions*, 2nd ed., Cambridge University Press, London, 1944.

66 Weber, H., Über eine Darstellung willkurlicher Funktionen durch Bessel'sche Funktionen, *Math. Ann.*, **6,** 146, 1873.

67 Weber, H., Über die Besselschen Funktionen und ihre Anwendung an die Theorie der Elektrischen Strome, *J. f. Math.*, **75,** 75, 1873.

Professor Ian N. Sneddon
Department of Mathematics
University of Glasgow
Glasgow G12 8QQ
Great Britain

Cz. Woźniak

Non-standard analysis and its application to mechanics

Introduction

In the last few years a new approach to some branches of mathematics has been developed by Robinson [1], Luxemburg [2] and others. This approach is referred to as non-standard analysis. The purpose of the paper is to apply the ideas of non-standard analysis to the Newtonian mechanics of finite systems of particles (of discrete systems) and to deduce some consequences of this non-standard theory of discrete systems.

The paper is divided into three sections. They concern, successively, non-standard analysis, discrete mechanics and the interaction between the concepts of non-standard analysis and those of discrete mechanics. In sections 1 and 2, we describe the known concepts needed for subsequent developments. The foundations and the interesting special case of the non-standard theory of discrete systems are given in Section 3. Some general conclusions concerning the applications of the non-standard approach to discrete mechanics end the paper.

The applications of non-standard analysis to mechanics expounded previously (*see*, for example, Robinson [1]) were confined to the special problems of continuum mechanics. Here we deal with particle mechanics, but we show that certain non-standard discrete systems can be described by standard equations of continuum mechanics and we prove that such passage from non-standard discrete mechanics to continuum mechanics is unique.

1 Tools from non-standard analysis

1.1 Some concepts of mathematical logic

To describe our main analytical tool, which is the notion of enlargement of a certain structure, we have to introduce first the concepts of the higher-order structure, the formal language, the L-structure and the non-standard model of L-structure. All statements expounded below are given without proof; the particulars can be found in Robinson [1] or Luxemburg [2].

Before we introduce the concept of the higher-order structure, we have to define the notion of the class of types. There is a set T of elements τ called types which are defined inductively: (1°) zero is a type, $0 \in T$; (2°) if $\tau_1, \ldots, \tau_n$ are types, $\tau_1, \ldots, \tau_n \in T$, then $(\tau_1, \ldots, \tau_n)$ is a type, $(\tau_1, \ldots, \tau_n) \in T$; (3°) the class of types T is the smallest class satisfying conditions (1°) and (2°).

By the higher-order structure (called the structure below) we mean a set $\mathcal{M} = (A_\tau)_{\tau \in T}$ indexed in T such that:

1. the set A_0 is a given set, the elements of which are called individuals or relations of the type 0;
2. the set A_τ for each $\tau = (0, \ldots, 0)$ is a given set of n-ary relations, $n \geqslant 1$, in the product $A_0 \times \ldots \times A_0$, the elements of which are called relations of the type $\tau = (0, \ldots, 0)$ (relations of the first order);
3. the set A_τ for each $\tau = (\tau_1, \ldots, \tau_n)$, $n \geqslant 1$, is a given set of n-ary relations in the product $A_{\tau_1} \times \ldots \times A_{\tau_n}$ (provided that $A_{\tau_1}, \ldots, A_{\tau_n}$ have been defined previously), the elements of which are said to be relations of the type $\tau = (\tau_1, \ldots, \tau_n)$(relations of higher order or relations between relations).

For example, $A_{(0)}$ is the given set of subsets of A_0; $A_{(\tau)}$ is a given set of subsets of relations of type τ (i.e., the set of subsets of A_τ).

The structure $\bar{\mathcal{M}} = (\bar{A}_\tau)_{\tau \in T}$ is called full if $\bar{A}_0$ is a given set of individuals and $\bar{A}_\tau$ is the set of all relations of the type τ for each $\tau \in T$. For the full structure, we have $\bar{A}_\tau \in \bar{A}_{(\tau)}$ for each $\tau \in T$.

Note that, for an arbitrary structure $\mathcal{M} = (A_\tau)_{\tau \in T}$, we can define the full structure $\bar{\mathcal{M}} = (\bar{A}_\tau)_{\tau \in T}$, setting $\bar{A}_0 \equiv A_0$. The relations belonging to A_τ will be called internal, while the relations of $\bar{\mathcal{M}}$ belonging to $\bar{A}_\tau - A_\tau$ are said to be external (with respect to the given structure $\mathcal{M}$). Since $\bar{A}_0 - A_0 = \varnothing$, there are no external individuals.

The next basic concept we introduce is that of the formal language. First of all, any formal language L includes a class of so-called atomic symbols such as constants (the set of all constants is assumed to be infinite), variables, relation symbols, type symbols, connectives, brackets and quantifiers. Using certain rules (cf., for example, Robinson [1]) we can construct from atomic symbols the well-formed formulae of L. Each well-formed formula is called a sentence iff every variable contained in it is under the scope of a quantifier. Within the formal language itself, any sentence of L is neither true nor false. In this paper, we shall not discuss the concept of language in a more precise way because it is not necessary to subsequent developments; we are not going to use the formal language in this paper. It must be stressed here that it is only the possibility of formulation of certain statements in the language L that is the important feature of our procedure (cf., Robinson [1], p. 60).

Now, let $\mathcal{M} = (A_\tau)_{\tau \in T}$ be a structure and let us denote by $M \equiv \bigcup A_\tau$ the set of all relations of $\mathcal{M}$. Let there also be given a certain formal language L and let C denote the set of all constants of L. If there is given a one-to-one correspondence $M \to C$ between all the relations of $\mathcal{M}$ and a certain subset of constants of L (i.e., a certain subset of C), then $\mathcal{M}$ will be called an L-structure (cf., Luxemburg [2], p. 22).

Let there be given a certain L-structure. Then a sentence of L is said to be defined in $\mathcal{M}$ iff all the constants contained in it denote relations of $\mathcal{M}$. Each sentence of L defined in $\mathcal{M}$ can be either true or false according to certain criteria which will not be given here (cf., Robinson [1]). The intuitive sense of these criteria is rather clear; roughly speaking, the sentence of L is true in $\mathcal{M}$ (provided the L-structure $\mathcal{M}$ is given) if it represents a certain relation of $\mathcal{M}$ in the formal language L (it is an atomic sentence) or it can be obtained from true atomic sentences by using the connectives, brackets or quantifiers consistent with their meaning (for example, the sentence 'not W' is true in $\mathcal{M}$ iff the sentence 'W' is not true in $\mathcal{M}$, etc.) More generally, the sentence of L is true in $\mathcal{M}$ (the L-structure $\mathcal{M}$ being given) if it represents a certain statement which can be verified by using the relations of $\mathcal{M}$.

1.2 Non-standard models and enlargements

By the non-standard model of the given L-structure, we shall mean such an L-structure $^*\mathcal{M} = (^*A_\tau)_{\tau = T}$ that each sentence of L which is true in $\mathcal{M}$ is also true in $^*\mathcal{M}$. It follows that each relation belonging to $\mathcal{M}$ has its corresponding relation belonging to $^*\mathcal{M}$, both being denoted in L by the same constant. The relations of $^*\mathcal{M}$ denoted in L by symbols which also denote certain relations of $\mathcal{M}$ will be called standard. In what follows we shall denote by A_τ the set of all standard relations belonging to $^*A_\tau$, $A_\tau \subset {}^*A_\tau$. This kind of notation identifies the relations of A_τ and their sets with the corresponding standard relations of $^*A_\tau$ and their sets, respectively. Moreover, we observe that each statement concerning the L-structure $\mathcal{M}$ also concerns the L-structure $^*\mathcal{M}$, provided it can be formulated in the formal language L.

From now on, we shall consider only full L-structures $\bar{\mathcal{M}} = (\bar{A}_\tau)_{\tau \in T}$ in which the set $\bar{A}_0$ of all individuals is infinite. Then it can be proved (cf., Robinson [1]) that there exists a non-standard model $^*\mathcal{M} = (^*A_\tau)_{\tau \in T}$ of $\bar{\mathcal{M}} = (\bar{A}_\tau)_{\tau \in T}$, called enlargement, which is the proper extension of $\bar{\mathcal{M}}$, i.e., $^*A_\tau - \bar{A}_\tau \neq \varnothing$ for each $\tau \in T$. The enlargement $^*\mathcal{M} = (^*A_\tau)_{\tau \in T}$ of the full L-structure $\bar{\mathcal{M}} = (\bar{A}_\tau)_{\tau \in T}$ need not be full. Denoting by $^*\bar{\mathcal{M}} = (^*\bar{A}_\tau)_{\tau \in T}$ the full structure based on the set $^*\bar{A}_0 \equiv {}^*A_0$ of individuals, we shall deal with the internal relations (belonging to $^*M \equiv \cup {}^*A_\tau$) as well as with the external relations (belonging to $\cup({}^*\bar{A}_\tau - {}^*A_\tau)$). Each standard relation is internal. Note that since all sentences which are true in $\bar{\mathcal{M}}$ are also true in $^*\mathcal{M}$, then in all statements 'transferred' in this way from $\bar{\mathcal{M}}$ to $^*\mathcal{M}$ we shall deal only with internal relations. No statement in which an external relation is involved can be obtained in this fashion. It must be stressed, however, that the external relations will appear in our consideration in the natural way, cf., below.

The existence of the enlargement $^*\mathcal{M} = ({}^*A_\tau)_{\tau \in T}$ of the given structure $\bar{\mathcal{M}} = (\bar{A}_\tau)_{\tau \in T}$ is all we need for the development of the non-standard theory of certain known theory which can be expressed in terms of the L-structure $\bar{\mathcal{M}}$.

1.3 Non-standard model of analysis

Following Robinson [1] we shall now outline an example of the enlargement which is called the non-standard model of analysis and which includes the non-standard theory of real numbers.

Let $\bar{\mathcal{M}} = (\bar{A}_\tau)_{\tau \in T}$ be a full structure in which $\bar{A}_0 \equiv R$, i.e., the set of all individuals (relations of the type 0) coincides with the set of all real numbers. Then there exists the L-structure $^*\mathcal{M} = ({}^*A_\tau)_{\tau \in T}$ which is the enlargement of the L-structure $\bar{\mathcal{M}} = (\bar{A}_\tau)_{\tau \in T}$. Let us denote $^*R \equiv {}^*A_0$ and let us refer to individuals of $^*\mathcal{M}$ as real numbers. The set R in $\bar{\mathcal{M}}$ and the set $^*R$ in $^*\mathcal{M}$ are denoted in L by the same constant as certain relation of the type (0); that is why the set $^*R$ is said to be the set of real numbers. At the same time, we shall put $R \subset {}^*R$, referring to elements of R in $^*\mathcal{M}$ as standard real numbers. Each such number is denoted in L by the same constant as the corresponding well-known real number in $\bar{\mathcal{M}}$. Moreover, the relations of order, addition, multiplication, division, etc., are defined in $^*\mathcal{M}$ because they are defined in $\bar{\mathcal{M}}$ (division by zero is not allowed in $^*\mathcal{M}$ because it is not allowed in $\bar{\mathcal{M}}$). Following Robinson [1], we can prove that to $^*R - R$ belong infinite real numbers (positive and negative) which are either greater or smaller than any standard real number. Thus, we conclude that to $^*R - R$ there belong also the inverses of infinite numbers which are called infinitesimal numbers. In the set of infinitesimal numbers, we include also the standard number zero. By forming the sums of standard numbers and infinitesimal numbers, we obtain the set of what will be called finite numbers (or near-standard numbers). We prove that each finite number has a uniquely defined standard part; the standard part of the finite real number a will be denoted by ^0a. At the same time, each finite number a can be uniquely represented in the form of $a = {}^0a + \mathrm{d}a$, where $\mathrm{d}a$ is the infinitesimal number.

The set of natural numbers as certain relation of the type (0) in $\mathcal{M}$ is denoted by a certain constant of L. The same constant has to denote a certain subset $^*N$ of $^*R$ (certain relation of the type (0) in $^*\mathcal{M}$). The elements of $^*N$ will be called natural numbers. Any natural number $n \in {^*N}$ which is denoted in L by the same constant as the well-known natural number in $\bar{\mathcal{M}}$ is said to be the standard natural number. The set of all standard natural numbers will be denoted by N, where $N \subset {^*N}$. We can prove (cf., Robinson [1], p. 50) that all natural numbers belonging to $^*N - N$ are infinite (greater than any standard real number) and that there does not exist the smallest infinite natural number.

In the foregoing considerations, we have dealt with different new relations, for example, relations of the type (0) such as R, N, $^*R - R$, $^*N - N$. The question arises whether these relations belong to $^*\mathcal{M}$, i.e., whether they are internal relations or not. It can be proved, cf. Robinson [1], that all relations mentioned above are external. As the relations which are internal but not standard (i.e., which belong to $^*\mathcal{M}$ but cannot be denoted by any constant which denotes certain relation of $\bar{\mathcal{M}}$), we can mention an arbitrary non-standard real number (it is a relation of the type 0): the subset $(a, b) \subset {^*R}$, where either a or b are not standard real numbers (it is a relation of the type (0)) and many others.

All known concepts of the analysis, such as continuity, differentiability, etc., that can be expressed in the formal language L are also meaningful in the non-standard model of analysis, i.e., in the L-structure $^*\mathcal{M} = (^*A_\tau)_{\tau \in T}$, $^*A_0 \equiv {^*R}$. The form of the suitable definitions and theorems related to the L-structure $^*\mathcal{M}$ is analogous to the corresponding definitions and theorems related to the L-structure $\bar{\mathcal{M}}$, provided no external relation is involved in the statements related to $^*\mathcal{M}$. It means that all statements concerning the L-structure $^*\mathcal{M}$ which are transferred from the L-structure $\bar{\mathcal{M}}$ can be formulated exclusively in terms of the relations belonging to $^*\mathcal{M}$, i.e., in terms of internal relations. Thus, the statements 'for all' and 'there exist' concerning $\bar{\mathcal{M}}$ have to be reformulated to the form 'for all internal' and 'there exist internal', respectively, when these statements concern $^*\mathcal{M}$, cf. Robinson [1], p. 49. For example, the well-known definition of the sum a of infinite sequence $a_0, a_1, \ldots$ (whenever that sum exists) within the non-standard model of analysis will be given by the following statement: 'the real number a is the sum of the infinite internal sequence $\{a_n\}$ of real numbers $a = \sum_{n=0}^{\infty} a_n$ if, for every positive ε in $^*R$, there exists a natural number ν in $^*N$ such that $|\sum_{n=0}^{k} a_n - a| < \varepsilon$ for all $k > \nu$'. Because all real numbers in $^*\mathcal{M}$ (all individuals of $^*\mathcal{M}$) are internal, we did not use the term 'internal real number' in the foregoing definition.

Let $f(x)$ be a standard real-valued function defined on $^*R$; for each $x \in R$, the corresponding value of f also belongs to R. In order that the standard real number $f'(x_0)$ be the derivative of f at the standard point

$x_0 \in {}^*R$ (according to the known definition transferred from $\bar{\mathcal{M}}$ to ${}^*\mathcal{M}$) it is necessary and sufficient that

$$\frac{f(x_0+h)-f(x_0)}{h} \simeq f'(x_0) \quad \text{or} \quad {}^0\left[\frac{f(x_0+h)-f(x_0)}{h}\right] = f'(x_0) \tag{1.1}$$

for all infinitesimal real numbers $h \neq 0$, where ${}^0[\,]$ denotes the standard part of the value in the brackets, provided the latter is finite (for proof, cf., Robinson [1], p. 68), and where $a \simeq b$ denotes that the difference $a-b$ is an arbitrary infinitesimal real number. In an analogous way, we can define the higher-order derivatives as well as partial derivatives of the standard functions.

Let a, b, where $a<b$, be two standard real numbers. Let us put $x_i = a + i(b-a)/\omega$, $i = 0, 1, \ldots, \omega$, where ω is a natural number (finite or infinite). Then, the sequence $x_0 = a$, x_i, $x_2, \ldots, x_\omega = b$ will represent a certain finite internal partition of the interval $(a, b) \subset {}^*R$. This internal partition is not standard if ω is a non-standard (infinite) natural number. Moreover, let $\xi_1, \xi_2, \ldots, \xi_\omega$ be an internal sequence such that $x_i \leqslant \xi_{i+1} < x_{i+1}$, $i = 0, 1, \ldots, \omega - 1$. Then, for any internal partition of $(a, b) \subset {}^*R$ in which ω is an infinite natural number, and for any internal sequence $(\xi_1, \ldots, \xi_\omega)$ defined above, we have

$$\sum_{i=1}^{\omega} f(\xi_i)(x_i - x_{i-1}) \simeq \int_a^b f(x)\,\mathrm{d}x \quad \text{or} \quad {}^0\left[\sum_{i=1}^{\omega} f(\xi_i)(x_i - x_{i-1})\right] = \int_a^b f(x)\,\mathrm{d}x, \tag{1.2}$$

provided that the integral exists (for proof, cf., Robinson [1] p. 72). In the same way, we can define the surface and volume integrals of standard functions in standard regions.

2 Newtonian discrete mechanics

2.1 Basic statements

Let us assume that in the Gallilean space-time there is fixed one inertial coordinate system. Then each point of the space-time is represented by the sequence (x^1, x^2, x^3, t), where the triples $x \equiv (x^1, x^2, x^3)$ of real numbers represent the points of the physical space and the real numbers t represent the time instants. Let us also introduce the space of forces, elements of which will be represented by the triples $f \equiv (f^1, f^2, f^3)$ of real numbers.

Now let us define an infinite set D of elements d and to each $d \in D$ let us assign the sequences of n positive real numbers $\{m_\mu\}$; real-valued functions $\{x_\mu(t)\}$ having derivatives up to second order and real-valued functions $\{f_\mu(t)\}$, $\mu = 1, \ldots, n$, are defined for each time instant t and interrelated by means of the condition

$$m_\mu \ddot{\mathbf{x}}_\mu(t) = \mathbf{f}_\mu(t), \qquad \mu = 1, \ldots, n. \tag{2.1}$$

Moreover, let us assign to each $d \in D$ the system of relations

$$\left.\begin{aligned}
&\mathbf{f}_\mu(t) = \mathbf{h}_\mu(t, \mathbf{x}_1, \ldots, \mathbf{x}_n, \dot{\mathbf{x}}_1, \ldots, \dot{\mathbf{x}}_n) \\
&\qquad + g_\mu(t, \rho(\mathbf{x}_\mu, \mathbf{x}_1), \ldots, \rho(\mathbf{x}_\mu, \mathbf{x}_n)) + \mathbf{r}_\mu(t), \qquad \mu = 1, \ldots, s \\
&\alpha_\pi(t, \mathbf{x}_1, \ldots, \mathbf{x}_n, \dot{\mathbf{x}}_1, \ldots, \dot{\mathbf{x}}_n) = 0, \qquad \pi = 1, \ldots, s < 3n, \\
&\sum_{\mu=1}^{h} \mathbf{r}_\mu \cdot \delta\mathbf{x}_\mu = 0 \text{ for each } \delta\mathbf{x}_\mu \text{ such that} \\
&\sum_{\mu=1}^{n} \frac{\partial \alpha_\pi}{\partial \dot{\mathbf{x}}_\mu} \cdot \delta\mathbf{x}_\mu = 0, \qquad \pi = 1, \ldots, s,
\end{aligned}\right\} \tag{2.2}$$

where $\mathbf{h}_\mu(\,)$, $\mathbf{g}_\mu(\,)$, $\mathbf{r}_\mu(\,)$, $\mu = 1, \ldots, n$ and $\alpha_\pi(\,)$, $\pi = 1, \ldots, s$, also satisfy certain conditions of regularity as well as the conditions

$$\sum_{\mu=1}^{n} \mathbf{g}_\mu(t, \cdot) = 0, \qquad \sum_{\mu=1}^{n} \mathbf{x}_\mu(t) \times \mathbf{g}_\mu(t, \cdot) = 0 \tag{2.3}$$

and where $\rho(\mathbf{x}, \mathbf{y})$ is a distance between points $\mathbf{x}$ and $\mathbf{y}$ in the physical space.†

If the foregoing conditions hold, then element $d \in D$ is said to be the discrete system or the system of material particles. We shall also assume that to each system of functions interrelated by Eqns (2.1)–(2.3) there is assigned one and only one element of D; the set D will then be called the set of all discrete systems. In the set D, we shall also define the relation of identity, putting $d_1 \equiv d_2$ iff all sequences of relations in Eqns (2.1)–(2.3) for d_1 can be obtained from the corresponding relations for d_2 by a certain permutation of the sequences $\{m_\mu\}$, $\{\mathbf{x}_\mu(t)\}, \ldots$, $\mu = 1, \ldots, n$, and by a certain permutation of the sequence $\{\alpha_\pi(\,)\}$, $\pi = 1, \ldots, s$.

The physical interpretation of Eqns (2.1)–(2.3) is obvious; they represent Newtonian mechanics of the sequence of n material particles referred to the fixed inertial coordinate system. The symbols m_μ, $\mathbf{x}_\mu(t)$, $\mathbf{f}_\mu(t)$, $\mathbf{h}_\mu(\,)$, $\mathbf{g}_\mu(\,)$, $\mathbf{r}_\mu(t)$, $\mu = 1, \ldots, n$, denote the mass, the motion, the resultant force, the external force, the force of the interparticle interaction and the reaction force due to the aholonomic constraints $\alpha_\pi(\,) = 0$, respectively, assigned to the μth particle of the given discrete system. To make our notations simpler, we have omitted from Eqns (2.1)–(2.3) the argument d as an argument of any function in these equations.

† All small bold-face letters denote the triples of real numbers or real-valued functions, while the italic-face letters denote the real numbers or real-valued functions. The dot between the bold face letters denotes the scalar product and the cross the vector product.

2.2 Formal theory

Now let there be given a full L-structure $\bar{\mathcal{M}} = (\bar{A}_\tau)_{\tau \in T}$, such that $R \subset \bar{A}_0$ and $D \subset \bar{A}_0$, the sets R, D being the sets of real numbers and discrete systems, respectively. The possibility of realization of all statements given in section 2.1 being within the formal language L is a very important feature of our approach. Thus, any statement in which the discrete system (the element of D) is involved and which is a consequence of axioms of the discrete mechanics formulated above, can be expressed in the form of a sentence of L which is true in the structure $\bar{\mathcal{M}} = (\bar{A}_\tau)_{\tau \in T}$, $R \subset \bar{A}_0$, $D \subset A_0$. The set of all such sentences will be referred to as the Newtonian discrete mechanics.

3 Non-standard theory of discrete systems

3.1 Non-standard model of discrete mechanics

Let the L-structure $^*\mathcal{M} = (^*A_\tau)_{\tau \in T}$ be an enlargement of the L-structure $\bar{\mathcal{M}} = (\bar{A}_\tau)_{\tau \in T}$ introduced in section 2.2. From certain theorems concerning the so-called partial structures (cf., Robinson [1], p. 44), we obtain $^*R \subset {}^*A_0$, $^*D \subset {}^*A_0$. At the same time, we have $R \subset {}^*R$ and $D \subset {}^*D$, cf., section 1. The elements of $^*D$ will be called discrete systems and the elements of D will be referred to as standard discrete systems. Because each statement of the Newtonian discrete mechanics (in its formal form, cf., section. 2.2) is true in the structure $\bar{\mathcal{M}}$, then it is also true in the structure $^*\mathcal{M}$, according to the definition of the enlargement $^*\mathcal{M}$ as the non-standard model of $\bar{\mathcal{M}}$. Thus, to each discrete system $d \in {}^*D$ we assign the internal sequences of positive real numbers m_μ and vector or scalar functions $\mathbf{x}_\mu(\,)$, $\mathbf{f}_\mu(\,)$, $\mathbf{h}_\mu(\,)$, $\mathbf{g}_\mu(\,)$, $\mathbf{r}_\mu(\,)$, $\alpha_\pi(\,)$, interrelated by Eqns (2.1)–(2.3) and satisfying the known conditions of regularity. All relations (2.1)–(2.3) are treated now within the non-standard model of analysis or, to more exact, as the relations belonging to the structure $^*\mathcal{M} = (^*A_\tau)_{\tau \in T}$, $^*D \subset {}^*A_0$, $^*R \subset {}^*A_0$. At the same time, to any system of internal relations of the form (2.1)–(2.3), there is assigned one element of $^*D$ (one discrete system) and in the set of all discrete systems $^*D$ the relation of identity is defined in the same way as before.

The set of statements given in section 2.1, together with all their consequences, expressed in the form of a set of sentences of the formal language L which are true in the L-structure $^*\mathcal{M} = (^*A_\tau)_{\tau \in T}$ (in the enlargement of the L-structure introduced in section 2.2), will be called the non-standard model of the Newtonian discrete mechanics.

The extension of classical discrete mechanics developed here enables us to construct new discrete systems (non-standard discrete systems, i.e., these belonging to the non-empty set $^*D - D$) which cannot be described by classical mechanics. They are the systems for which at least one of the relations (2.1)–(2.2) is non-standard. As an example of a non-standard discrete system, we can take the system in which n is an infinite natural number, in which masses m_μ are infinitesimal, etc. An example of such system will be detailed in the last section. However, we do not assume that each non-standard discrete system must have a physical meaning, and describe the motion of a certain real material system.

3.2 Topologies in the physical space

Within the non-standard model of discrete mechanics, we deal with the physical space $^*E = (^*R)^3$ in which two kinds of topologies can be defined, cf., Robinson [1] p. 106. First, we shall define the Q-topology by specifying as the base the set of all open Q-balls B, where $B \equiv \{\mathbf{y} \mid \rho(\mathbf{x}, \mathbf{y}) < r\}$, point $\mathbf{x} \in {}^*E$ is a centre and $r \in {}^*R$, $r > 0$, is the radius of the ball B. Second, we shall define the S-topology defining the set of all open S-balls, $S \equiv \{\mathbf{y} \mid {}^\circ\rho(\mathbf{x}, \mathbf{y}) < {}^\circ r\}$, where ${}^\circ\rho$, ${}^\circ r$ denote the standard parts of ρ, r, respectively. It follows that we can introduce functions, defined on certain subsets of $^*E$, which are discrete in the Q-topology (Q-discrete functions) but they can be continuous in the S-topology (S-continuous functions). For the details (which are not necessary for our further considerations) the reader is referred to Robinson [1].

Within the non-standard model of discrete mechanics, we can uniquely define the local interactions between material particles even if the collisions of particles are neglected (we did not introduce the concept of separate particle before, but it can be easily done). Such interactions cannot be defined in classical discrete mechanics. The interactions between the μth particle and the discrete system $d \in {}^*D$ will be called local (or S-local) if the function $\mathbf{g}_\mu(t, \rho(\mathbf{x}_\mu, \mathbf{x}_1), \ldots, \rho(\mathbf{x}_\mu, \mathbf{x}_n))$ takes a value different from zero only if at least one of the distances is infinitesimal: ${}^\circ\rho(\mathbf{x}_\mu, \mathbf{x}_\nu) = 0$ and $\rho(\mathbf{x}_\mu, \mathbf{x}_\nu) > 0$, and if the value of the function $\mathbf{g}_\mu(t, \rho(\mathbf{x}_\mu, \mathbf{x}_1), \ldots, \rho(\mathbf{x}_\mu, \mathbf{x}_n))$ does not depend on the distances $\rho(\mathbf{x}_\mu, \mathbf{x}_\nu)$, which are not infinitesimal. The local interactions can only take place in non-standard discrete systems; moreover, this concept is closely related to the existence of two kinds of topologies in the physical space. For the details, the reader is referred to Woźniak [7].

3.3 From discrete to continuum mechanics

We shall now see that within the non-standard theory of discrete systems we can construct, in an unambiguous manner, certain material continua.

The term 'continuum' will be understood here in the sense of the S-topology in the physical space $^*E$.

Let $^*\Omega$ be a certain standard regular region in the physical space $^*E$, i.e., the region which is denoted in the formal language L by the same constant as a certain regular region Ω in E. It follows that all points $\mathbf{X} \in {}^*\Omega$ are finite and have uniquely defined standard parts $^\circ\mathbf{X} \in \Omega \subset {}^*\Omega$. Let us introduce the internal fine partition of $^*\Omega$ into volume elements with volumes $dv_R \equiv dX^1\, dX^2\, dX^3$, and centres $\mathbf{X}_A$, $A = 1, \ldots, \omega$, where ω is an infinite natural number. The set of all points $\mathbf{X}_A$, $A = 1, \ldots, \omega$, is Q-discrete, i.e., for each $\mathbf{X}_A$ there exists a Q-ball with the centre $\mathbf{X}_A$ which does not contain any other point of this set. The set of points under consideration will be also called S-continuous in the region $^*\Omega$. This means that each S-ball in the region $^*\Omega$ contains points of this set. Let us assign to each $\mathbf{X}_A$ certain discrete subsystem of n_A material particles with masses $m_\gamma(\mathbf{X}_A)$, $\gamma = 1, \ldots, n_A$, and let $\mathbf{X}_A + \mathbf{d}_\gamma(\mathbf{X}_A)$, $\gamma = 1, \ldots, n_A$, where $\mathbf{d}_\gamma(\mathbf{X}_A)$ are fixed infinitesimal vectors, be a reference configuration of an arbitrary discrete subsystem. Each subsystem under consideration will be called the material grain; we assume that the correspondence between grains and points $\mathbf{X}_A$, $A = 1, \ldots, \omega$, is one to one. Moreover, let us also assume that each grain undergoes only homogeneous deformations, i.e., motion of each grain is given by $\mathbf{X} = \boldsymbol{\psi}(\mathbf{X}_A, t) + \mathbf{F}(\mathbf{X}_A, t)\mathbf{d}_\gamma(\mathbf{X}_A)$, where $\boldsymbol{\psi} \equiv (\psi^k(\mathbf{X}_A, t))$, $\mathbf{F} \equiv (F^k_{\cdot\alpha}(\mathbf{X}_A, t))$ are internal vector and matrix functions, respectively, assigned to each grain.† We also assume that $\sum_\gamma m_\gamma(\mathbf{X}_A)\mathbf{d}_\gamma(\mathbf{X}_A) = 0$ for $A = 1, \ldots, \omega$ and that $\det \mathbf{F}(\mathbf{X}_A, t) > 0$. The interparticle interactions $\mathbf{g}_\mu(t, \cdot)$ will be defined for each grain independently and have the potential $\pi(\mathbf{X}_A, |\mathbf{F}(\mathbf{X}_A, t)\mathbf{d}_\gamma(\mathbf{X}_A)|)$.

To describe the discrete system under consideration, i.e., the discrete system made of all grains, let us also denote by $\mathbf{h}_\gamma(\mathbf{X}_A, \cdot)$ the external force acting at the γth particle of the Ath grain, $\gamma = 1, \ldots, n_A$, $A = 1, \ldots, \omega$, and let us assume that the motion of different grains is restricted by the holonomic constraints

$$\gamma_s(t, \boldsymbol{\psi}(\mathbf{X}_1, t), \ldots, \boldsymbol{\psi}(\mathbf{X}_\omega, t), \mathbf{F}(\mathbf{X}_1, t), \ldots, \mathbf{F}(\mathbf{X}_\omega, t)) = 0 \qquad s = 1, \ldots, S, \quad (3.1)$$

where $\gamma_s(\cdot)$ are differentiable internal functions and S is a finite or infinite natural number. The reaction force due exclusively to the constraints (3.1) and acting at the γth particle of the Ath grain will be denoted by $\bar{\mathbf{r}}_\gamma(\mathbf{X}_A, t)$; we do not introduce a seperate notation for the reaction force maintaining the homogeneous deformations of grains.

Now, from the general relations (2.1), (2.2) and from the foregoing assumptions, we can obtain the system of governing relations for the

† The capital bold-face letters denote 3×3 matrices; indices k, α run over the sequence 1, 2, 3.

discrete system under consideration. The calculations are simple and will not be given here, cf., Woźniak [4, 5]. To write down the basic system of relations we shall first define the following internal functions which describe each grain:

$$\left.\begin{aligned}
&\rho_R(\mathbf{X}_A) \equiv \sum_{\gamma=1}^{n_A} m_\gamma(\mathbf{X}_A)/\mathrm{d}v_R, \qquad \mathbf{b}_R(\mathbf{X}_A, \cdot) \equiv \sum_{\gamma=1}^{n_A} \mathbf{h}_\gamma(\mathbf{X}_A, \cdot)/\mathrm{d}v_R,\\
&\mathbf{d}_R(\mathbf{X}_A, t) \equiv \sum_{\gamma=1}^{n_A} \bar{\mathbf{r}}_\gamma(\mathbf{X}_A, t)/\mathrm{d}v_R,\\
&\qquad\qquad \mathbf{T}_R(\mathbf{X}_A, t) \equiv \sum_{\gamma=1}^{n_A} \bar{\mathbf{r}}_\gamma(\mathbf{X}_A, t) \otimes \mathbf{d}_\gamma(\mathbf{X}_A)/\mathrm{d}v_R,\\
&\sigma(\mathbf{X}_A, \mathbf{F}^{\mathrm{T}}\mathbf{F}) \equiv \pi(\mathbf{X}_A, \cdot) \Big/ \sum_{\gamma=1}^{n_A} m_\gamma(\mathbf{X}_A),\\
&\qquad\qquad \mathbf{J}_R(\mathbf{X}_A) \equiv \sum_{\gamma=1}^{n_A} m_\gamma(\mathbf{X}_A)\, \mathbf{d}_\gamma(\mathbf{X}_A) \otimes \mathbf{d}_\gamma(\mathbf{X}_A)/\mathrm{d}v_R.
\end{aligned}\right\} \quad (3.2)$$

Using (3.2) we obtain, finally,

$$\rho_R(\mathbf{X}_A)\ddot{\boldsymbol{\psi}}(\mathbf{X}_A, t) = \mathbf{b}_R(\mathbf{X}_A, t, \boldsymbol{\psi}(\mathbf{X}_A, t), \dot{\boldsymbol{\psi}}(\mathbf{X}_A, t), \mathbf{F}(\mathbf{X}_A, t), \dot{\mathbf{F}}(\mathbf{X}_A, t)) + \mathbf{d}_R(\mathbf{X}_A, t), \quad (3.3a)$$

$$\mathbf{T}_R(\mathbf{X}_A, t) = \rho_R(\mathbf{X}_A)\frac{\partial\sigma(\mathbf{X}_A, \mathbf{F}^{\mathrm{T}}\mathbf{F})}{\partial\mathbf{F}} + \mathbf{J}_R(\mathbf{X}_A)\ddot{\mathbf{F}}^{\mathrm{T}}(\mathbf{X}_A, t), \qquad A = 1, \ldots, \omega, \quad (3.3b)$$

$$\sum_{A=1}^{\omega} [\mathbf{d}_R(\mathbf{X}_A, t) \cdot \delta\boldsymbol{\psi}(\mathbf{X}_A, t) + \mathbf{T}_R(\mathbf{X}_A, t) \cdot \delta\mathbf{F}(\mathbf{X}_A, t)]\, \mathrm{d}v_R = 0, \quad (3.3c)$$

where the relation (3.3c) has to be satisfied by any $\delta\boldsymbol{\psi}(\mathbf{X}_A, t)$, $\delta\mathbf{F}(\mathbf{X}_A, t)$ consistent with the equations of constraints (3.1).

Equations (3.1)–(3.3) represent the governing relations of a certain class of discrete systems which are non-standard. The subscript 'R' indicates that the corresponding density is related to the reference configuration of the system. Now we confine ourselves to a certain subclass of the non-standard discrete systems under consideration. To this end, we introduce first the concept of the quasi-regular discrete function defined on the set $\{\mathbf{X}_1, \ldots, \mathbf{X}_\omega\} \subset {}^*\Omega$.

Let $f(\mathbf{X}_A)$ be an arbitrary function in Eqns (3.3) defined on the set $\{\mathbf{X}_1, \ldots, \mathbf{X}_\omega\} \subset {}^*\Omega$, all other arguments of this function being fixed. Suppose that $f(\mathbf{X}_A)$ are finite for each $A = 1, \ldots, \omega$, and let $f(\mathbf{X}_A) \simeq f(\mathbf{X}_B)$ for each pair $(\mathbf{X}_A, \mathbf{X}_B)$ such that ${}^0\rho(\mathbf{X}_A, \mathbf{X}_B) = 0$. Then we can define the function $f(\mathbf{X})$, $\mathbf{X} \in \Omega \subset {}^*\Omega$, which is said to be a standard part of $f(\mathbf{X}_A)$ (cf. Robinson [1], p. 115, where the symbol f is replaced by ${}^0f$), putting $f(\mathbf{X}) \equiv f(\mathbf{X}_A)$ for each $\mathbf{X}_A$ such that ${}^0\rho(\mathbf{X}, \mathbf{X}_A) = 0$. The function $f(\mathbf{X})$ defined above can be identified with the function ${}^0f(\mathbf{X})$ defined in the region Ω of the physical space E. Thus, there exists a standard function defined in the region $^*\Omega$ of the space $^*E$, which will be denoted by the same symbol $^0f(\mathbf{X})$, $\mathbf{X} \in {}^*\Omega$. Let the condition $^0f(\mathbf{X}_A) = f(\mathbf{X}_A)$ hold for $A = 1, \ldots, \omega$ and let the function $^0f(\mathbf{X})$ satisfy certain conditions of regularity (in $^*\Omega$). Then the function $f(\mathbf{X}_A)$, $A = 1, \ldots, \omega$, will be called the discrete quasi-regular function. For example, if the standard function $^0f(\mathbf{X})$ is continuous or differentiable, then the non-standard function $f(\mathbf{X}_A)$, $A = 1, \ldots, \omega$, will be called discrete quasi-continuous or discrete quasi-differentiable, respectively. We can also say that the non-standard function $f(\mathbf{X}_A)$ defined on the set $\{\mathbf{X}_1, \ldots, \mathbf{X}_\omega\}$, Q-discrete and S-continuous in $^*\Omega$, $^*\Omega$ being the standard regular region in $^*E$, is called quasi-regular, if there exists a regular standard function $^0f(\mathbf{X})$, $\mathbf{X} \in {}^*\Omega$, such that $f(\mathbf{X}_A) = {}^0f(\mathbf{X}_A)$ for $A = 1, \ldots, \omega$.

Now we investigate the subclass of non-standard discrete systems, governed by Eqns (3.1), (3.3), in which all functions are finite, the standard part of $\mathbf{J}_R$ is equal to zero (cf., Woźniak [5]), and all other functions are sufficiently quasi-smooth discrete functions. Then, taking the standard parts of Eqns (3.3), we obtain

$$\rho_R(\mathbf{X})\ddot{\boldsymbol{\psi}}(\mathbf{X}, t) = \mathbf{b}_R(\mathbf{X}, t, \boldsymbol{\psi}(\mathbf{X}, t), \dot{\boldsymbol{\psi}}(\mathbf{X}, t), \mathbf{F}(\mathbf{X}, t), \dot{\mathbf{F}}(\mathbf{X}, t)) + \mathbf{d}_R(\mathbf{X}, t), \tag{3.4a}$$

$$\mathbf{T}_R(\mathbf{X}, t) = \rho_R(\mathbf{X}) \frac{\partial \sigma(\mathbf{X}, \mathbf{F}^{\mathrm{T}}\mathbf{F})}{\partial \mathbf{F}}, \qquad \mathbf{X} \in \Omega, \tag{3.4b}$$

$$\int_\Omega [\mathbf{d}_R(\mathbf{X}, t) \cdot \delta\boldsymbol{\psi}(\mathbf{X}, \mathrm{t}) + \mathbf{T}_R(\mathbf{X}, t) \cdot \delta\mathbf{F}(\mathbf{X}, t)]\, dv_R = 0. \tag{3.4c}$$

Under the conditions given above, we can also assume that the standard part of Eqns (3.1) reduces to the form

$$\boldsymbol{\gamma}(\mathbf{X}, t, \boldsymbol{\psi}(\mathbf{X}, t), \mathbf{F}(\mathbf{X}, t), \nabla\boldsymbol{\psi}(\mathbf{X}, t), \nabla\mathbf{F}(\mathbf{X}, t), \ldots, \nabla^k\boldsymbol{\psi}(\mathbf{X}, t), \nabla^l\mathbf{F}(\mathbf{X}, t)) = \mathbf{0}, \qquad \mathbf{X} \in \Omega \tag{3.5}$$

where $\nabla = (\partial/\partial X^\alpha)$, $\alpha = 1, 2, 3$, $\nabla^m \equiv \nabla(\nabla^{m-1})$, $\mathbf{X} \equiv (X^\alpha)$, and $\boldsymbol{\gamma}$ is a certain k-dimensional differentiable vector function. Equation (3.4c) has to hold for any $\delta\boldsymbol{\psi}(\mathbf{X}, t)$, $\delta\mathbf{F}(\mathbf{X}, t)$ admissible by Eqn (3.5).

Note that, from the non-standard equations (3.1) and (3.3), under certain restrictions imposed on the class of functions in these equations, we have obtained Eqns (3.4), (3.5) of the classical mathematical analysis. The latter will be a starting point for the further investigations, which lead to the definition (within classical analysis) of what will be called a generalized elastic continuum. It must be stressed that the passage given here from a certain subclass of the non-standard discrete system to the 'standard' generalized continuum is unique and has nothing in common with the extrapolation procedures used in the pseudo-continuum approaches.

3.4 Generalized elastic continua

Let $B_R \subset \Omega$ be a fixed regular region and let Eqns (3.5) in $\Omega - \bar{B}_R$ reduce to the form $\nabla\boldsymbol{\psi}(\mathbf{X}, t) - \mathbf{F}(\mathbf{X}, t) = \mathbf{0}$, while in B_R they are given by

$$\boldsymbol{\gamma}(\mathbf{X}, t, \boldsymbol{\psi}, \mathbf{F}, \nabla\boldsymbol{\psi}, \nabla\mathbf{F}, \ldots, \nabla^k\boldsymbol{\psi}, \nabla^k\mathbf{F}) = 0, \qquad \mathbf{X} \in B_R. \tag{3.6}$$

Let $\bar{\mathbf{T}}_R$ be a boundary value of $\mathbf{T}_R$ in $\Omega - \bar{B}_R$ on the surface ∂B_R, $\mathbf{n}_R$ be the unit vector normal to ∂B_R and let us denote $\mathbf{p}_R \equiv \bar{\mathbf{T}}_R \mathbf{n}_R$. Let us also define the field $\mathbf{t}_R$ on ∂B_R by means of the following kinetic boundary condition:

$$\mathbf{t}_R + \mathbf{p}_R = \mathbf{0} \quad \text{on} \quad \partial B_R. \tag{3.7}$$

Then, from Eqn (3.4c), we obtain the relation

$$\oint_{\partial B_R} \mathbf{t}_R \cdot \delta\boldsymbol{\psi}\, \mathrm{d}s_R + \int_{B_R} (\mathbf{d}_R \cdot \delta\boldsymbol{\psi} + \mathbf{T}_R \cdot \delta\mathbf{F})\, \mathrm{d}v_R = 0, \tag{3.8}$$

which has to hold for any $\delta\boldsymbol{\psi}$, $\delta\mathbf{F}$ consistent with Eqn (3.6). At the same time, the following field equations have to hold in B_R:

$$\rho_R \ddot{\boldsymbol{\psi}} = \mathbf{b}_R + \mathbf{d}_R, \qquad \mathbf{T}_R = \rho_R \frac{\partial\sigma}{\partial\mathbf{F}} \quad \text{in} \quad B_R. \tag{3.9}$$

Equations (3.6)–(3.9) describe the so-called generalized hyperelastic continuum and B_R is a region in the physical space occupied by this continuum in the reference configuration. All governing equations mentioned above belong to classical analysis, but the fields involved can be interpreted within the non-standard theory of discrete systems, i.e., by means of Eqns (3.2) and the conditions given in sections 3.3. We have obtained here a new interpretation of the Piola–Kirchhoff stress tensor $\mathbf{T}_R$, strain energy function σ, etc., and we have also arrived at new fields: the field $\mathbf{d}_R$, which will be called the density of internal interactions, and the field $\mathbf{t}_R$, which is said to be the density of boundary interactions. The kinematic fields $\mathbf{F}$, $\boldsymbol{\psi}$ also have clear physical interpretation as the local deformation of material grains and its motion, cf., section 3.3.

The interesting special cases of the generalized elastic continuum will be obtained by specifying the form of Eqns (3.6). Using (3.6)–(3.9), we can easily prove the following statements.

1°. If Eqn (3.6) has the form $\mathbf{F} - \nabla\boldsymbol{\psi} = \mathbf{0}$, then we arrive at the well-known equations of non-linear elasticity, obtaining from (3.8) that $\mathbf{d}_R = \operatorname{Div} \mathbf{T}_R$ in B_R and $\mathbf{T}_R = \bar{\mathbf{T}}_R \mathbf{n}_R$ on ∂B_R.

2°. If Eqn (3.6) has the form $\mathbf{F} - \nabla\boldsymbol{\psi} = \mathbf{0}$, $\boldsymbol{\alpha}(\mathbf{X}, t, \boldsymbol{\psi}, \nabla\boldsymbol{\psi}, \ldots, \nabla^k\boldsymbol{\psi}) = \mathbf{0}$, $\boldsymbol{\alpha}$ being a known differentiable vector function, then we arrive at the equations of the non-linear elasticity with the simple (for $k = 1$) or non-simple (for $k > 1$) kinematic constraints $\alpha(\cdot) = 0$.

3°. If Eqn (3.6) is the identity (i.e., if there are no restrictions of the form (3.6)), then we obtain $\mathbf{t}_R = \mathbf{0}$, $\mathbf{d}_R = \mathbf{0}$ and $\mathbf{T}_R = \mathbf{0}$. This case represents the *S*-continuum of non-interacting material grains.
4°. If Eqn (3.6) is independent of $\mathbf{F}$, then we arrive at $\mathbf{T}_R = \mathbf{0}$. The dynamics of such a system of grains is independent of its material properties.
5°. If Eqn (3.6) has the form $\mathbf{F} - \nabla\boldsymbol{\psi} = \mathbf{0}$, $\boldsymbol{\alpha}(\mathbf{X}, t, \mathbf{F}) = \mathbf{0}$, $\boldsymbol{\alpha}$ being the known differentiable vector function, then we arrive at the theory of elasticity with constraints imposed on the state of stress.

The special cases mentioned above have found many applications in the different problems of engineering mechanics, cf., Woźniak [6] and related papers.

3.5 Conclusions

The *Q*-discrete and *S*-continuous system of elastic grains, which was analysed in sections 3.3, 3.4, represents only one example of the applications of the non-standard theory of discrete systems. However, some conclusions of a more general character can be also formulated.

1. The non-standard approach to discrete mechanics does not change the meaning of the laws of mechanics, but only enables us to apply these laws to a wider class of objects (discrete systems) than the classical approach.
2. Within the non-standard theory of discrete systems, we can analyse new discrete systems (which were called non-standard) which cannot be described by classical discrete mechanics (for example, discrete *S*-continuous systems) and we can investigate new kinds of problems (for example, local interactions in discrete systems).
3. The non-standard theory of discrete systems gives us new interpretations of some concepts of continuum mechanics (for example, the concept of the stress tensor) as well as the concept of the material continuum itself.

In this contribution we have confined ourselves to some simple applications of the non-standard analysis to mechanics; other applications will be treated in subsequent papers.

References

1 Robinson, A., *Non-Standard Analysis.* North-Holland, Amsterdam and London, 1974.
2 Luxemburg, W. A. J., A general theory of monads, in *Proceedings of the*

International Symposium on the Applications of Model Theory to Algebra, Analysis and Probability, California Instutute of Technology, 1967, (W. A. J. Luxemburg, ed.), Holt, Rinehart and Winston, New York, 1969.

3 *Contributions to Non-Standard Analysis*, North-Holland, Amsterdam and London, 1972.

4 Woźniak, Cz., Non-standard approach to the theory of elasticity, *Bull. Acad. Polon. Sci., Sér. Sci. Techn.*, **24,** 229–240, 1976.

5 Woźniak, Cz., On the non-standard formulation of mechanics, *Arch. Mech. Warszawa,* **29,** 593–605, 1977.

6 Woźniak, Cz., On non-standard continuum mechanics, *Bull. Acad. Polon. Sci., Sér. Sci. Techn.*, **24,** 23–37, 1976.

7 Woźniak, Cz., Lecture Notes on the Non-Standard Methods in Mechanics, CISM, Udine, October 5–13, 1978.

Professor Czesław Woźniak,
Institute of Mechanics,
University of Warsaw,
Pałac Kultury i Nauki, p. 937,
Warszawa,
Poland